Aquatic Ecosystems and Biodiversity

Aquatic Ecosystems and Biodiversity

Editor: Ellie Ward

STATES
ACADEMIC PRESS
www.statesacademicpress.com

States Academic Press,
109 South 5th Street,
Brooklyn, NY 11249, USA

Visit us on the World Wide Web at:
www.statesacademicpress.com

ISBN: 978-1-63989-056-9 (Hardback)

Cataloging-in-Publication Data

Aquatic ecosystems and biodiversity / edited by Ellie Ward.
p. cm.
Includes bibliographical references and index.
ISBN 978-1-63989-056-9
1. Aquatic ecology. 2. Aquatic biodiversity. 3. Aquatic biology. 4. Aquatic habitats. I. Ward, Ellie.
QH541.5.W3 A68 2022
577.6--dc23

Table of Contents

Preface VII

Chapter 1 **Trace metal distribution in pelagic fish species** 1
Imane Afandi, Sophia Talba, Ali Benhra, Samir Benbrahim, Rachid Chfiri, Maylis Labonne, Hicham Masski, Raymond Laë, Luis Tito De Morais, Mohammed Bekkali and Fatima Zohra Bouthir

Chapter 2 **Determination of amino acid and fatty acid composition of goldband goatfish [*Upeneus moluccensis* (Bleeker, 1855)] fishing** 16
Güntekin Doğan and Ömer Osman Ertan

Chapter 3 **Bioactivity of aqueous and organic extracts of sea cucumber *Holothuria leucospilota* (Brandt 1835) on pathogenic Candida and Streptococci** 30
Motahereh Ghadiri, Soodabeh Kazemi, Behrooz Heidari and Mehdi Rassa

Chapter 4 **α-Melanocyte-stimulating hormone directly increases the plasma calcitonin level and involves calcium metabolism in goldfish** 43
Takahiro Ikari, Yuki Kobayashi, Yoichiro Kitani, Toshio Sekiguchi, Masato Endo, Akira Kambegawa, Kiyoshi Asahina, Atsuhiko Hattori, Yoshiaki Tabuchi, Thumronk Amornsakun, Kanta Mizusawa, Akiyoshi Takahashi and Nobuo Suzuki

Chapter 5 **Survival time of juvenile pen shell *Atrina pectinata* (Bivalvia: Pinnidae) in hyposaline water** 53
Takeo Kurihara, Shoji Nakano, Yukihiko Matsuyama, Kazumasa Hashimoto, Katsumasa Yamada, Atsushi Ito and Masaei Kanematsu

Chapter 6 **Characterization of carotenoid pigments from bacterial symbionts of soft-coral *Sarcophyton* sp.** 64
Lia Kusmita, Erlita Verdia Mutiara, Handung Nuryadi, Petrick Ariska Pratama, Awang Surya Wiguna and Ocky Karna Radjasa

Chapter 7 **Cypermethrin induced stress and changes in growth of freshwater fish *Oreochromis niloticus*** 73
Rajib Majumder and Anilava Kaviraj

Chapter 8 **Annual variations in hepatic antioxidant defenses and lipid peroxidation in a temperate fish, common bream *Abramis brama* (L.)** 85
Alexey A. Morozov, Grigorii M. Chuiko and Victoria V. Yurchenko

Chapter 9 **Effects of dietary protein and essential amino acid deficiencies on growth, body composition and digestive enzyme activities of silvery-black porgy (*Sparidentex hasta*)** 94
Mansour Torfi Mozanzadeh, Morteza Yaghoubi, Jasem G. Marammazi, Omid Safari and Enric Gisbert

Chapter 10 **Food-deprivation-induced suppression of pituitary-testicular-axis in the tilapia *Oreochromis mossambicus*** 105
R. P. Pikle, R. M. Jatiger and C. B. Ganesh

Chapter 11 **Adenosine monophosphate degradation and inosinic acid accumulation in the shrimp *Penaeus japonicus*** 116
Hiroko Seki, Kaori Nakazato and Naoko Hamada-Sato

Chapter 12 **Chironomidae larvae associated with *Eichhornia azurea* leaf detritus: decomposition community structure and colonization dynamics** 132
Carolina Vieira da Silva and Raoul Henry

Chapter 13 **Effect of dietary soybean meal and multienzyme on intestine histology of beluga sturgeon (*Huso huso*)** 146
M. Sohrabnezhad, M. Sudagar and M. Mazandarani

Chapter 14 **Remarkably high ingestion ratio of acidic food in juvenile marble goby, *Oxyeleotris marmorata*** 156
Chui-Fen Teoh, Leong-Seng Lim and Gunzo Kawamura

Chapter 15 **A metabolic study in hepatopancreas of *Litopenaeus vannamei* response to white spot syndrome virus** 162
Hao Wu, Jianlong Zhang, Yongxing He, Junfang Zhou, Jizhou Yan and Min Jiang

Chapter 16 **Assessment of bacterial communities in skin ulceration in cultured sea cucumber *Apostichopus japonicus* (Selenka)** 169
Yi Yang, Yuchun Li and Zhenlin Liang

Chapter 17 **Determining the growth and mortality parameters of longtail tuna (*Thunnus tonggol* Bleeker, 1851) using length frequency data in coastal waters** 177
Mehran Yasemi, Alireza Nazari Bajgan and Mehran Parsa

Chapter 18 **Induction of DNA damage in the peripheral blood of zebrafish (*Danio rerio*) by an agricultural organophosphate pesticide, monocrotophos** 186
Avelyno H. D'Costa, S. K. Shyama, M. K. Praveen Kumar and Trisha M. Fernandes

Chapter 19 **Effects of salinity on the growth and mucous cells of the abalone *Haliotis diversicolor* Reeve, 1846** 195
Lota Alcantara Creencia and Tadahide Noro

Chapter 20 **Locus-specific EPIC-PCR primers for four distinct calmodulin genes of the Pacific bluefin tuna *Thunnus orientalis*** 206
Seinen Chow, Nobuaki Suzuki, Yoji Nakamura, Motoshige Yasuike, Kenji Saitoh and Takashi Yanagimoto

Chapter 21 **Endocrine disrupting effects of 4-nonylphenol on plasma vitellogenin, reproductive system and histology in koi carp (*Cyprinus carpio*)** 213
Parisa Amaninejad, Homayoun Hosseinzadeh Sahafi, Mehdi Soltani and Seyed Pezhman Hosseini Shekarabi

Permissions

List of Contributors

Index

Preface

Aquatic ecosystem refers to the ecosystem of a waterbody and encompasses both marine and freshwater ecosystems. Marine ecosystems encompass oceanic ecosystems as well as ecosystems of salt marshes, estuaries, lagoons, mangroves, and coral reefs. It may differ both in terms of salinity and biodiversity from freshwater ecosystems, which encompass wetland, river, lake, and pond ecosystems. Many abiotic factors determine the biodiversity of such ecosystems, such as salinity, nutrient levels, the amount of dissolved oxygen, etc. Moreover, biotic factors such as predation by coastal herbivores, including mammals, snails, and geese as well as mechanisms of competition and mutualism influence the biodiversity of aquatic ecosystems. This book attempts to understand aquatic ecosystems and aquatic biodiversity and how different factors impact them. Different approaches, evaluations, methodologies and advanced studies on these domains have been included herein. This book is a resource guide for experts as well as students.

This book is a comprehensive compilation of works of different researchers from varied parts of the world. It includes valuable experiences of the researchers with the sole objective of providing the readers (learners) with a proper knowledge of the concerned field. This book will be beneficial in evoking inspiration and enhancing the knowledge of the interested readers.

In the end, I would like to extend my heartiest thanks to the authors who worked with great determination on their chapters. I also appreciate the publisher's support in the course of the book. I would also like to deeply acknowledge my family who stood by me as a source of inspiration during the project.

Editor

1

Trace metal distribution in pelagic fish species from the north-west African coast (Morocco)

Imane Afandi · Sophia Talba · Ali Benhra · Samir Benbrahim · Rachid Chfiri · Maylis Labonne · Hicham Masski · Raymond Laë · Luis Tito De Morais · Mohammed Bekkali · Fatima Zohra Bouthir

Abstract In the current study, ten elements contents (Fe, Zn, Mn, Cu, Cr, Co, Ni, Cd, Pb and Hg) have been measured in muscle and liver of four pelagic fish species (*Engraulis encrasicolus*, *Sardina pilchardus*, *Scomber japonicus* and *Trachurus trachurus*) from the north-west African coast (South Atlantic Moroccan coast), collected during summer and autumn seasons (July and December 2013, respectively). Significant differences in metal contents were found between the different species ($p < 0.05$). Metals levels were also much higher in liver than those recorded in muscle tissues. The concentrations of Fe, Zn, Cd, Co, Cu and Pb were significantly higher in mackerel liver ($p < 0.05$).While, in muscle, anchovy presents a higher content of Mn, Cu, Cr, Ni and Pb. A high level of cadmium was recorded in liver of the different species which can be attributed to an anthropogenic source (phosphate industry) and to natural sources (upwelling activities). The main concentration of toxic elements (Cd, Pb and Hg) recorded in the four edible muscles of pelagic fish species, under study, were below the established values by the European Commission Regulations and show that their effect on the consumers health can be considered as negligible.

Keywords South Atlantic coast of Morocco · Pelagic fish · Trace metals · Bioaccumulation · Upwelling

Introduction

The metal pollution of the marine ecosystem has always been admitted as a serious environmental issue (Blakas et al. 1982; Tariq et al. 1991). Among environmental pollutants, metals are of particular concern; due to their potential toxic effect and ability to bioaccumulate in aquatic compartment (Censi et al. 2006). Trace metals are naturally and lowly present in seawater but their concentration levels have increased due to anthropogenic pollutants (Kargın et al. 2001). Fish and other aquatic organisms are persistently exposed to

---

I. Afandi (✉) · M. Bekkali
Faculty of Sciences Ain Chock, Hassan II University, Maarif, B. P. 5366, 20100 Casablanca, Morocco
e-mail: imaneafandi@hotmail.com

I. Afandi · S. Talba · A. Benhra · S. Benbrahim · R. Chfiri · H. Masski · F. Z. Bouthir
National Institute of Halieutic Research, Street Sidi Abderrahmane Club equestre Ould Jmel, 20100 Casablanca, Morocco

S. Talba
Faculty of Sciences Ben M'sik, Avenue du Cdt Driss Lharti, 20700 Casablanca, Morocco

M. Labonne · R. Laë · L. Tito De Morais
UMR LEMAR 6539, IUEM, Place copernic, Plouzané, 29820 Brest, France

chemicals in polluted and contaminated waters (Burger et al. 2002) and are exposed to accumulate considerable amounts of toxic elements from their living environment (Suhaimi et al. 2005). Thus, accumulation occurs in the tissues of fish and can be transferred to higher trophic levels via food chain. Therefore, it is important to evaluate the concentration of toxic elements in fish, because it is representing a potential health risks to the fish organisms and to the higher trophic level organism like humans that potentially consume them (Li et al. 2015; Vu et al. 2017). Actually, numerous studies have been carried out on trace metals contamination of different species of edible fish (Prudente et al. 1997; Kucuksezgin et al. 2001; Lewis et al. 2002; El Morhit et al. 2009; Chahid et al. 2014; Afandi et al. 2015; Diop et al. 2016).

The Moroccan coast, with the Canary Current upwelling ecosystem (CanC), constitute one of the four main eastern boundaries upwelling ecosystems (EBUEs) of the world, which give rise to highly productive ecosystems and fisheries (Pauly and Christensen 1995). In this area, the primary production exceeds 300 g cm^{-2} $year^{-1}$ (Carr 2002; Carr and Kearns 2003). Based on this high production, various commercially and socio-economically important fisheries take place in this area such as small pelagic fishes (Boëly and Fréon 1979). Small pelagic fishes are the major target in the fisheries in the Northwest coast of Africa. Over the last decade, total catch has been fluctuating around an average of 1.8 million tones (FAO 2012). However, this fishing potential has experienced a decrease in the recent decades. Beyond the impact of fishing activities (Kifani et al. 2008), the Moroccan coast is under threat from the impacts of anthropogenic pollution (UNEP 1999, 2006) and global changes that will affect the water resources either directly (Biswas 2008), by potentiating effects of contaminants (Parry et al. 2007; Couillard et al. 2008), through variations in ocean dynamics and upwelling activities.

In this study, four pelagic fish species including (*Engraulis encrasicolus* (anchovy), *Sardina pilchardus* (sardine), *trachurus trachurus* (horse mackerel) and *Scomber japonicus* (mackerel)) were selected. These fish species are of particular importance due to their intermediate position in the food web but also for their abundance. They are of crucial economic and biological importance, especially as they are the most consumed species by the Moroccan population. All these species have a gregarious behavior; however, they represent different spatial distribution and diet behaviors. Anchovy is found in the North Atlantic, North and Mediterranean Sea. It lives in coastal waters up to 150 m of depth. It is a planktonophagous species that mainly feeds on phytoplankton (diatoms, dinoflagellates…) and zooplankton (copepods, radiolaria, eggs, larvae…) (INRH 2013). In the Atlantic Ocean, sardines are distributed from the North Sea to the Mauritanian coast and can even reach the waters of Senegal (Fréon and Stéquert 1979). It lives in coastal waters up to 120 m of depth. It is a planktonophagous species whose diet consists largely of phytoplankton. Mackerel has a cosmopolitan distribution including warm and temperate waters of the Atlantic, Indian and Pacific Oceans and adjacent seas. It is essentially a near coastal species, with a vertical distribution ranging between 0 and 300 m of depths. Mackerel show a diet that varies according to its stage of life, from a zoo planktonophagous during larval and post-larval stage to piscivory as adult (small pelagic fish, particularly anchovies). The European horse mackerel is distributed mainly on the continental shelf of the Northeast Atlantic. Along the West African coast, this species is distributed over the coast at over 300 m of depth, with preference for the deepest areas of the continental shelf (FAO 2001). His diet is mainly composed of decapods (shrimps), juvenile anchovies, myctophids and carangids (FAO 1983; COPACE 1984).

Thus, the purpose of this work (involved in the framework of EPURE project research program) is to determine and compare the concentrations of Zn, Fe, Cu, Pb, Cd, Cr, Co and Hg in the liver and muscle of four small pelagic fish species from the south Atlantic Moroccan coast during two periods of upwelling phenomena, in July season of strong upwelling and December season of weak upwelling. The objective is also to describe the differences of the concentrations of trace metals between species, between upwelling seasons and to compare the results with guidelines set down by the European Commission Regulation (European Commission Regulation (EC) 2008, 2014, 2015) for the safe consumption limits of fish and the other studies.

Materials and methods

Sampling

Four pelagic marine fish (*Engraulis encrasicolus, Sardina pilchardus, Scomber japonicus* and *Trachurus trachurus*) were collected on cruises of the R/V ANTEA in summer and autumn season (July and December 2013). The samples were caught with a pelagic trawl at 27 locations in southern Moroccan coastal waters (22–30°N) (Fig. 1). The four species were sampled according to their availability and presence in sufficient quantity. After collection, the total length (cm) and body weight (g) of fishes were measured (Table 1). The dissection of samples was done on board, then three replicate, composed of tissues from five individuals, of muscle and liver were removed with ceramic knife and frozen at − 20 °C. In the laboratory, samples were freeze dried for 72 h at − 55 °C. Then each dried fish sample was homogenized by mixing in rotating glass bottle, and stored until analysis.

Samples preparation and analytical method

The digestion procedure is carried out according to the Association of Official Analytical Chemists (AOAC) method: 2000. The total mercury levels were determined by an Automated Mercury Analyzer, Aula 254 (Gold trap). The analyses of the other metals were performed on an inductively coupled plasma mass spectrometer (ICP-Q-MS X-Series 2, Thermo Scientific) in Ocean Spectrometry Pole at IUEM (Brest, France). Typically, 400 μL of the archive solution were added to 800 μL of 0.80 M HNO_3 (Merck®-Ultrapur Grade) spiked with in (1 ppb) into acid-washed 13 mm o.d. polypropylene tube with rounded bottom (VWR). Samples were introduced to a PFA-ST nebulizer/via a modified SC-Fast introduction system consisting of an SC-2 auto sampler, a six-port valve and a vacuum rinsing pump.

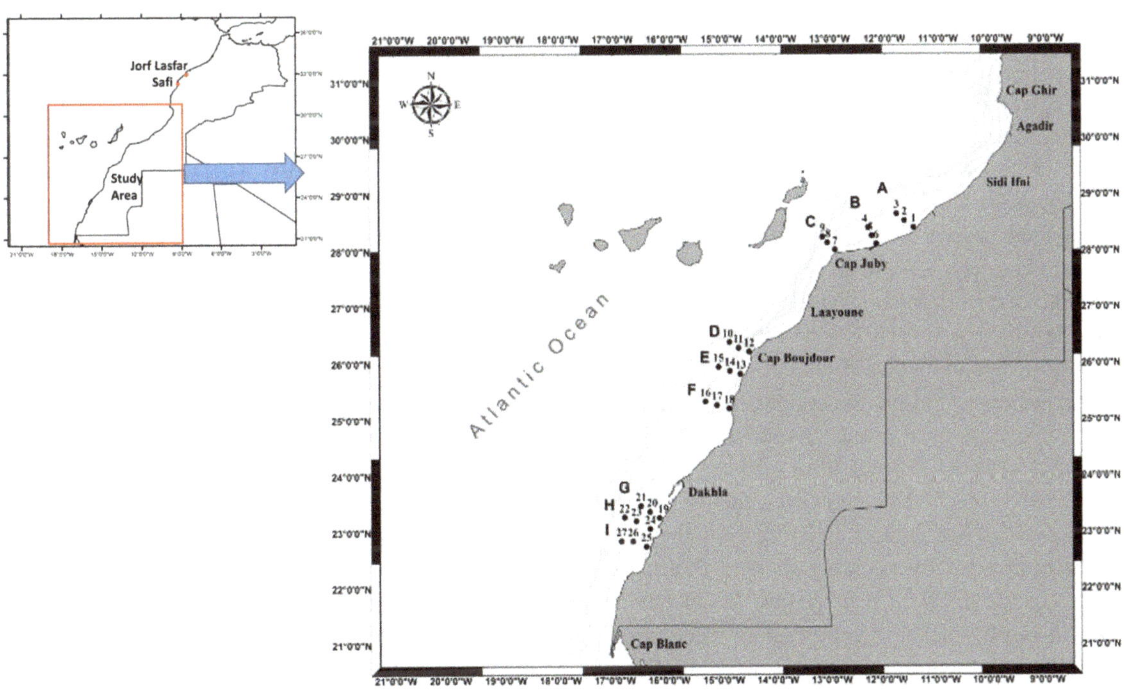

Fig. 1 Study area and sampling location

Table 1 The mean length and weight of fish species

Species	Length (cm)	Weight (g)
Anchovy (*Engraulis encrasicolus*)	13.5 ± 1.15	19.3 ± 5.05
Sardine (*Sardina pilchardus*)	19 ± 2.40	43.7 ± 14.40
Mackerel (*Scomber japonicus*)	23 ± 5.50	65.3 ± 15.80
Horse Mackerel (*Trachurus trachurus*)	21.5 ± 4.50	92.4 ± 40.60

Quality assurance and statistical analysis

Stock standard solutions were prepared gravimetrically from mono and multielemental certified standards (Plasma Cal, SCP Science). A 7-point external standard curve was prepared by serial dilutions and analyzed every 30 samples. Indium and rhodium were added as internal standards.

Precision (degree of reproducibility) and accuracy (degree of veracity) of our procedure were controlled through analyses of a certified reference materials Dorm-4 (fish protein, National Research Council, Canada), using the same digest procedures used for samples. Repeated measurements of these reference materials (every 15 samples) yielded a precision (relative standard deviation) between 8 and 15%. Accuracy was extremely good with a variation ranging from 5 to 8% between our measurements and the certified values.

All statistical calculations were carried out with XLSTAT for Windows (Version 2015.1.03.15945). Statistical significance was defined at 95% ($p < 0.05$). The whole data were subjected to statistical analysis. The inter-species and seasonal differences in element concentrations between fish were performed with a one-way ANOVA, followed by post hoc Turkey tests. Pearson's correlation coefficients were used to examine relationships between the elements in the muscle and liver of the fish species.

Results

The concentration of ten metals elements measured in liver and muscle tissues are reported for each species in Tables 2 and 3. The results show that whatever the species, metals were more concentrated in the livers than in muscles. We found significant variation in the element concentrations in liver and muscle tissues (one-way ANOVA test, $p < 0.05$). Considering the average concentration of metals in all pelagic species, the metals in livers are present in the following order: Fe > Zn > Cd > Cu > Mn > Co > Ni > Hg > Cr > Pb. However, the sequence of metal content in muscle present the following order Fe > Zn > Cu > Mn > Ni > Cr > Cd > Hg > Pb–Co. The order of metal accumulation in both organs showed that Cd and Co are much more accumulated in liver than in muscle. Moreover, for all species, the sequence of toxic trace element was as follows: Cd > Hg > Pb in liver and muscle tissues.

In both organs, iron showed the highest metal level in studied fish, with the highest concentration recorded in mackerel tissues (1158.1 ± 514.8 in liver and 68.52 ± 30.7 μg g^{-1} d.w. in muscle). Following iron, zinc showed the second highest levels. In liver of mackerel, Zn represent the highest level (129.65 ± 22.15 μg g^{-1} d.w.), however, in muscle the highest concentration of Zn was observed in anchovy and mackerel (36.9 ± 7.5 and 37.5 ± 10.9 μg g^{-1} d.w., respectively). The maximum content of Cd and Cu was observed in liver of mackerel (50.07 ± 20.3; 28.30 ± 6.37 μg g^{-1} d.w., respectively). In muscle Cd content was high in mackerel (0.198 ± 0.3 μg g^{-1} d.w.), however, the maximum level of Cu, was observed in anchovy (6.18 ± 1.6 μg g^{-1} d.w.). Levels of toxic metals in livers show that Cd and Pb are more accumulated by mackerel; however, Hg is more accumulated by anchovy (0.49 ± 0.00 μg g^{-1} d.w.). The levels of Pb showed a weak variation in muscles of the different fish species. In spite its small size, anchovy had Cd levels slightly higher of those found in horse mackerel. Furthermore, mercury level was significantly high in muscle of horse mackerel ($p < 0.05$). Concentrations of Ni were high in liver of anchovy and horse mackerel (0.521 ± 0.28; 0.437 ± 0.52 μg g^{-1} d.w., respectively) and in muscle of anchovy (0.33 ± 0.28 μg g^{-1} d.w.). Co present the same behavior as Cd with high level in liver of mackerel (1.59 ± 0.8 μg g^{-1} d.w.) and muscle of anchovy and mackerel (0.064 ± 0.013; 0.068 ± 0.037 μg g^{-1} d.w., respectively). Results of concentration of Mn showed that the highest concentrations are registered in liver and muscle of anchovy (4.95 ± 0.67, 2.0 ± 0.3 μg g^{-1}

Table 2 Mean (± SD) concentration of metals in liver of different fish species (μg g^{-1} Dry weight)

Species	Fe	Zn	Co	Ni	Mn	Cr	Cu	Pb	Cd	Hg
Liver										
Engraulis encrasicolus										
$n = 6$										
Mean ± SD	$432.15^{bb} \pm 57.86$	$104.70^{bcab} \pm 15.10$	$0.392^{bcbc} \pm 0.14$	$0.521^{aa} \pm 0.28$	$4.951^{aa} \pm 0.67$	$0.088^{aa} \pm 0.04$	$15.522^{bb} \pm 2.07$	$0.078^{bb} \pm 0.02$	$13.009^{bcbc} \pm 5.67$	$0.49^{aa} \pm 0.00$
Min–Max	356.63–513.82	86.78–122.76	0.23–0.61	0.28–0.90	4.42–6.28	0.03–0.14	12.55–17.86	0.05–0.11	5.46–19.28	0.48–0.5
Sardina pilchardus										
$n = 31$										
Summer	606.53 ± 150.66	92.80 ± 15.84	0.22 ± 0.07	0.26 ± 0.22	3.30 ± 0.93	0.149 ± 0.16	18.26 ± 6.16	0.08 ± 0.04	7.45 ± 2.69	0.11 ± 0.13
Autumn	301.540 ± 78.50	88.03 ± 3.22	0.21 ± 0.02	0.15 ± 0.09	4.92 ± 0.61	0.118 ± 0.06	19.48 ± 1.80	0.04 ± 0.03	6.23 ± 3.13	0.05 ± 0.00
Mean ± SD	$559.605^{bb} \pm 180$	$92.064^{cb} \pm 14.67$	$0.221^{cc} \pm 0.06$	$0.242^{aa} \pm 0.21$	$3.611^{bb} \pm 1.09$	$0.143^{aa} \pm 0.15$	$18.431^{bb} \pm 5.73$	$0.073^{bb} \pm 0.04$	$7.215^{cc} \pm 2.77$	$0.10^{cb} \pm 0.129$
Min–Max	224.83–944.79	64.39–125.46	0.13–0.38	0.04–0.83	2.05–5.59	0.02–0.71	10.75–32.58	0.01–0.16	2.99–13.19	0.020–0.486
Scomber japonicus										
$n = 37$										
Summer	1085.11 ± 441.52	129.97 ± 22.86	1.45 ± 0.73	0.42 ± 0.34	3.81 ± 1.03	0.095 ± 0.07	27.95 ± 6.31	0.21 ± 0.09	48.16 ± 19.42	0.21 ± 0.09
Autumn	2254,10 ± 12.34	125,21 ± 5.91	3.71 ± 0.02	0.82 ± 0.08	5.21 ± 0.14	0.128 ± 0.13	34.05 ± 5.70	0.25 ± 0.11	78.74 ± 9.50	N.D.
Mean ± SD	$1158.172^{aa} \pm 514.8$	$129.659^{aa} \pm 22.15$	$1.591^{aa} \pm 0.8$	$0.447^{aa} \pm 0.34$	$3.887^{bab} \pm 1.05$	$0.098^{aa} \pm 0.08$	$28.305^{aa} \pm 6.37$	$0.218^{aa} \pm 0.09$	$50.072^{aa} \pm 20.3$	$0.210^{bab} \pm 0.09$
Min–Max	372.91–2262.82	94.39–182.46	0.17–3.72	0.13–1.58	2.50–6.27	0.04–0.43	16.78–40.96	0.09–0.43	6.72–85.45	0.092–0.342
Trachurus trachurus										
$n = 30$										
Summer	1105.77 ± 679.10	112.64 ± 35.72	0.88 ± 0.42	0.45 ± 0.56	3.79 ± 0.94	0.122 ± 0.10	24.49 ± 12.57	0.05 ± 0.03	21.17 ± 12.64	0.13 ± 0.08
Autumn	930,20 ± 65.71	112,26 ± 5.09	1.29 ± 0.80	0.39 ± 0.06	4.59 ± 1.42	0.061 ± 0.02	16.82 ± 9.49	0.09 ± 0.02	34.99 ± 17.13	0.30 ± 0.00
Mean ± SD	$1077.681^{aa} \pm 623.83$	$112.575^{ba} \pm 32.55$	$0.938^{bb} \pm 0.49$	$0.437^{aa} \pm 0.52$	$3.927^{bab} \pm 1.05$	$0.111^{aa} \pm 0.09$	$22.958^{bab} \pm 12.27$	$0.058^{bb} \pm 0.03$	$23.557^{bb} \pm 14.19$	$0.155^{bcb} \pm 0.094$
Min–Max	477.63–3182.56	68.30–210.71	0.35–2.04	0.03–2.66	2.58–6.80	0.04–0.48	9.04–67.94	0.003–0.12	8.84–65.35	0.034–0.302

Mean metal concentration of different species from all sites sharing a common letter for a particular metal are not significantly different $p < 0.05$

N.D. no determined

Table 3 Mean (± SD) concentration of metals in muscle of different fish species (μg g^{-1} Dry weight)

Species	Fe	Zn	Co	Ni	Mn	Cr	Cu	Pb	Cd	Hg
Muscle										
Engraulis encrasicolus										
$n = 6$										
Mean ± SD	$62.43^{abab} \pm 5.38$	$36.92^{aba} \pm 7.50$	$0.064^{abab} \pm 0.013$	$0.332^{aa} \pm 0.280$	$2.002^{aa} \pm 0.380$	$0.27^{aa} \pm 0.442$	$6.18^{aa} \pm 1.61$	$0.066^{aa} \pm 0.037$	$0.195^{aba} \pm 0.072$	$0.052^{bb} \pm 0.011$
Min–Max	56.39–68.56	29.53–51.26	0.04–0.07	0.11–0.69	1.48–2.61	0.05–1.17	4.04–8.72	0.02–0.12	0.12–0.29	0.044–0.060
Sardina pilchardus										
$n = 31$										
Summer	65.79 ± 34.00	30.66 ± 11.00	0.05 ± 0.13	0.16 ± 0.27	1.24 ± 0.50	0.12 ± 0.13	4.94 ± 3.31	0.04 ± 0.07	0.07 ± 0.04	0.03 ± 0.01
Autumn	62.87 ± 9.53	37.54 ± 4.54	0.06 ± 0.01	0.23 ± 0.22	1.38 ± 0.22	0.07 ± 0.02	4.46 ± 0.81	0.02 ± 0.01	0.03 ± 0.01	0.05 ± 0.01
Mean ± SD	$66.10^{aa} \pm 31.1$	$31.91^{ba} \pm 10.64$	$0.049^{bcab} \pm 0.022$	$0.172^{aa} \pm 0.255$	$1.26^{bb} \pm 0.462$	$0.11^{ba} \pm 0.121$	$4.86^{aa} \pm 3.027$	$0.035^{ba} \pm 0.033$	$0.067^{ba} \pm 0.067$	$0.037^{bb} \pm 0.015$
Min–Max	31.42–141.84	15.74–60.09	0.02–0.12	0.02–1.29	0.21–2.35	0.03–0.66	1.74–14.32	0.0008–0.13	0.01–0.33	0.016–0.061
Scomber japonicus										
$n = 37$										
Summer	62.00 ± 25.83	35.00 ± 8.73	0.06 ± 0.11	0.12 ± 0.11	0.57 ± 0.58	0.11 ± 0.11	3.41 ± 1.26	0.03 ± 0.33	0.20 ± 0.03	0.09 ± 0.04
Autumn	125.14 ± 11.00	59.66 ± 9.00	0.14 ± 0.01	0.33 ± 0.07	0.75 ± 0.09	0.15 ± 0.01	5.88 ± 0.51	0.05 ± 0.00	0.19 ± 0.04	0.14 ± 0.00
Mean ± SD	$68.52^{aa} \pm 30.7$	$37.50^{aa} \pm 10.93$	$0.068^{aa} \pm 0.037$	$0.140^{aa} \pm 0.124$	$0.591^{cc} \pm 0.554$	$0.11^{ba} \pm 0.109$	$3.60^{bab} \pm 1.38$	$0.035^{ba} \pm 0.028$	$0.198^{aa} \pm 0.318$	$0.094^{bb} \pm 0.042$
Min–Max	27.68–155.38	21.07–68.68	0.03–0.15	0.01–0.42	0.16–3.51	0.02–0.52	1.32–6.50	0.01–0.16	0.04–1.39	0.040–0.162
Trachurus trachurus										
$n = 30$										
Summer	33 87 ± 10.23	17.92 ± 6.17	0.04 ± 0.09	0.16 ± 0.18	0.48 ± 0.18	0.14 ± 0.09	2.07 ± 0.61	0.03 ± 0.15	0.07 ± 0.02	0.24 ± 0.15
Autumn	54.00 ± 9.00	36.00 ± 16.52	0.04 ± 0.02	0.11 ± 0.09	0.64 ± 0.11	0.06 ± 0.01	2.47 ± 0.80	0.04 ± 0.02	0.10 ± 0.08	0.15 ± 0.07
Mean ± SD	$37.07^{bb} \pm 12.5$	$20.78^{cb} \pm 10.57$	$0.037^{cb} \pm 0.019$	$0.148^{aa} \pm 0.163$	$0.513^{cc} \pm 0.175$	$0.13^{ba} \pm 0.089$	$2.13^{cb} \pm 0.64$	$0.029^{ba} \pm 0.022$	$0.077^{ba} \pm 0.143$	$0.221^{aa} \pm 0.137$
Min–Max	12.75–67.55	9.78–51.92	0.01–0.10	0.01–0.69	0.29–0.94	0.02–0.43	1.17–3.87	0.004–0.08	0.004–0.77	0.064–0.430

Mean metal concentration of different species from all sites sharing a common letter for a particular metal are not significantly different $p < 0.05$

d.w., respectively). Finally, Cr represent the higher concentrations in liver of Sardine and horse mackerel (0.143 ± 0.15 µg g^{-1} d.w) however, in muscle anchovy represent the maximum content (0.27 ± 0.4 µg g^{-1} d.w). The concentrations of Fe, Zn, Cd, Co, Cu and Pb were significantly higher in mackerel liver ($p < 0.05$). Whereas Mn, Cu, Cr, Ni and Pb were higher in muscle of anchovy (Tables 2, 3).

The mean concentration of metals in tissues of sardine, mackerel and horse mackerel during summer and autumn seasons (anchovy was not represented due to unavailability during warm season) are giving in Figs. 2 and 3 and Tables 2 and 3. During both seasons, the results showed that the concentration of metals were higher in liver than in muscle. Therefore, there was no clear seasonal pattern in metal concentrations (Figs. 2, 3). The results showed that, indeed, sardine concentrations of Fe, Pb were significantly higher ($p < 0.05$) during summer season, whereas Mn were significantly higher ($p < 0.05$) during autumn season. For mackerel, concentrations of Fe, Co and Cd were significantly higher ($p < 0.05$) during autumn season. For horse mackerel only Pb concentrations that were significantly higher ($p < 0.05$) during autumn season. In muscle, concentrations of Fe, Co Ni, Cu, Zn were significantly higher ($p < 0.05$) during autumn season for mackerel. However, only Zn and Fe were significantly higher ($p < 0.05$) during autumn season for horse mackerel. No significant differences were observed for the other elements studied here.

Inter-elemental relationships in fish tissues (inter-metal correlation) were assessed by the mean of Person's correlation coefficient. Tables 4 and 5 show values of correlation coefficients among metal concentrations. A high significant positive correlation was observed in tissue samples for the trace metal pairs of Fe and (Zn, Cu, Ni, Co), Zn and (Cu, Co), Mn and (Cu, Cr, Ni), Cu and (Ni, Co) in liver. In addition, a high significant correlations were found in muscle between Fe and (Zn, Cu, Co, Cd), Zn and (Cu, Co, Cd), Cu and (Co, Pb, Cd); Co and (Pb, Cd) and Pb–Cd.

Discussion

In this study, the investigation of trace metal contamination levels in fish from the southern Atlantic Moroccan coast showed the presence of differences of contamination between species and considered organs, as well as

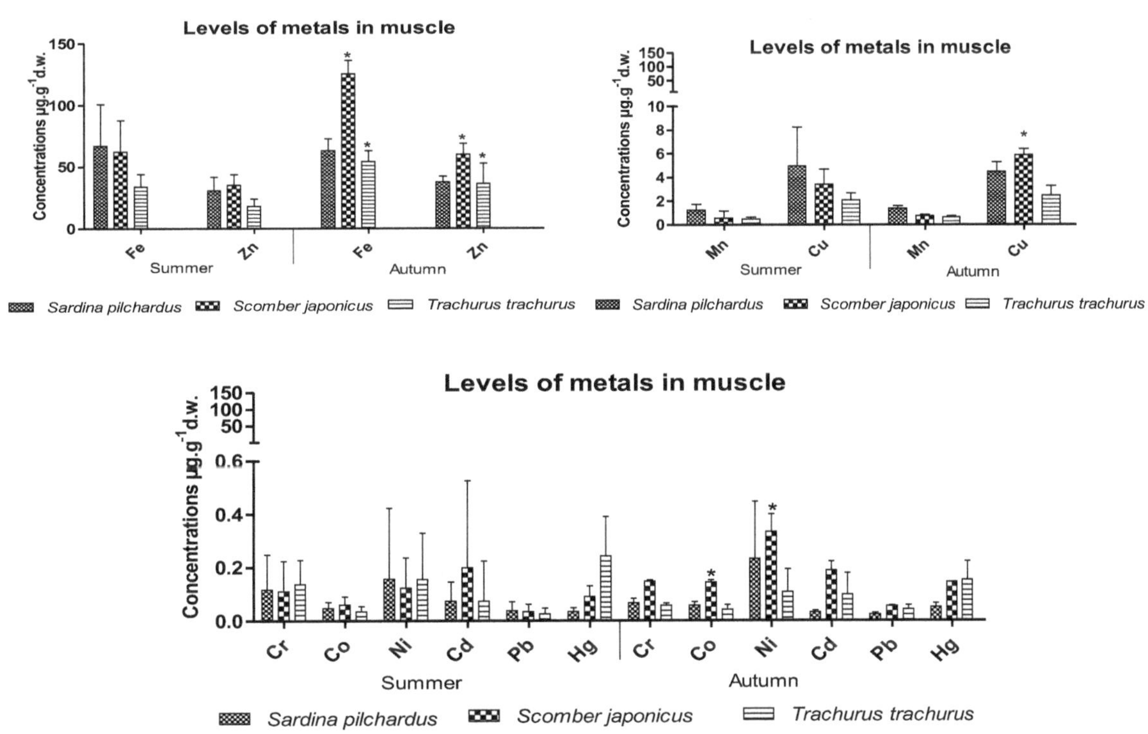

Fig. 2 Seasonal variation in mean metals concentrations (± SD) in muscle of the three pelagic fish species (asterisk denote significant differences between season)

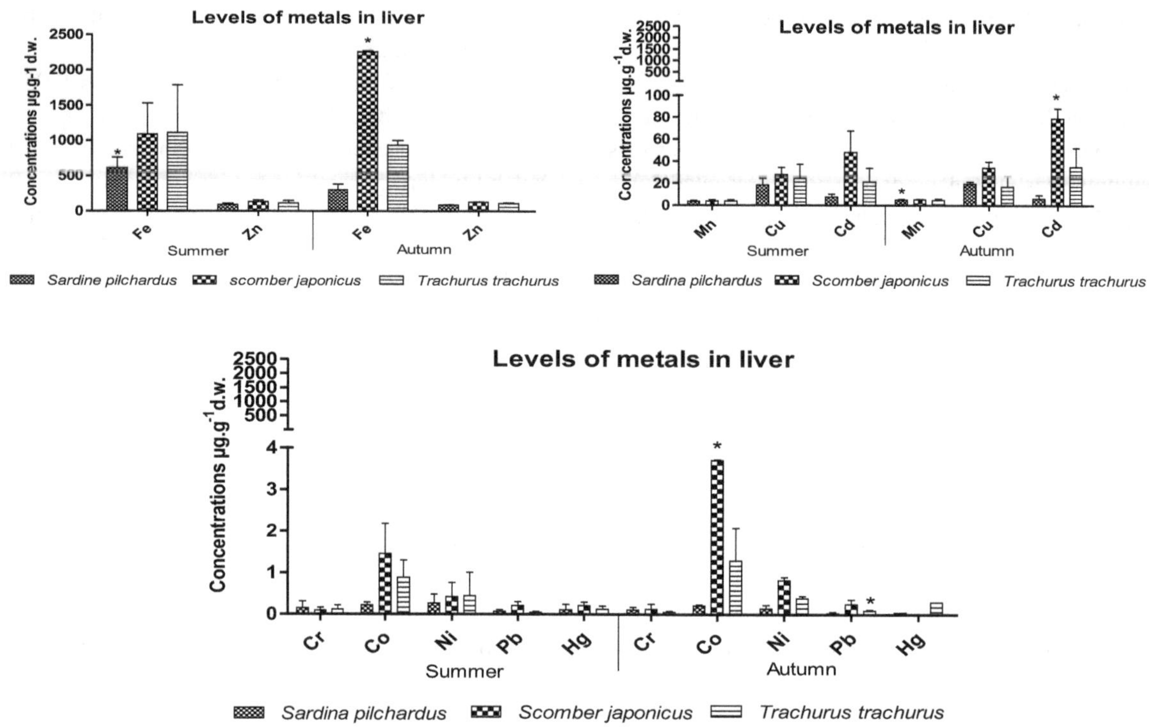

Fig. 3 Seasonal variation in mean metals concentrations (± SD) in liver of the three pelagic fish species (asterisk denote significant differences between season)

Table 4 Pearson product-moment correlation coefficients between trace metals in the muscle tissue of fishes

Metal	Fe	Zn	Mn	Cr	Cu	Ni	Co	Pb	Cd	Hg
Fe	**1**									
Zn	**0.727**	**1**								
Mn	**0.473**	**0.362**	**1**							
Cr	0.325	0.158	**0.428**	**1**						
Cu	**0.848**	**0.656**	**0.659**	**0.458**	**1**					
Ni	**0.614**	**0.452**	**0.515**	**0.792**	**0.720**	**1**				
Co	**0.753**	**0.644**	0.263	0.329	**0.672**	**0.481**	**1**			
Pb	**0.396**	0.335	**0.463**	0.279	**0.532**	**0.394**	0.329	**1**		
Cd	0.178	0.073	− 0.051	0.193	0.030	0.087	0.177	0.077	**1**	
Hg	− 0.313	− **0.400**	− **0.365**	0.074	− 0.339	0.004	− 0.117	− 0.151	− 0.034	**1**

Bold value indicate significant levels at $p < 0.05$

very significant variations between the elements of metals analyzed. Two groups of elements were determined for all the data obtained. It is well known that some of the trace elements control important biological processes by facilitating the binding of molecules to their receptor sites on cell membrane, e.g., Fe, Cu, Zn, Mn, Cr, Ni and Co, representing a group of essential elements to metabolic activities and physiological functions which are present in all animal tissues. However, other metals are known to be potentially toxic, i.e., Hg, Pb and Cd, and may be of natural or anthropogenic origin, representing a group of toxic elements.

Metal concentrations in fish organs

The results demonstrated that, whatever the species, elements show higher concentrations in liver than in muscle tissue (Tables 2, 3) as reported in other studies (Diop et al. 2016). Metal accumulation in fish organs

Table 5 Pearson product-moment correlation coefficients between trace metals in the liver tissue of fishes

Variables	Fe	Zn	Mn	Cr	Cu	Ni	Co	Pb	Cd	Hg
Fe	**1**									
Zn	**0.787**	**1**								
Mn	0.137	**0.510**	**1**							
Cr	− 0.105	0.013	0.226	**1**						
Cu	**0.661**	**0.708**	0.267	0.199	**1**					
Ni	0.234	**0.463**	**0.395**	− 0.029	0.197	**1**				
Co	**0.686**	**0.668**	0.198	− 0.165	**0.735**	**0.398**	**1**			
Pb	**0.472**	**0.492**	− 0.191	− 0.227	**0.675**	0.169	**0.631**	**1**		
Cd	**0.687**	**0.651**	0.033	− 0.233	**0.747**	0.205	**0.764**	**0.876**	**1**	
Hg	0.173	0.335	**0.457**	− 0.161	0.280	0.217	0.311	0.264	0.363	**1**

Bold value indicate significant levels at $p < 0.05$

relay on the physiological role of the organs (Uysal et al. 2008). Liver represent an organ with high metabolic activities and accumulate more elements than organ with lower metabolic activities like muscle (Ploetz et al. 2007). These organs with high metabolic activities contain metal-binding proteins involved in the reduction of trace metal functions, like metallothioneins in the liver, allowing them to accumulate a significant higher level of element than the muscle tissue (Roesijadi 1996; JariC et al. 2011). In our study, all metals concentrations in all fish species show the following trend, liver > muscle.

Metal concentrations in different fish species

In our study, the results showed that fish exhibited wide inter-specific variations in metals accumulation in both organs. The liver tissue of predatory fish (mackerel and horse mackerel) contained significantly higher levels of Fe, Co, Cu, Zn, Cd (Table 2), due to their feeding at the higher trophic levels, therefore, by the bioaccumulation and biomagnification of the elements through the food chain. Many studies attributed high metal accumulation to the feeding habit of the fish. Mackerel and horse mackerel feed on fish, crustaceans and cephalopods. Therefore, there is an evidence for metal bioaccumulation and biomagnification; however, evidence for Cd biomagnifications is inconsistent (Vieira et al. 2011).

In the edible muscles of the pelagic species, anchovy was found to have significant higher content of Cr, Mn, Ni, Cu and Pb. Actually; it is well known that metal concentrations in fish tissues are related to metal contents in environment. Thus, this finding in anchovy could be linked to feeding on zooplankton since it is able to accumulate a high metal content. Topping (1973) have also reported that fish species feeding on plankton contain higher concentration of trace metals than bottom feeding fish. Gundogdu et al. (2016) reported the same results with high metals content in muscle of *E. encrasicolus* from Black sea than content found in muscle of *T. trachurus* and *M. barbatus*. The reason for high metal concentrations in small fish could also be due to the metal complex with the mucus in skin that is impossible to be purified completely from fish tissue before the analysis. Thus, for small fish the skin may be an important site for the uptake of metals due to their high surface area to body ratio.

Nevertheless, the bioaccumulation process of metals in organisms can also influenced by biotic factors, defining for each species and for each development stage, the ecological characteristics (habitat. diet) and the structural and functional properties of the biological barriers that separate living organisms from their environment and controlling metal absorption. (Boudou and Ribeyre 1997; Andres et al. 2000; Barron 2003).

Metal accumulation patterns

The general order of monitored metal bioaccumulation in pelagic fish was Fe > Zn > Cd > Cu > Mn > Co > Ni > Hg > Cr > Pb in liver and Fe > Zn > Cu > Mn > Ni > Cr > Cd > Hg > Pb–Co in muscle tissue. We observed that the essential metals showed the same order of magnitude, however, only the toxic elements presented different order depending on fish environment. In comparison with data reported for pelagic fish from other locations, the observed amount of sequences are different from that indicated by Copat et al. (2013) in muscle of pelagic fish (muscle of anchovy, mackerel and horse mackerel) from Eastern Mediterranean sea (Zn > Mn > Ni > Cr > Pb > Cd), Canli and Atli (2003) in sardine tissues (Fe > Zn > Pb > Cu > Cr > Cd), Ersoy and Çelik (2009) in muscle (Fe > Zn > Cu > Mn > Ni > Pb > Cr > Cd) and liver of mackerel (Fe > Zn > Cu > Mn > Cd > Pb > Ni > Cr) both from the Mediterranean sea, and Bilandžić et al. (2011) in muscle tissue of sardine and mackerel from Adriatic Sea (Cu > Hg > Pb > Cd). From these observations, we have noted that cadmium element is the less accumulated metal, at the opposition of our results, which show a large accumulation of cadmium especially in the liver. Cadmium is particularly concentrated in liver of the four studied fish species (Table 2). Data from a previous study showed that cadmium level in liver of various species is less than 10 mg/kg d.w., and more often less than 1 mg/kg d.w. (Romeo et al. 1999; Henry et al. 2004; Diop et al. 2016). Cadmium were already founded at high levels in fish and mollusks from West African coastal areas such as the Mauritanian water (Romeo et al. 1999; Sidoumou et al. 2005), Moroccan coast (Benbrahim et al. 2006; Maanan 2008; El Morhit et al. 2013) or Senegalese coast (Bodin et al. 2013; Diop et al. 2016). These high concentrations of Cd were attributed to natural origins, such as upwelling of deep waters, which takes place along the Western Atlantic coast (Romeo et al. 1999). Auger et al. (2015), has demonstrated that coastal upwelling in Western Atlantic coast, represent the main natural origin of cadmium enrichment of marine ecosystems, whereas the main anthropogenic origin is the phosphate industry effluents. Actually, Morocco is the second phosphate producer in the world. About 27 million tons of phosphate ores are extracted and processed annually to produce phosphoric acid and phosphorus-based fertilizers ("Annual Report, Office Chérifien des Phosphates, Maroc," 2012; Gaudry et al., 2007). The phosphoric acid is produced, from the chemical reaction between sulphuric acid and phosphate ores. During the reaction, calcium sulphate (phosphogypsum), is produced and spread out into the ocean (Auger et al. 2015).The phosphogypsum resultant from production of phosphoric acid and phosphorus-based fertilizers contain a significant abundance of crustal elements such as heavy metal, especially cadmium. Thus, the phosphate industry releases a large quantity of Cd off the coast of Morocco around Safi and Jorf-Lasfar (33°N) (Gaudry et al. 2007) (Fig. 1). However, wind-inducing upwelling of intermediate water along WN African Coast and driving to a re-injection of element (Such as Cadmium) in the surface layer (Auger et al. 2015). In the same study, it has been demonstrated that potential phytoplankton Cd-uptake and Cd-bioaccumulation coincide with the upwelling enrichment and dispersion patterns in waters. This result can explain the high Cd content in muscle of anchovy which feed directly on plankton. The concentrations of Cd in liver and muscle of marine fish were in the same order of those reported in Mauritanian coast (Sidoumou et al. 2005; Romeo et al. 1999; Diop et al. 2016) and higher than those reported in pelagic fish from Moroccan coast (in central and south zone) (Chahid et al. 2014; Afandi et al. 2015) and in Portuguese water (Vieira et al. 2011).

However, metals such as Fe, Zn, Mn, Cr, Ni, Co and Cu are essential elements with useful biological functions. Thus, most organisms have biochemical mechanisms control the amount of these elements into their cells. A number of studies had investigated metal contents in small pelagic fish tissues comparing with those studies, levels of Cu, Mn, Cr, Ni, Fe, Zn Pb and Cd reported in our study are in agreement with those found in muscle and liver of sardine from the upwelled water of Senegalese coast (Diop et al. 2016) except for Cr, Fe, Pb and Cd contents in liver which were higher than our results. Also, El Morhit et al. (2013) reported metal content in muscle of sardine collected from South Atlantic coast of Morocco, which were in same order of magnitude than our results but with higher content of Ni. However, lower metal contents were reported in muscle of anchovy (Cu, Fe, and Zn), sardine tissues (Cd, Cu, Co, Fe, Ni, Pb, and Zn) and in muscle of *scombrus* (Cd, Co, Cu, Fe, Pb, and Zn) from Marmara Sea (Ergül and Aksan 2013). As stated above, upwelled waters, where the different studied species were collected, may be naturally rich in metal element. Metal content of fish from Mauritanian coast showed a lower concentration of Cd and Cu in sardine, mackerel and horse mackerel, and of Zn in sardine and mackerel, however, higher content of Zn were reported in muscle of horse mackerel (Romeo et al. 1999). Therefore, difference in metal accumulation can be controlled by numerous factors, corresponding to the physicochemical characteristics of the aquatic biotopes and to their natural or anthropogenic variations, influencing metals bioavailability, via the chemical speciation reactions, and their transfer and bioaccumulation behavior, in relation to the adaptive responses to the main physiological functions (e.g., respiration, osmo regulation, nutrition) (Boudou and Ribeyre 1997; Andres et al. 2000; Barron 2003).

Cd, Pb and Hg are non-essential elements with high toxic potential even at low concentrations. Legal thresholds are not available for essential elements in Europe. The European Community proposed threshold values of metal concentrations in fish muscle only for nonessential metals (e.g., Cd, Pb and Hg). The threshold values are expressed as μg g^{-1} wet weight and are 0.3 for Pb, 0.5 for Hg and 0.05 for Cd in fishery products. For physiological reasons, some species accumulate this elements more than others and for these species a higher acceptable limit applies for Cd (0.1 μg g^{-1} wet weight for mackerel and 0.25 μg g^{-1} wet weight for sardine and anchovy (European Commission (EC) 2008, 2014, 2015). For an easier comparison, our results in dry weight have been converted into μg g^{-1} wet weight using a conversion factor $CF = \frac{Cd}{Cw} = \frac{100}{100-\%H}$ of four between wet and dry mass (Cresson et al. 2017), where Cd and Cw represent the concentrations expressed to dry and wet mass, respectively, and %H is the percentage of humidity in wet tissues determined according to the Association of Official Analytical Chemists (AOAC) method: 2000. The Cd concentrations in muscle tissues were 0.05 μg g^{-1} in anchovy and mackerel, and were 0.02 μg g^{-1} in sardine and horse mackerel. For Pb, the concentrations were 0.01 μg g^{-1} in sardine, mackerel and horse mackerel muscles, however, the concentration were 0.02 μg g^{-1} in anchovy. The Hg concentrations in muscle tissues of anchovy, sardine, mackerel and horse mackerel were 0.013 μg g^{-1}, 0.009 μg g^{-1}, 0.024 μg g^{-1} and 0.055 μg g^{-1}, respectively. The overall Cd, Pb and Hg levels in the muscle tissues of all the specimens analyzed were well under the proposed limit values. Results of cadmium contents for Sardine are comparable with those reported by Diop et al. (2016) from Senegalese Coast and higher with those reported by Chahid et al. (2013) and Vieira et al.(2011). However, these authors reported higher levels of lead and mercury in the same species. For mackerel, results found in our study present also higher cadmium contents and lower levels of Pb than mackerel from Atlantic Ocean in Portuguese waters, and Mediterranean Sea (Vieira et al. 2011; Falcó et al. 2006). Moreover, lower content of cadmium were reported in horse mackerel from Adriatic Sea and Atlantic Sea (Storelli 2008; Vieira et al. 2011) and higher content of Hg and Pb (Storelli 2008; Vieira et al. 2011; Mendil et al. 2010) (Table 6).

Seasonal concentration patterns

In our study area, there was no clear seasonal pattern in metal contents in fish muscles. In liver tissue, metals concentrations (except Mn and Cu) show a decrease between summer and autumn seasons for sardine, this can probably be attributed to the upwelling intensity, which is greater during the summer season and lower during the autumn season (Makaoui et al. 2005). It is well known that upwelling phenomena contribute to metals bioavailability in environment. Bruland and Franks (1983) reported in the coastal waters of California, elevated levels of Cd, Cu and Zn, explaining that it could be a result from the upwelling of nutrient-rich waters. The same result was also reported by Roméo and Gnassia-Barelli (1988) for Cd, Fe and Zn on the Mauritanian coast. Whereas in mackerel's liver, seasonal variation of all metals concentrations (except Zn) show an increase between summer and autumn seasons for mackerel. However, in horse mackerel liver no consistent patterns were recorded. Therefore, the seasonal variations of metals concentrations measured in our study could depend upon several factors such as growth and reproductive cycles and changes in water temperature (Ersoy and Çelik 2009). These variations can also be reported to the water circulation along the coast and upwelling intensity. Additionally, the differences noted in seasonal patterns between species could be due to variations in feeding habits, in spatial migration, feeding and intrinsic factors such as different rates of physiological process and uptake.

Inter-elemental relationships in fish tissues (inter-metal correlation) (Tables 4, 5) shows significant correlations among the elements (Fe, Zn, Mn, Cr, Cu, Ni, Co and Pb) in muscle and (Fe, Zn, Mn, Cu, Co, Pb and Cd) in liver. The correlation between the different elements observed in the same tissue of fish can be explained by the similar accumulation behavior of trace elements in the fishes and their interactions (Kojadinovic et al. 2007).

Conclusion

Based on the analyses of fish samples, trace metal concentrations in fish from the south Atlantic Moroccan coast were broadly comparable to those found in similar national and international studies. In this work, there

Table 6 Comparison heavy metal concentrations (μg g^{-1}) in fish with values taken from the open literature

Fish species	Region	Fe	Zn	Mn	Cu	Co	Cr	Ni	Cd	Pb	Hg	References
Sardine	Atlantic sea (ww)								0.0017–0.0151	0.0029–0.056	0.0116–0.028	Vieira et al. 2011
Sardine	Mediterranean sea (ww)								0.002–0.01	0.01–0.08	0.07–0.09	Falcó et al. 2006
Sardine	Senegalese Coast (dw)	(L) 599.8	(L) 90.23	(L) 3.84	(L) 18.19	–	(L) 1.244	(L) 0.394	(L) 18.11	(L) 0.249	–	Diop et al. 2016
		(M) 72.3	(M) 20.92	(M) 1.87	(M) 4.96		(M) 0.248	(M) 0.126	(M) 0.051	(M) 0.023		
Sardine	Marmara Sea (dw)	(L) 110	(L) 55.4		(L) 3.55	(L) 0.09	(L) 0.32	(L) 0.36	(L) 0.07	(L) 0.09	(L) 0.02	Ergül and Aksan 2013
		(M) 39.7	(M) 30.8		(M) 3.03	(M)0.04	(M) 0.37	(M) 0.05	(M) 0.05	(M) 0.10	(M) 0.11	
Anchovy		(M) 26 .3	(M) 24.5		(M) 2.7	(M) 0.05	(M) 0.33	(M) 0.06	(M) 0.02	(M) 0.03	(M) 0.11	
Mackerel		(M) 33.9	(M) 18		(M) 2.05	(M) 0.04	(M) 0.72	(M) 0.27	(M) 0.01	(M) 0.01	(M) 1.31	
Sardine	Atlantic coast Morocco (ww)	(M) 11.67–15.90	(M) 3.54–12.70	(M) 0.02–0.18	(M) 0.01–1.10		(M) 0.001–0.02	(M) 1.10–2.8	(M) 0.01–0.23	(M) 0.001–0.008		El Morhit et al. 2013
Horse Mackerel	Adriatic sea (ww)								0.01–0.03	ND–0.06	0.16–2.41	Storelli 2008
Horse Mackerel	Atlantic sea (ww)								0.003–0.0141	0.0031–0.0215	0.038–0.3371	Vieira et al. 2011
Horse Mackerel	Black sea (ww)								0.22	0.64		Mendil et al. 2010
Mackerel	Mediterannean sea (ww)								0.003–0.01	0.01–0.02	0.06–0.15	Falcó et al. 2006
Sardine	Atlantic sea Morocco (ww)								0.0058	0.02	0.038	Chahid et al. 2014
Sardinella	Mauritanian Coast (dw)		1.6		2.8				0.02		0.09	Romeo et al. 1999
Mackerel			32		1.7				0.03		ND	
Horse Mackerel			42		1.6				0.04		0.03	
Anchovy	Moroccan Atlantic coast (dw)	432.15 (L)	104.7 (L)	4.95 (L)	15.52 (L)	0.39 (L)	0.088 (L)	0.52 (L)	13.0 (L)	0.078 (L)	0.49(L)	Present study
		62.42 (M)	36.19 (M)	2.00 (M)	6.17 (M)	0.06(M)	0.269 (M)	0.33 (M)	0.195 (M)	0.066 (M)	0.052 (M)	
Sardina		559.6 (L)	104.7 (L)	3.611 (L)	18.4 (L)	0.221(L)	0.143 (L)	0.242 (L)	7.215 (L)	0.073 (L)	0.10(L)	
		66.09 (M)	31.9 (M)	1.263(M)	4.86 (M)	0.049 (M)	0.109 (M)	0.172 (M)	0.067 (M)	0.035 (M)	0.037 (M)	
Mackerel		1158.1(L)	129.6 (L)	3.88 (L)	28.30 (L)	1.59 (L)	0.098 (L)	0.447 (L)	50.07 (L)	0.218 (L)	0.21(L)	
		68.52 (M)	37.49 (M)	0.58 (M)	3.6 (M)	0.068 M)	0.113 (M)	0.140 (M)	0.198 (M)	0.035 (M)	0.094 (M)	
Horse Mackere		1077.6 (L)	112.57(L)	3.92 (L)	22.95 (L)	0.93(L)	0.111 (L)	0.437 (L)	23.55 (L)	0.058 (L)	0.15(L)	
		37.07 (M)	20.78 (M)	0.51 (M)	2.13 (M)	0.037 (M)	0.125 (M)	0.148 (M)	0.077 (M)	0.029 (M)	0.22(M)	

dw dry weight, *ww* wet weight

was no clear seasonal pattern in metal contents in fish muscles. However, we observed, a significant bioaccumulation behavior between species and organs. A high level of cadmium was recorded in liver of the different species which can be attributed to anthropogenic sources (phosphate industry) and to natural sources (upwelling activities). The toxic elements (Cd, Pb and Hg) recorded in edible muscle of the four pelagic fish species, were below the limit value established by the European Commission Regulation (European Commission (EC) 2008, 2014, 2015) and show that consumption of fish from the south Moroccan Atlantic coast cannot have a risk for health of consumers. However, the coastal zone and its components represent an expanding dynamic environment, which continues to be a key and a promising sector for the development of Morocco, and therefore, constant monitoring of the marine ecosystems in the southern coast sea is required by the concerned agency due to the increase of anthropogenic activities for safe supply of fish.

Acknowledgements This work was supported financially by ANR EPURE project (Trace-metal Element climatic Perturbations, Upwelling and Resources). We would like to express our gratitude to the project coordinators and the crew that participated in the survey on aboard the R/V Antea and to the engineers of the Pole Spectrometry Ocean Brest (PSO, IUEM, Brest, France) for their help running ICP-MS measurements. A special thanks go to Ms Zineb Zine El Abidine for editing English language of the whole text.

References

Afandi I, Chfiri R, Benhra A, Benbrahim S, Bekkali M, Bouthir FZ (2015) Heavy metal in four fish species from the Atlantic coast of Morocco. Int J Adv Res 3:368–376

Andres S, Ribeyre F, Tourencq JN, Boudou A (2000) Interspecific comparison of cadmium and zinc contamination in the organs of four fish species along a polymetallic pollution gradient (Lot River, France). Sci Total Environ 248:11–25

Annual Report, Office Chérifien des Phosphates, Maroc (2012). http://www.ocpgroup.ma/finance/rapports-publies/rapport-annuel

AOAC (2000) Official methods of analysis of AOAC international. Animal feed.Chap.4 p.5. http://webpages.icav.up.pt/PTDC/CVTNUT/4294/2012/AOAC%202000.pdf

Auger PA, Machu E, Gorgues T, Grima N, Waeles M (2015) Comparative study of potential transfer of natural and anthropogenic cadmium to plankton communities in the North-West African upwelling. Sci Total Environ 505:870–888

Barron MG (2003) Bioaccumulation and bioconcentration in aquatic organisms. In: Hoffman DJ, Rattner BA, Burton GA Jr, Cairns J Jr (eds) Handbook of ecotoxicology, 2nd edn. Lewis Publishers, Boca Raton, pp 877–892

Benbrahim S, Chafik A, Chfiri R, Bouthir FZ, Siefeddine M, Makaoui A (2006) Etude des facteurs influençant la répartition géographique et temporelle de la contamination des côtes atlantiques marocaines par les métaux lourds: cas du mercure, du plomb et du cadmium. Mar Life 16:37–47

Bilandžić N, Đokić M, Sedak M (2011) Metal content determination in four fish species from the Adriatic Sea. Food Chem 124:1005–1010

Biswas AK (2008) Current directions: integrated water resources development a second look. Water Int 33:274–278

Blakas IT, Tugrel S, Salhogln I (1982) Trace metal levels in fish and crustaceans from Northeastern Mediterranean coastal waters. Mar Environ Res 6:281–289

Bodin N, N'Gom-Kâ R, Kâ S, Thiaw O, De Morais LT, Le Loc'h F, Chiffoleau J-F (2013) Assessment of trace metal contamination in mangrove ecosystems from Senegal, West Africa. Chemosphere 90:150–157

Boëly T, Fréon P (1979) Les ressources pélagiques côtières, In : Les ressources halieutiques de l'Atlantique centre-est. I. Les ressources du Golfe de Guinée de l'Angola à la Mauritanie, FAO Document Technique sur les Pêches 186 (1):13-78

Boudou A, Ribeyre F (1997) Mercury in the food web: accumulation and transfer mechanisms. Met Ions Biol Syst 34:289

Bruland KW, Franks RP (1983) Mn, Ni, Cu, Zn and Cd in the western North Atlantic. Trace metals in sea water. Springer, New York, pp 395–414

Burger J, Gaines KF, Boring CS, Stephens WL, Snodgrass J, Dixon C, McMahon M, Shukla S, Shukla T, Gochfeld M (2002) Metal levels in fish from the Savannah River potential hazards to fish and other receptors. Environ Res 89:85–97

Canli M, Atli G (2003) The relationships between heavy metal (Cd, Cr, Cu, Fe, Pb, Zn) levels and the size of six Mediterranean fish species. Environ Pollut 121:129–136

Carr ME (2002) Estimation of potential productivity in Eastern Boundary Currents using remote sensing. Deep Sea Res Part II 49:59–80

Carr ME, Kearns EJ (2003) Production regimes in four Eastern Boundary Current systems. Deep Sea Res Part II 50:3199–3221

Censi P, Spoto SE, Saiano F, Sprovieri M, Mazzola S, Nardone G, Di Geronimo SI, Punturo R, Ottonello D (2006) Heavy metals in coastal water systems. A case study from the northwestern Gulf of Thailand. Chemosphere 64:1167–1176

Chahid A, Hilali M, Benlhachimi A, Bouzid T (2013) Contents of cadmium, mercury and lead in fish from the Atlantic sea (Morocco) determined by atomic absorption spectrometry. Food Chem 147:357–360

Chahid A, Hilali M, Benlhachemi A, Kadmiri IM, Bouzid T (2014) Concentrations of heavy metals in muscle, liver and gill of *Sardinapilchardus* (Walbaum, 1792): risk assessment for the consumers. J Environ Occup Sci 3:47–52

European Commission Regulation (E.C.) (2008) No 629/2008 of 2 July 2008 amending Regulation (EC) No 1881/2006 setting maximum levels for certain contaminants in foodstuffs. Official Journal of the European Union. 5.L.173/6

European Commission Regulation (E.C.) (2014) No 488/2014 of 12 May 2014 amending Regulation (EC) no 1881/2006 as regards the maximum levels for cadmium in foodstuffs. Official Journal of the European Union. 5. L 138/75

European Commission Regulation (E.C.) (2015) No 2015/1005 of 25 June 2015 amending Regulation (EC) no 1881/2006 as regards the maximum levels for lead in foodstuffs. Official Journal of the European Union. 5. L 161/9

COPACE RI (1984) Les pêches dans l'Atlantique Centre-Est Recherches soviétiques sur les chinchards et les maquereaux de la zone nord du COPACE

Copat C, Arena G, Fiore M, Ledda C, Fallico R, Sciacca S, Ferrante M (2013) Heavy metals concentrations in fish and shellfish from eastern Mediterranean Sea: consumption advisories. Food Chem Toxicol 53:33–37

Couillard Y, Cattaneo A, Gallon C, Courcelles M (2008) Sources and chronology of fifteen elements in the sediments of lakes affected by metal deposition in a mining area. J Paleolimnol 40:97–114

Cresson P, Travers-Trolet M, Rouquette M, Timmerman C, Giraldo C, Lefebvre S, Ernande B (2017) Underestimation of chemical contamination in marine fish muscle tissue can be reduced by considering variable wet: dry weight ratios. Mar Pollut Bull 123(1–2):279–285

Diop M, Howsam M, Diop C, Cazier F, Goossens JF, Diouf A, Amara R (2016) Spatial and seasonal variations of trace elements concentrations in liver and muscle of round Sardinelle (*Sardinellaaurita*) and Senegalese sole (*Soleasenegalensis*) along the Senegalese coast. Chemosphere 144:758–766

El Morhit M, Fekhaoui M, Elie P, Girard P, Yahyaoui A, El abidi A, Jbilou M (2009) Heavy metals in sediment, water and the European glass eel, *Anguilla anguilla* (Osteichthyes: Anguillidae) from Loukkos river estuary (Morocco, eastern Atlantic). Cybium 33:219–228

El Morhit M, Belghity D, El Morhit A (2013) Metallic contamination in muscle of three fish species in the southern Atlantic coast the Laâyoune (morocco). Larhyss J 15:21–34

Ergül HA, Aksan S (2013) Evaluation of non-essential element and micronutrient concentrations in seafood from the Marmara and Black Seas. J Black Sea Mediterr Environ 19(3):312–331

Ersoy B, Çelik M (2009) Essential elements and contaminants in tissues of commercial pelagic fish from the Eastern Mediterranean Sea. J Sci Food Agric 89:1615–1621

Falcó G, Llobet JM, Bocio A, Domingo JL (2006) Daily intake of arsenic, cadmium, mercury, and lead by consumption of edible marine species. J Agric Food Chem 54:6106–6112

FAO (1983) Rapport du deuxième groupe de travail *ad hoc* sur les chinchards et les maquereaux de la zone Nord du CO PACE. Nouadhibou, COPACE/PACE/SERIES/83/27

FAO (2001) Rapport du groupe de travail de la FAO surl'évaluation des petits pélagiques au large de l'Afrique NordOccidentale. FAO 657:1–133

FAO (2012) Science and management of small Pelagics. FAO Fish Aquac Proc 18:1–606

Fréon P, Stéquert B (1979) Note sur la présence de *Sardina pilchardus* (Walb.) au Sénégal: étude de la biométrie et interprétation. Cybium 6:65–90

Gaudry A, Zeroual S, Gaie-Levrel F, Moskura M, Boujrhal F-Z, El Moursli RC, Guessous A, Mouradi A, Givernaud T, Delmas R (2007) Heavy metals pollution of the atlantic marine environment by the moroccan phosphate industry, as observed through their bioaccumulation in *Ulva lactuca*. Water Air Soil Pollut 178:267–285

Gundogdu A, Culha ST, Kocbas F, Culha M (2016) Heavy metal accumulation in muscles and total bodies of *Mullus barbatus, Trachurus trachurus* and *Engraulis encrasicolus* captured from the coast of Sinop, Black Sea. Pak J Zool 48:25–34

Henry F, Amara R, Courcot L, Lacouture D, Bertho ML (2004) Heavy metals in four fish species from the French coast of the Eastern English Channel and Southern Bight of the North Sea. Environ Int 30:675–683

INRH (2013) Etat des Stocks et des Pêcheries Marocaines. Rapport technique 1p

JariĆ I, VisnjiĆ-JeftiĆ Ź, CvijanoviĆ G, GaĆiĆ Z, JovanoviĆ L, SkoriĆ S, Lenhardt M (2011) Determination of differential heavy metal and trace element accumulation in liver, gills, intestine and muscle of sterlet (*Acipenserruthenus*) from the Danube River in Serbia by ICP-OES. Microchem J 98:77–81

Kargın F, Dönmez A, Çoğun HY (2001) Distribution of heavy metals in different tissues of the shrimp *Penaeussemiculatus* and *Metapenaeusmonocerus* from the Iskenderun Gulf, Turkey: seasonal variations. Bull Environ Contam Toxicol 66:102–109

Kifani S, Masski H, Faraj A (2008) The need of an ecosystem approach to fisheries: the Moroccan upwelling-related resources case. Fish Res 94:36–42

Kojadinovic J, Potier M, Corre ML, Cosson PR, Bustamante P (2007) Bioaccumulation of trace elements in pelagic fish from the Western Indian Ocean. Environ Pollut 146:548–566

Kucuksezgin F, Altay O, Uluturhan E, Kontas A (2001) Trace metal and organochlorine residue levels in red mullet (Mullusbarbatus) from the Eastern Aegean, Turkey. Water Res 35:2327–2332

Lewis MA, Scott GI, Bearden DW, Quarles RL, Moore J, Strozier ED, Sivertsen SK, Dias AR, Sanders M (2002) Fish tissue quality in near-coastal areas of the Gulf of Mexico receiving point source discharges. Sci Total Environ 284:249–261

Li PF, Zhang J, Xie HJ, Liu C, Liang S, Ren YG, Wang WX (2015) Heavy metal bioaccumulation and health hazard assessment for three fish species from Nansi Lake, China. Bull Environ Contam Toxicol 94:431–436

Maanan M (2008) Heavy metal concentrations in marine molluscs from the Moroccan coastal region. Environ Pollut 153:176–183

Makaoui A, Orbi A, Hilmi K, Zizah S, Larissi J, Talbi M (2005) L'upwelling de la côte atlantique du Maroc entre 1994 et 1998. CR Geosci 337:1518–1524

Mendil D, Demirci Z, Tuzen M, Soylak M (2010) Seasonal investigation of trace element contents in commercially valuable fish species from the Black sea, Turkey. Food Chem Toxicol 48:865–870

Parry ML, Canziani OF, Palutikof JP, van der Linden PJ, Hanson CE (2007) Climate Change 2007: impacts, adaptation and vulnerability. Contribution of Working Group II to the Fourth Assessment Report of the Intergovernmental Panel on Climate Change. Cambridge University Press, Eds., Cambridge, UK

Pauly D, Christensen V (1995) Primary production required to sustain globalfisheries. Nature 374:255

Ploetz M, Fitts BE, Rice TM (2007) Differential accumulation of heavy metals in muscle and liver of a marine fish, (King Mackerel, Scomberomorus cavalla Cuvier) from the Northern Gulf of Mexico, USA. Bull Environ Contam Toxicol 78:124–127

Prudente M, Kim EY, Tanabe S, Tatsukawa R (1997) Metal levels in some commercial fish species from Manila Bay, the Philippines. Mar Pollut Bull 34:671–674

Roesijadi G (1996) Metallothionein and its role in toxic metal regulation. Comp Biochem Physiol Part C 113:117–123

o M, Gnassia-Barelli M (1988) *Donaxtrunculum* and *Venus verrucosa* as bioindicators of trace metal concentrations in Mauritanian coastal waters. Mar Biol 99:223–227

Romeo M, Siau Y, Sidoumou Z, Gnassia-Barelli M (1999) Heavy metal distribution in different fish species from the Mauritania coast. Sci Total Environ 232:169–175

Sidoumou Z, Gnassia-Barelli M, Siau Y, Morton V, Romeo M (2005) Distribution and concentration of trace metals in tissues of different fish species from the Atlantic Coast of Western Africa. Bull Environ Contam Toxicol 74:988–995

Storelli MM (2008) Potential human health risks from metals (Hg, Cd, and Pb) and polychlorinated biphenyls (PCBs) via seafood Roméonsumption: estimation of target hazard quotients (THQs) and toxic equivalents (TEQs). Food Chem Toxicol 46:2782–2788

Suhaimi F, Wong SP, Lee VLL, Low LK (2005) Heavy metals in fish and shellfish found in local wet markets. Singap J Prim Ind

Tariq J, Jaffar M, Moazzam M (1991) Concentration correlations between major cations and heavy metals in fish from the Arabian Sea. Mar Pollut Bull 22:562–565

Topping G (1973) Heavy metals in fish from Scottish waters. Aquaculture 1:373–377

UNEP (1999) Overview of land-based sources and activities affecting the marine, coastal and associated freshwater environment in the West and Central African Region. UNEP Regional Seas Reports and Studies 171

UNEP (2006) Marine and coastal ecosystems and human wellbeing: a synthesis report based on the findings of the Millennium Ecosystem Assessment. UNEP 76

Uysal K, Emre Y, Köse E (2008) The determination of heavy metal accumulation ratios in muscle, skin and gills of some migratory fish species by inductively coupled plasma-optical emission spectrometry (ICP-OES) in Beymelek Lagoon (Antalya/Turkey). Microchem J 90:67–70

Vieira C, Morais S, Ramos S, Delerue-Matos C, Oliveira MBPP (2011) Mercury, cadmium, lead and arsenic levels in three pelagic fish species from the Atlantic Ocean: intra-and inter-specific variability and human health risks for consumption. Food Chem Toxicol 49:923–932

Vu CT, Lin C, Yeh G, Villanueva MC (2017) Bioaccumulation and potential sources of heavy metal contamination in fish species in Taiwan: assessment and possible human health implications. Environ Sci Pollut Res 24:19422–19434

Zuraini A, Somchit MN, Solihah MH, Goh YM, Arifah AK, Zakaria MS, Somchit N, Rajion MA, Zakaria ZA, Mat Jais AM (2006) Fatty acid and amino acid composition of three local Malaysian *Channa* spp. fish. Food Chem 97:674–678

2

Determination of amino acid and fatty acid composition of goldband goatfish [*Upeneus moluccensis* (Bleeker, 1855)] fishing from the Gulf of Antalya (Turkey)

Güntekin Doğan · Ömer Osman Ertan

Abstract In this study, we aimed to determine the basic food components, fatty acids and amino acids, and variations in these components with months in goldband goatfish (*Upeneus moluccensis*) that fishing from Gulf of Antalya. As a result of the analyzes, the crude fat values were determined between 1.43 and 3.78%, and the crude protein values were determined between 20.79 and 22.16%. The most abundant fatty acids were determined: palmitic acid (C16:0), stearic acid (C18:0), palmitoleic acid (C16:1c9), oleic acid (C18:1c9), linoleic acid (C18:2n-6), eicosatrienoic acid (C20:3n-3), arachidonic acid (C20:4n-6), eicosapentaenoic acid (C20:5n-3), docosapentaenoic acid (C22:5n-6), and docosahexaenoic acid (C22:6n-3). The most abundant amino acids were determined lysine and leucine, aspartic acid, glutamic acid, alanine, and glycine. The differentiations of essential nutrient components, fatty acids, and amino acids were found generally significant ($P < 0.05$).

Keywords Goldband goatfish · *Upeneus moluccensis* · Fatty acids · Amino acids · Food components · Catching season

Introduction

Seafood is a highly valuable food source because of rich in protein, amino acid, unsaturated fatty acid (especially omega-3), and vitamin components which are necessary for healthy and balanced nutrition (Gulyavuz and Unlusayın 1999; Simsek et al. 2009). The taste of fish meat is closely related to the protein and fat content, and also the seasonal variations of these components are important determinant of both consumer choice and quality of the processed product. Depending on the season, water temperature and nutrients in the environment affect the biochemical composition of fish meat (Kuzu 2005). In many studies, the relationship between seasonal changes and the chemical composition of fish meat has been shown (Ersoy 2006; Kandemir and Polat 2007; Polat et al. 2009; Ozogul et al. 2011). In a study, it was emphasized that the chemical composition of different fish species depends on variables such as seasonal change, migration, sexual maturity, and nutrient cyclicity (Kuzu 2005).

G. Doğan (✉)
Suleyman Demirel University, Isparta, Turkey
e-mail: guntekin32@hotmail.com

O. O. Ertan
Department of Fisheries Basic Sciences, Egirdir Fisheries Faculty, Suleyman Demirel University, Isparta, Turkey

Scientific studies in various fields reveal that nutrients and nutritional habits play an important role in some diseases that people are confronted with, and the necessity of more conscious nutrition is needed. A number of studies have been undertaken to elucidate the positive effect of fish consumption on human health and to investigate the therapeutic properties of eicosapentaenoic acid (EPA) and docosahexaenoic acid (DHA) found in seafood (Broughton et al. 1997; Haris 1997; Conquer et al. 2000; Norrish et al. 2000; Ozkan and Koca 2006; Ebbesson et al. 2008). EPA and DHA must be taken as essential fatty acids from the outside for a healthy life (Canbulat and Ozcan 2008).

Amino acids are determinants of quality in fish and crustaceans (Ruiz-Capillas and Moral 2001). Most amino acids such as glutamic acid, aspartic acid, alanine, and glycine are important, because they are responsible for the taste and flavor (Ruiz-Capillas and Moral 2004). Sea food is important source of protein because of the high content of aspartic acid, glutamic acid, lysine, arginine, and leucine (Rosa and Nunes 2003; Erkan and Ozden 2007). In fish, 50–80% of the non-protein nitrogenous compounds are amino acids and significant amounts of these are proline, arginine, lysine, alanine, histidine, glutamic acid, and taurine (Ruiz-Capillas and Moral 2001; Ozden 2005). The most consumed and delicious portions of fish are muscle tissue and they contain plenty of flavor amino acids. Amino acids provide tissue healing and growth (Oluwaniyi et al. 2010).

Although some studies have been published (Celik et al. 1999; Ersoy 2006; Simsek et al. 2009; Oksuz et al. 2011) about the fatty acid and the basic food components of the goldband goatfish (*Upeneus moluccensis*), but no information was found about the amino acid composition and seasonal study about the fatty acid and the basic food components of the goldband goatfish. In this study, the seasonal variation of the basic food components, amino acids, and fatty acids in goldband goatfish which were caught during the fishing season in Gulf of Antalya and have economic importance was investigated.

Materials and methods

Providing and preservation of samples

Sampling were conducted monthly during the catching season (September, October, November, December, January, February, March, and April). The goldband goatfish (*Upeneus moluccensis*) samples were purchased sufficient quantity at the time of reaching on port as the freshest state and brought to the Food Processing Laboratory of Egirdir Fisheries Faculty under cold chain conditions in 2 h. First, total length and weight measurements were determined and viscera were removed. The samples to be used in the determination of nutrient components were packed in the amount required for each analysis and stored at -80 ± 1 °C. 32 samples were used for each month. Female and male fish sample numbers were not considered.

Determination of the basic food components

Moisture analyzes of the samples were determined triplicate ($n = 3$) by an automatic moisture analyzer (AND MX-50 Moisture Analyzer, Japan). Crude ash analyzes were made triplicate ($n = 3$) according to AOAC (2002a), 920.153 Method. Crude protein analyzes were made duplicate ($n = 2$) according to AOAC (2010), 960.52 Micro Kjeldahl Method in the Turkey Scientific and Technical Research Institute Marmara Research Center (TUBITAK MAM).

Crude fat analyzes were made triplicate ($n = 3$) according to Lovell (1981). The glass containers to be used in the analyzes were first fixed weight. 5 g of sample placed in the tube were homogenized in a homogenizer (WiseTis HG-15D, Korea) with chloroform (Riedel–de–Haen, 24216), methanol (Riedel–de–Haen, 24229), and distilled water. The tubes were centrifuged (Sigma 2–16 K, Germany) at 3000 rpm for 10 min. At the end of the centrifugation, the methanol and water phase at the top, the fish meat phase in the middle, and the chloroform phase containing fat at the bottom were formed in the tubes. 1 mL was pipetted from the bottom chloroform phase and placed in fixed weight glass containers and evaporated in a hot water bath (WiseBath, Korea) for 1 h at 80° C. At last cooled to room temperature in the desiccator and weighed on a precision scale and the amount of crude fat was determined as %.

Determination of fatty acids

Fatty acid analyzes were made duplicate ($n = 2$) in Suleyman Demirel University (SDU) Experimental and Observational Student Research and Application Center. Fat extractions were made according to Bligh and Dyer (1959), methyl ester analyzes of fatty acids were made according to AOAC (2002b), 996.06 Method using gas chromatography (GC, Perkin Elmer Auto System XL, USA). After identifying 37 standard locations, Supelco 18919 F.A.M.E. Mix C4-C24, was injected into the GC. The fatty acids of the samples were determined in the reference assay conditions of AOAC 996.06.

Determination of amino acids

The amino acid analyzes of the samples were made duplicate ($n = 2$) in TUBITAK MAM according to Dimova (2003), and Gheshlaghi et al. (2008), using the high-performance liquid chromatography (HPLC) method. This method is based on the reading of ultra fast liquid chromatography–ultraviolet (UFLC-UV) detector by derivatization with phenyl isothiocyanate and acetonitrile:methanol:triethylamine solution after acidic hydrolysis applied to disassociate the constituent proteins into amino acid components. Tryptophan is completely disappeared as a result of acid hydrolysis; for that reason, tryptophan analyzes were made by base hydrolysis method. The sulfur-containing amino acids immediately expose to degradation when hydrolyzed with a strong acid solution, so sulfur-containing amino acids did not determine. Totally, 17 amino acids (aspartic acid, glutamic acid, serine, glycine, arginine, histidine, threonine, lysine, alanine, proline, leucine, isoleucine, tyrosine, phenylalanine, valine, methionine, and tryptophan) were determined as mg/100 g.

Statistical analyzes

The data, obtained in the study, were subjected to analysis of variance (F test) using the SPSS 16.0 program. Averages of significant variance sources were compared using the Duncan multiple comparison test as significance level $P = 0.05$.

Results and discussion

Average length and weight values

The monthly average of total length values was determined between 13.15 ± 0.12 cm and 14.89 ± 0.20 cm, and the monthly average weight values were determined between 24.98 ± 1.31 cm and 33.67 ± 1.34 g (Table 1). We intended that purchased fish should be about the same length and weight each month when sampling is being done. However, it was observed that the average length and weight values of fish purchased in September, February, and April were higher than the other months, depending on the fishing month, fishing area, feeding regime, and spawning season.

Proximate analyzes

Moisture

In our study, the moisture values of *U. moluccensis* samples were found between 78.74 ± 0.32 and 80.30 ± 0.45%, and there was no significant ($P > 0.05$) difference detected in monthly moisture change rates (Table 2).

The moisture content of seafood varies greatly with species, sex, and age. It is generally the moisture contents found between 70 and 85% in fish. This ratio decreases to 50–60% in some fishes (Gulyavuz and Unlusayın 1999). In the previous studies, the moisture contents of *Mullus barbatus* were determined 59.75% by Guner et al. (1998), 79.71% by Kalogeropoulos et al. (2004), and 79.00% by Gumus et al. (2009). Erkan et al. (2010b) determined the humidity value 70.25% in *Mullus surmuletus.* Oksuz et al. (2011), determined the humidity values 79.41% in *Upeneus moluccensis* and 73.14% in *M. surmuletus.* Ersoy (2006), found the

Table 1 Monthly average lengths and weights of *U. moluccensis*

	Sep.	Oct.	Nov.	Dec.	Jan.	Feb.	Mar.	Apr.	*P* values ($P < 0.05$)
Length (cm)	14.61 ± 0.20^{a}	13.15 ± 0.12^{c}	13.86 ± 0.19^{b}	$13.59 \pm 0.22b^{c}$	$13.39 \pm 0.11b^{c}$	14.89 ± 0.20^{a}	13.65 ± 0.17^{bc}	14.53 ± 0.14^{a}	0.000
Weight (g)	33.56 ± 1.12^{a}	25.19 ± 0.67^{c}	28.48 ± 1.15^{b}	24.98 ± 1.31^{c}	25.74 ± 0.59^{bc}	33.67 ± 1.34^{a}	26.79 ± 1.03^{bc}	33.31 ± 0.94^{a}	0.000

Different lowercase letters in the same line indicate significant differences ($P < 0.05$)

Table 2 Montly changes in the basic nutritional components of *U. moluccensis* (%)

	Sep.	Oct.	Nov.	Dec.	Jan.	Feb.	Mar.	Apr.	*P* values ($P < 0.05$)
Moisture	79.22 ± 0.52^{a}	78.97 ± 1.26^{a}	78.74 ± 0.32^{a}	78.89 ± 1.24^{a}	79.12 ± 0.50^{a}	79.75 ± 0.24^{a}	80.30 ± 0.45^{a}	79.90 ± 0.21^{a}	0.744
Crude ash	1.93 ± 0.09^{a}	1.82 ± 0.10^{ab}	1.86 ± 0.05^{ab}	1.61 ± 0.11^{b}	1.67 ± 0.07^{ab}	1.68 ± 0.08^{ab}	1.59 ± 0.11^{b}	1.65 ± 0.09^{b}	0.132
Crude fat	1.45 ± 0.01^{d}	3.08 ± 0.16^{b}	2.00 ± 0.06^{c}	3.54 ± 0.04^{a}	2.10 ± 0.12^{c}	2.12 ± 0.12^{c}	2.08 ± 0.09^{c}	3.78 ± 0.09^{a}	0.000
Crude protein	21.35 ± 0.04^{c}	21.50 ± 0.00^{bc}	21.22 ± 0.28^{cd}	21.60 ± 0.16^{bc}	22.16 ± 0.03^{a}	21.88 ± 0.07^{ab}	20.79 ± 0.10^{d}	20.82 ± 0.13^{d}	0.001

Different lowercase letters in the same line indicate significant differences ($P < 0.05$)

content of moisture in *U. moluccensis* between 75.79 and 78.27% in September, December, March, and May, and he indicated that these changes originated from the fish diet. Polat et al. (2009), found the changes in moisture content of *M. barbatus* in Autumn, Winter, and Spring seasons 74.13–75.21 and 73.84%, respectively, and they indicated that this changes originated from the feeding regime. Ozogul et al. (2011) investigated the changes in moisture content of *Upeneus pori* and *M. barbatus* seasonally. They determined the moisture contents between 75.31 and 79.08% in *U. pori* and between 76.23 and 76.92% in *M. barbatus*.

The moisture values were compared with the previous studies, it was observed that similar results were obtained in some studies (Kalogeropoulos et al. 2004; Gumus et al. 2009; Oksuz et al. 2011), and we found higher values than some studies (Guner et al. 1998; Erkan et al. 2010b; Polat et al. 2009; Tulgar and Berik 2012). We considered that the moisture value differentiations depend on species, season, nutritional regime, size, spawning season, sex, and catching area.

Crude ash

In our study, the crude ash was found between 1.59 ± 0.11 and 1.93 ± 0.09%. The highest crude ash was determined in September, and the lowest crude ash was determined in March (Table 2).

Gulyavuz and Unlusayın (1999) stated that the amount of inorganic matter in fish meat is about 1–2%. Aquatic organisms take inorganic materials from food and water, and store them in skeletal tissue and other members. There are many factors affecting the amount of inorganic matter such as season, biological difference (species, length, age, sex, and sexual maturity), nutrition, and environmental conditions (water chemistry, temperature, and salinity) (Cakli 2007). In the previous studies, the crude ash contents of *M. barbatus* were found 0.90% by Guner et al. (1998) and 1.45% by Gumus et al. (2009). Erkan et al. (2010b) determined the crude ash content 1.10% in *M. surmuletus*. Oksuz et al. (2011) found the crude ash content 1.10% in *U. moluccensis* and 1.60% in *M. surmuletus* and they were indicated that the difference is due to the mineral content of the species. Ersoy (2006) found the crude ash contents in the *U. moluccensis*, caught in September, December, March, and May, 1.29, 1.35, 1.24, and 1.38%, respectively. Polat et al. (2009) found that the crude ash contents of *M. barbatus* were 1.11% in Autumn, 1.09% in Winter, and 1.24% in Spring and they determined that seasonal variations were significant ($P < 0.05$). Ozogul et al. (2011) found the crude ash contents of *U. pori* between 1.42 and 1.65% and the crude ash contents of *M. barbatus* between 1.23 and 1.66%.

When we compared our values with the previous studies, we found that crude ash values are higher than the previous studies due to species, nutritional regime, and hunting area. In some studies (Polat et al. 2009; Oksuz et al. 2011; Tulgar and Berik 2012), seasonal variations were found generally significant ($P < 0.05$). We also found monthly variations generally significant ($P < 0.05$) and we thought that the monthly differences in our study depend on the spawning season, size, nutritional status, and catching area.

Crude fat

In our study, it was determined that the crude fat contents changed between 1.45 ± 0.01 and 3.78 ± 0.09%. The highest crude fat content was determined in April, before the spawning season, and the lowest crude fat was determined in September, after spawning season. The crude fat values were detected significantly ($P < 0.05$) higher in April and December and significantly ($P < 0.05$) lower in September than the other months (Table 2). We thought that the distinction in crude fat contents depends mainly on the spawning season, age, sex, and nutritional density.

In fish, the amount of fat may be below 1% and may exceed 30% according to the species and biological status and the fish are grouped according to the fat content (Cakli 2007). In addition, the fat content changes depending on the age, sex, catching season, nutritional status, and living environment (Gulyavuz and Unlusayın 1999).

In the previous studies, the crude fat contents of *M. barbatus* were determined 9.72% by Guner et al. (1998), 3.52% by Kalogeropoulos et al. (2004), and 1.75% by Gumus et al. (2009), and they considered them as low-fat fish. Celik et al. (1999) found that crude fat content was 6.67% in *M. barbatus*, 2.67% in *M. surmuletus,* and 4.91% in *U. moluccensis*. Erkan et al. (2010b) determined the crude fat content 14.46% in *M. surmuletus*. Oksuz et al. (2011) found that crude fat content was 4.35% in *U. moluccensis* and 10.38% in *M.*

surmuletus in their study. They noted that although the species belong to the same family and the same habitat, they may have distinct fat content. Ersoy (2006) investigated the nutrient composition of *U. Moluccensis* caught in September, December, March, and May and she determined the crude fat contents as 3.26, 0.64, 2.20, and 0.51%, respectively. Polat et al. (2009), in their study, determined the crude fat content of *M. barbatus* 3.68% in Spring, 5.33% in Winter, and 5.76% in Autumn. They thought that the fall in crude fat content during the spring season was to be due to the supply of raw energy needed for egg development during gonadogenesis formation, which reached the highest point just before the spawning. Ozogul et al. (2011) studied seasonally changes in crude fat content of *U. pori* and *M. barbatus* in their study. They determined the crude fat contents of *U. pori* between 1.07 and 2.10% and the crude fat contents of *M. barbatus* between 1.07 and 3.00%. They argued that the changes in fat rates are due to the season of feeding and spawning.

When we compared the fat content with the previous studies, we found similar results in some studies (Celik et al. 1999; Kalogeropoulos et al. 2004; Ersoy 2006; Gumus et al. 2009; Ozogul et al. 2011) and higher results in some studies (Guner et al. 1998; Erkan et al. 2010b; Oksuz et al. 2011). In general, significant ($P < 0.05$) variations were determined in seasonal studies (Polat et al. 2009; Tulgar and Berik 2012). We also found monthly variations generally significant ($P < 0.05$). We determined the *U. moluccensis* in the low-fat fish class according to the Cakli (2007).

Crude protein

In our study, we determined the crude protein content between 20.79 and 22.16%. The highest crude protein was determined in January, and the lowest crude protein was determined in March and April when the spawning period was near. Crude protein was detected in January significantly ($P < 0.05$) higher than other months, while crude protein values were detected in March and April significantly ($P < 0.05$) lower (Table 2). We thought that the differences in crude protein are mainly due to spawning season and nutritional density.

Gulyavuz and Unlusayın (1999) reported that the protein content of fish meat is about 14–20%, and this values changes with age, gender, feeding environment, spawning, and migration season. Cakli (2007) also reported that protein contents of aquatic organisms are about 11–25%. In the previous studies, the crude protein contents of *M. barbatus* were determined 16.3% by Guner et al. (1998), 14.79% by Kalogeropoulos et al. (2004), and 14.84% by Gumus et al. (2009). Celik et al. (1999) found crude protein contents 21.32% in *M. barbatus,* 19.56% in *M. surmuletus,* and 20.32% in *U. moluccensis*. Erkan et al. (2010b) determined the crude protein content 14.07% in *M. surmuletus*. Ersoy (2006) investigated the nutrient composition of the *U. moluccensis* fillets in September, December, March, and May and determined crude protein contents as 19.32, 21.00, 18.75, and 19.36%, respectively. She stated that the protein contents changed depending on the catching season and the feeding density. Polat et al. (2009) found that the crude protein content of *M. barbatus* was 20.43% in Spring, 17.90% in Winter, and 18.97% in Autumn. The amount of crude protein in the spring, before spawning period, was significantly higher than other seasons ($P < 0.05$). Ozogul et al. (2011) found that the crude protein contents of *U. pori* were between 17.68 and 21.27% and the crude protein contents of *M. barbatus* were between 18.25 and 20.52% in their studies. They found the lowest protein contents in both species in Spring.

When we compared the our protein values with the previous studies, similar results were appeared in some studies (Celik et al. 1999; Ersoy 2006) and lower protein values were appeared in some studies (Guner et al. 1998; Kalogeropoulos et al. 2004; Gumus et al. 2009; Tulgar and Berik 2012). In seasonal studies (Ersoy 2006; Polat et al. 2009; Ozogul et al. 2011; Tulgar and Berik 2012), generally significantly ($P < 0.05$) different values were determined as our monthly study. When we examined the crude protein values of the *U. moluccensis* in our study, we concluded that it was an important protein source in all months (although the protein contents decreased in March and April when the spawning season approached).

Fatty acids

In our study, it was observed that the monthly variations of fatty acid values of *U. moluccensis* samples are generally significant ($P < 0.05$). We found that the most abundant fatty acids were palmitic acid and stearic acid as saturated fatty acids, palmitoleic acid and oleic acid as monounsaturated fatty acids, eicosatrienoic acid, arachidonic acid, EPA, and DHA as polyunsaturated fatty acids (Table 3).

Table 3 Montly changes in fatty acid contets of *U. moluccensis* (% total fatty acids)

Fatty acids	Sep.	Oct.	Nov.	Dec.	Jan.	Feb.	Mar.	Apr.	P values ($P < 0.05$)
C14:0	2.43 ± 0.11^{b}	2.08 ± 0.16^{bc}	2.88 ± 0.02^{a}	2.87 ± 0.15^{a}	2.06 ± 0.12^{bc}	2.16 ± 0.08^{bc}	2.18 ± 0.14^{bc}	1.81 ± 0.10^{c}	0.001
C15:0	0.89 ± 0.02^{ab}	0.73 ± 0.04^{c}	1.00 ± 0.01^{a}	1.00 ± 0.04^{a}	0.99 ± 0.05^{a}	0.88 ± 0.01^{b}	0.91 ± 0.05^{ab}	0.90 ± 0.02^{ab}	0.004
C16:0	20.46 ± 0.32^{c}	24.33 ± 0.67^{a}	24.10 ± 0.05^{a}	24.86 ± 0.55^{a}	21.12 ± 0.59^{bc}	21.28 ± 0.08^{bc}	20.58 ± 0.028^{c}	22.07 ± 0.09^{b}	0.000
C17:0	1.96 ± 0.01^{c}	1.70 ± 0.02^{d}	2.10 ± 0.05^{b}	2.30 ± 0.02^{a}	2.04 ± 0.03^{bc}	2.03 ± 0.02^{bc}	2.37 ± 0.05^{a}	2.00 ± 0.02^{bc}	0.000
C18:0	8.89 ± 0.04^{b}	7.09 ± 0.07^{e}	7.40 ± 0.10^{d}	8.29 ± 0.03^{c}	8.53 ± 0.02^{c}	9.16 ± 0.18^{a}	8.82 ± 0.02^{b}	7.38 ± 0.03^{d}	0.000
C21:0	0.51 ± 0.02^{c}	0.48 ± 0.01^{cd}	0.44 ± 0.01^{de}	0.40 ± 0.02^{e}	0.34 ± 0.01^{f}	0.60 ± 0.02^{b}	0.65 ± 0.01^{a}	0.65 ± 0.03^{a}	0.000
∑SFA	35.14 ± 0.40^{a}	36.41 ± 0.80^{bc}	37.92 ± 0.09^{b}	39.72 ± 0.74^{a}	35.08 ± 0.77^{a}	36.11 ± 0.22^{a}	35.51 ± 0.07^{a}	34.81 ± 0.13^{a}	0.001
C14:1	0.16 ± 0.01^{de}	0.23 ± 0.01^{a}	0.18 ± 0.01^{cd}	0.20 ± 0.01^{bc}	0.22 ± 0.2^{ab}	0.15 ± 0.01^{ef}	0.12 ± 0.00^{f}	0.19 ± 0.01^{bcd}	0.000
C16:1	3.38 ± 0.05^{d}	7.33 ± 0.12^{a}	6.70 ± 0.12^{ab}	5.52 ± 0.15^{c}	5.64 ± 0.41^{c}	4.58 ± 0.03^{d}	5.11 ± 0.25^{cd}	6.31 ± 0.19^{b}	0.000
C17:1	0.85 ± 0.11^{b}	0.67 ± 0.03^{c}	1.22 ± 0.03^{a}	1.09 ± 0.01^{a}	0.82 ± 0.02^{bc}	1.11 ± 0.01^{a}	1.10 ± 0.02^{a}	0.77 ± 0.06^{bc}	0.000
C18:ln9c	13.37 ± 0.05^{d}	19.10 ± 0.05^{a}	18.20 ± 0.26^{b}	17.64 ± 0.49^{b}	19.13 ± 0.01^{a}	17.70 ± 0.29^{b}	14.91 ± 0.09^{c}	19.09 ± 0.07^{a}	0.000
C20:1	0.61 ± 0.04^{de}	0.90 ± 0.07^{b}	0.66 ± 0.01^{de}	0.82 ± 0.02^{bc}	1.27 ± 0.01^{a}	0.67 ± 0.03^{de}	0.73 ± 0.05^{cd}	0.61 ± 0.02^{e}	0.000
∑MUFA	18.37 ± 0.07^{f}	28.23 ± 0.14^{a}	26.96 ± 0.41^{b}	25.27 ± 0.34^{c}	27.08 ± 0.43^{b}	24.21 ± 0.27^{d}	21.97 ± 0.23^{e}	26.97 ± 0.06^{b}	0.000
C18:2n6r	0.45 ± 0.01^{d}	1.12 ± 0.10^{a}	0.69 ± 0.01^{c}	0.53 ± 0.05^{d}	0.96 ± 0.01^{b}	0.68 ± 0.01^{c}	0.84 ± 0.01^{b}	0.88 ± 0.03^{b}	0.000
C18:2n6c	1.17 ± 0.02^{b}	1.67 ± 0.01^{a}	1.12 ± 0.02^{c}	1.06 ± 0.00^{d}	1.02 ± 0.00^{d}	1.17 ± 0.02^{b}	1.05 ± 0.01^{d}	1.18 ± 0.03^{b}	0.000
C18:3n6	0.58 ± 0.00^{cd}	0.37 ± 0.03^{d}	0.53 ± 0.01^{cd}	0.64 ± 0.01^{c}	0.96 ± 0.01^{b}	0.72 ± 0.07^{c}	1.15 ± 0.17^{b}	1.39 ± 0.03^{a}	0.000
C18:3n3	0.56 ± 0.06^{c}	0.54 ± 0.02^{c}	0.61 ± 0.01^{bc}	0.74 ± 0.02^{a}	0.66 ± 0.02^{abc}	0.55 ± 0.01^{c}	0.71 ± 0.08^{ab}	0.57 ± 0.01^{c}	0.028
C20:3 n6	0.70 ± 0.01^{a}	0.11 ± 0.01^{d}	0.10 ± 0.01^{d}	0.06 ± 0.01^{d}	0.21 ± 0.03^{c}	0.74 ± 0.06^{a}	0.78 ± 0.01^{a}	0.53 ± 0.02^{b}	0.000
C20:3 n3	4.14 ± 0.07^{b}	2.79 ± 0.05^{ef}	2.85 ± 0.01^{e}	2.68 ± 0.05^{fg}	3.46 ± 0.05^{d}	4.36 ± 0.04^{a}	3.84 ± 0.05^{c}	2.60 ± 0.02^{g}	0.000
C20:4 n6	5.40 ± 0.08^{d}	6.40 ± 0.09^{a}	6.48 ± 0.01^{a}	6.50 ± 0.08^{a}	6.06 ± 0.07^{b}	4.55 ± 0.02^{e}	5.77 ± 0.06^{c}	4.62 ± 0.05^{e}	0.000
C20:5 n3	2.02 ± 0.06^{b}	1.69 ± 0.09^{cd}	1.61 ± 0.04^{de}	1.45 ± 0.02^{e}	1.88 ± 0.10^{bc}	1.78 ± 0.01^{cd}	2.05 ± 0.06^{b}	2.33 ± 0.07^{a}	0.000
C22:5 n6	1.94 ± 0.03^{a}	1.14 ± 0.04^{e}	1.22 ± 0.03^{e}	1.34 ± 0.02^{d}	1.42 ± 0.04^{d}	1.91 ± 0.02^{ab}	1.83 ± 0.05^{b}	1.70 ± 0.02^{c}	0.000
C22:6 n3	26.09 ± 0.47^{a}	15.15 ± 0.37^{e}	17.11 ± 0.46^{cd}	17.39 ± 0.24^{c}	17.07 ± 0.65^{cd}	20.83 ± 0.01^{b}	18.09 ± 0.03^{c}	15.96 ± 0.08^{de}	0.000
∑PUFA	43.05 ± 0.62^{a}	30.98 ± 0.72^{d}	32.32 ± 0.55^{cd}	32.39 ± 0.37^{cd}	33.70 ± 0.97^{c}	37.29 ± 0.18^{b}	36.11 ± 0.02^{b}	31.76 ± 0.08^{d}	0.000
TOTAL	96.53 ± 0.16^{b}	95.50 ± 0.27^{c}	97.15 ± 0.06^{a}	97.32 ± 0.03^{a}	95.82 ± 0.24^{c}	97.54 ± 0.13^{a}	93.53 ± 0.19^{d}	93.48 ± 0.15^{d}	0.000
Unidentified	3.48 ± 0.16^{c}	4.50 ± 0.27^{b}	2.86 ± 0.06^{d}	2.68 ± 0.03^{d}	4.19 ± 0.24^{b}	2.47 ± 0.13^{d}	6.48 ± 0.19^{a}	6.53 ± 0.15^{a}	0.000
∑UNSFA	61.42 ± 0.56^{a}	59.21 ± 0.58^{b}	$59.28 \pm 0,14^{b}$	57.66 ± 0.71^{c}	60.78 ± 0.54^{a}	61.50 ± 0.09^{a}	58.08 ± 0.25^{bc}	58.73 ± 0.02^{bc}	0.001
∑UNSFA/ESFA	1.75 ± 0.04^{a}	1.63 ± 0.05^{b}	1.56 ± 0.01^{bc}	1.45 ± 0.05^{c}	1.73 ± 0.06^{a}	1.70 ± 0.02^{a}	1.64 ± 0.01^{b}	1.69 ± 0.01^{3}	0.004
EPA + DHA	28.11 ± 0.53^{a}	16.84 ± 0.46^{e}	18.72 ± 0.50^{cd}	18.84 ± 0.26^{cd}	18.95 ± 0.75^{cd}	22.61 ± 0.00^{b}	20.14 ± 0.09^{c}	18.29 ± 0.01^{d}	0.000
n3PUFA	32.81 ± 0.53^{a}	20.17 ± 0.52^{f}	22.18 ± 0.52^{de}	22.26 ± 0.32^{de}	23.07 ± 0.82^{d}	27.52 ± 0.04^{b}	24.69 ± 0.05^{c}	21.46 ± 0.05^{ef}	0.000
n6PUFA	10.24 ± 0.09^{d}	10.81 ± 0.21^{b}	10.14 ± 0.03^{de}	10.13 ± 0.05^{de}	10.63 ± 0.16^{bc}	9.77 ± 0.14^{e}	11.42 ± 0.07^{a}	10.30 ± 0.04^{cd}	0.000
n6/n3	0.31 ± 0.00^{e}	0.54 ± 0.01^{a}	0.46 ± 0.01^{bc}	0.46 ± 0.01^{c}	0.46 ± 0.01^{bc}	0.36 ± 0.01^{d}	0.46 ± 0.01^{bc}	0.48 ± 0.00^{b}	0.000

Table 3 continued

Fatty acids	Sep.	Oct.	Nov.	Dec.	Jan.	Feb.	Mar.	Apr.	P values ($P < 0.05$)
∑PUFA/ESFA	1.23 ± 0.04^{a}	$0.85 = 0.04^{d}$	0.85 ± 0.02^{d}	0.82 ± 0.03^{d}	0.96 ± 0.05^{bc}	1.03 ± 0.00^{b}	1.02 ± 0.01^{b}	0.91 ± 0.00^{cd}	0.000

Different lowercase letters in the same line indicate significant differences ($P < 0.05$)

∑*SFA* total saturated fatty acids, ∑*MUFA* total monounsaturated fatty acids, ∑*PUFA* total polyunsaturated fatty acids, ∑*UNSFA* total unsaturated fatty acids

The beneficial effects of fish fat on human health are known. Polyunsaturated fatty acids (especially omega-3) have been recognized as essential components of human food for the prevention of illness and a healthy life. There is a serious deficiency about intake of ω-3 polyunsaturated fatty acid (PUFA) in human beings, for that reason to consume more foods that containing these fatty acids is recommended (Erkan 2013). In the previous studies, Guner et al. (1998) found the major fatty acids in *M. barbatus* were palmitic acid, palmitoleic acid, stearic acid, oleic acid, EPA, docosapentaenoic acid (DPA, C22:5n-6) and DHA. They stated that the amount of crude fat and fatty acids must be taken into account when determining the nutritional value of fish. Kalogeropoulos et al. (2004) found the major fatty acids in *M. barbatus* were palmitic acid, palmitoleic acid, oleic acid, EPA, and DHA. Polat et al. (2009) determined that the predominant fatty acids were palmitic acid, stearic acid, oleic acid, palmitoleic acid, EPA, and DHA in *M. barbatus* in all seasons, and they reported that fatty acid components were varied with season and the amount of DHA + EPA and affects consumer preferences positively. Ozogul et al. (2011) found that the fatty acids of *U. pori* and *M. barbatus* were seasonally changed by a number of ecological and biological effects, and EPA and DHA values were high in all seasons. Oksuz et al. (2011) found that the most abundant fatty acids in *U. moluccensis* and *M. surmuletus* were palmitic acid, stearic acid, oleic acid, palmitoleic acid, EPA, and DHA. They reported the amount of omega-3 more than the amount of omega-6 in both species important for human health. Erkan (2013) studied omega-3 fatty acid distribution of consumed aquatic products in Turkey and found that the most abundant fatty acids in fish fat were palmitic acid, stearic acid, palmitoleic acid, oleic acid, EPA, and DHA.

When we compared with the previous studies, it was seen that similar results were detected the most abundant fatty acids. It was also found that the unsaturated fatty acid values were between 57.66 and 61.50% (Table 3) and these values were determined similar to the previous studies. It has been determined that the fatty acid content of fish fat changes in a number of studies (Gamez-Meza et al. 1999; Luzia et al. 2003; Kuzu 2005; Polat et al. 2009; Ozogul et al. 2011). Seasonal conditions, age, size, catching area, and spawning period have been shown as the reason for the change. In our study, we were also considered that the above-mentioned factors influence the distinctions in fatty acid amounts.

Amino acids

In our study, 17 amino acids were detected in the *U. moluccensis* samples. The amino acid contents of the samples changed monthly. We found that the most abundant amino acids were lysine and leucine as essential amino acids, aspartic acid, glutamic acid, alanine, and glycine as non-essential amino acids. When the total amino acid values were examined, it was determined that the highest amount of amino acid was found in December, and the lowest amount of amino acids were found in near spawning months (April and September). Monthly changes in the amounts of amino acids were found generally significant ($P < 0.05$) (Table 4). We thought that the monthly changes are caused by seasonal conditions, age, size, catching area, spawning season, and feeding conditions.

Fish are known to have high protein content and the most abundant amino acids are glutamic acid, aspartic acid, and lysine in aquatic organisms (Ozden and Erkan, 2008). In the previous studies, Kim and Lall (2000) investigated the amino acid contents of three kinds of marine flounder fish (*Hippoglossus hippoglossus*, *Pleuronectes ferruginea*, and *Paralichthys olivaceus*) in their study and found that the most abundant amino acids are aspartic acid, glutamic acid, glycine, leucine, and lysine. Zuraini et al. (2006) studied the amino acid contents of *Channa striats*, *Channa micropeltes*, and *Channa lucius*. They determined that the most abundant amino acids are glutamic acid, aspartic acid, and lysine, and they indicated that these species are rich about essential amino acids for human health and evolution. Ozden and Erkan (2008) examined the amino acid contents of three cultured species (*Dicentrarchus labrax*, *Sparus aurata*, and *Dentex dentex*) and determined that aspartic acid, glutamic acid, and lysine were the most abundant amino acids. They reported that aspartic acid and glutamic acid are important in enzyme solubility and protecting ionic character in enzyme activity. Adeyeye (2009), in his study, examined the amino acid contents of *Clarias anguillaris*, *Oreochromis niloticus*, and *Cynoglossus senegalensis* species, and he identified that glutamic acid, aspartic acid, and leucine were most common amino acids. He reported that these species are valuable amino acid sources. Erkan et al. (2010a) examined the amino acid content of *Trachurus trachurus* and found that the most abundant amino acids were aspartic acid, glutamic acid, and lysine. Erkan et al. (2010b) studied about the amino acid contents of *Engraulis encrasicolus*, *Pomatomus saltatrix*, *Sarda sarda*, *M. surmelutus*, and *Merlangius merlangus* and

Table 4 Montly changes in amino acid contents of *U. moluccensis* (mg/l00 g)

Amino acids	Sep.	Oct.	Nov.	Dec.	Jan.	Feb.	Mar.	Apr.	P values ($P < 0.05$)
Met.	500.5 ± 8.5^{e}	549.5 ± 2.5^{d}	512.5 ± 6.5^{e}	676.0 ± 6.0^{a}	603.5 ± 2.5^{c}	634.0 ± 2.0^{b}	507.0 ± 1.0^{e}	422.5 ± 0..5^{f}	0.000
Phe.	714.5 ± 12.5^{e}	826.5 ± 7.5^{c}	747.0 ± 12.0^{d}	974.0 ± 11.0^{a}	910.5 ± 7.5^{b}	901.5 ± 5.5^{b}	746.5 ± 4.5^{d}	585.5 ± 0.5^{f}	0.000
Lys.	257[illegible].0 ± 59.0^{e}	2933.5 ± 31.5^{c}	2744.5 ± 51.5^{d}	4638.0 ± 68.0^{a}	2259.5 ± 22.5^{f}	3821.5 ± 42.5^{b}	2688.0 ± 21.0^{de}	2621.5 ± 11.5^{de}	0.000
Val.	767.5 ± 13.5^{e}	872.0 ± 5.0^{c}	817.5 ± 12.5^{d}	1045.5 ± 10.5^{a}	957.0 ± 7.0^{b}	953.5 ± 16.0^{b}	726.0 ± 10.0^{f}	611.0 ± 2.0^{g}	0.000
Leu.	1249.0 ± 21.0^{e}	1424.5 ± 23 5^{c}	1343.0 ± 3.0^{d}	1769.0 ± 16.0^{a}	1621.5 ± 23.5^{b}	1622.0 ± 7.0^{b}	1308.5 ± 3.5^{d}	1069.5 ± 0.5^{f}	0.000
lie.	907.5 ± 16.5^{e}	1047.0 ± 9.0^{c}	990.5 ± 11.5^{d}	1286.0 ± 10.0^{a}	1200.0 ± 10.0^{b}	1180.5 ± 4.5^{b}	915.0 ± 6.0^{e}	771.0 ± 1.0^{f}	0.000
Thr.	849.0 ± 14.0^{e}	1096.5 ± 4.5^{c}	1106.5 ± 13.5^{c}	1038.5 ± 4.5^{d}	1250.0 ± 6.0^{a}	1205.5 ± 2.5^{b}	695.0 ± 2.0^{g}	774.0 ± 0.0^{f}	0.000
Tip.	226.0 ± 2.0^{abc}	232.0 ± 1.0^{ab}	241.5 ± 1.5^{a}	212.0 ± 3.0^{c}	239.5 ± 9.5^{a}	232.0 ± 6.0^{ab}	233.0 ± 2.0^{ab}	221.0 ± 4.0^{bc}	0.023
Arg.	491.0 ± 8.0^{g}	439.5 ± 10.5^{h}	547.0 ± 16.0^{f}	898.0 ± 9.0^{e}	1254.5 ± 4.5^{d}	1609.5 ± 4.5^{a}	1287.5 ± 5.5^{c}	1474.5 ± 3.5^{b}	0.000
His.	575.5 ± 16.5^{d}	750.5 ± 5.5^{b}	766.5 ± 8.5^{b}	841.0 ± 7.0^{a}	708.5 ± 4.5^{c}	751.0 ± 4.0^{b}	435.5 ± 1.5^{f}	499.0 ± 0.0^{e}	0.000
Ala.	971.5 ± 15.5^{de}	1085.0 ± 3.0$^{[illegible]}$	996.5 ± 12.5^{d}	1325.5 ± 6.5^{a}	1101.0 ± 4.0^{c}	1208.5 ± 7.5^{b}	957.5 ± 6.5^{e}	809.0 ± 1.0^{f}	0.000
Asp.	766.5 ± 12.5^{g}	1591.5 ± 8.50^{c}	1133.5 ± 14.5^{f}	1862.5 ± 8.5^{b}	1112.0 ± 3.0^{f}	2221.5 ± 12.5^{a}	1335.5 ± 4.5^{e}	1476.0 ± 6.0^{d}	0.000
Glu.	1422.0 ± 23.0^{g}	2023.5 ± 8.5$^{[illegible]}$	1668.5 ± 18.5^{e}	2709.0 ± 11.0^{a}	1484.5 ± 4.5^{f}	2637.0 ± 10.0^{b}	1864.5 ± 4.5^{d}	1992.5 ± 6.5^{c}	0.000
Tyr.	565.5 ± 15.5^{e}	648.5 ± 8.5$^{[illegible]}$	601.0 ± 10.0^{d}	756.5 ± 9.5^{a}	711.0 ± 7.0^{b}	720.0 ± 6.0^{b}	564.5 ± 4.5^{e}	458.0 ± 1.0^{f}	0.000
Gly.	1006.0 ± 18.0^{d}	1070.0 ± 4.0^{c}	951.5 ± 11.5^{e}	1276.0 ± 7.0^{a}	948.5 ± 3.5^{e}	1137.5 ± 3.5^{b}	943.5 ± 4.5^{e}	796.0 ± 4.0^{f}	0.000
Ser.	679 5 ± 13.5^{f}	877.5 ± 3.5^{e}	819.5 ± 9.5^{d}	966.0 ± 5.0^{b}	785.0 ± 4.0^{e}	1045.0 ± 3.0^{a}	621. ± 1.0^{g}	620.0 ± 0.0^{g}	0.000
Pro.	711 5 ± 9.5^{e}	728.5 ± 0.5^{d}	649.0 ± 7.0^{f}	873.0 ± 4.0^{b}	785.0 ± 2.0^{c}	997.5 ± 3.5^{a}	733.5 ± 1.5^{d}	604.5 ± 0.5^{g}	0.000
Total	14,974.0 ± 275.0^{e}	18,196.0 ± 137.0^{b}	16,636.0 ± 217.0^{c}	23,146.5 ± 190.5^{a}	17,931.5 ± 125.5^{b}	22,878.0 ± 129.0^{a}	16,562 ± 81.0^{c}	15.805.5 ± 7.5^{d}	0.000

Different lowercase letters in the same line indicate significant differences ($P < 0.05$)

they found that lysine, leucine, arginine, glutamic acid, and aspartic acid were the most abundant amino acids. Oluwaniyi et al. (2010) studied about the amino acid contents of *Clupea harengus*, *Scomber scombrus*, *Trachurus trachurus,* and *Urophycis tenuis* species. They reported that the most abundant amino acids are glutamic acid, aspartic acid, lysine, and leucine. Zhao et al. (2010) studied about the amino acid content of *Pampus punctatissimus*, an important fishery source in China, and found that glutamic acid, lysine, leucine, and aspartic acid were the most abundant amino acids. Ozden and Erkan (2011) studied the amino acid contents of some fishery products and found that the amino acid values changes according to species. Peng et al. (2013), identified that the most abundant amino acids of in tuna fish (*Thunnus albacares* and *Thunnus obesus*) were aspartic acid, glutamic acid, and lysine. They stated that the named fish are a good source of nutrients in terms of glutamic acid that required for cell proliferation. Kaya et al. (2014) investigated the amino acid content of, *Salmo trutta forma fario* that hunting and growing on the farms, and they found that glutamic acid, aspartic acid, and lysine values were higher than other amino acids. They found that the differences in total amino acid values were generally significant ($P < 0.05$) by species and by age, and they stated that the distinction in amino acid values was due to the spawning season, the feeding conditions, and the living area. Baki et al. (2015), compared the amino acid contents of natural and cultured sea bass (*Dicentrarchus labrax*) in their study, and they found that the most abundant amino acids were aspartic acid, glutamic acid, leucine, and lysine. Suseno (2015) reported that the most abundant amino acids were glutamic acid, arginine, leucine, and lysine in tuna fish (*Thunnus* sp.). Salma et al. (2016) found that the most abundant amino acids found in mackerel (*Scomber scombrus*) were glutamic acid, aspartic acid, and lysine.

We did not detect any study about amino acid contents of *U. moluccensis*. For that reason, we studied the work done with different types of fish and we found similar results for the most abundant amino acids. When we compared with the previous studies, we thought that the different results may be due to seasonal, biological differences, nutritional, and environmental conditions as well as methods used for amino acid determination. In addition, Kaya et al. (2014) found significant differences in monthly amino acid values in their studies ($P < 0.05$) like us.

Oluwaniyi et al. (2010) compared the essential amino acid contents of *C. harengus, S. scombrus, T. trachurus,* and *U. tenuis* with the standard amino acid values of FAO/WHO (1991). They determined that the essential amino acid values of the stated species were higher than the standard amino acid values, and according to these values, indicated species are a good source of amino acid. Zhao et al. (2010) compared the essential amino acid content of *P. punctatissimus* with the reference amino acid values of FAO/WHO/UNU (1985) and determined that the essential amino acid values of *P. punctatissimus* were higher than the reference values. Peng et al. (2013) compared the amino acid content of *T. albacares* and *T. obesus* with the standard amino acid values of FAO/WHO (1991) and found that the essential amino acid values of these species were higher than the standard amino acid values. They reported that the the essential amino acids of stated species were above the level required by humans by age. In our study, were compared the our 8-month averages of amino acid values in *U. moluccensis* with the average amino acid values of FAO (2013), and we determined that *U. moluccensis* has significant portion of the required amount of essential amino acids (Table 5).

Table 5 Comparison of the essential amino acid contets of *U. moluccensis* with the reference essential amino acid requirement according to FAO (2013), age groups

Amino acids (mg/g)	Reference amino acid values (FAO 2013)			Average amino acid content of *U. moluccensis* for 8 months
	Infant (0–6 month)	Child(6 month–3 year)	Older child, adolescent, adult	
His.	21	20	16	6.7
Iie.	55	32	30	10.4
Leu.	96	66	61	14.3
Lys.	69	57	48	30.3
Met.+Cys.	33	27	23	5.5
Phe.+Tyr.	94	52	41	14.3
Thr.	44	31	25	10.0
Trp.	17	8.5	6.6	2.3
Val.	55	43	40	8.4

Compared with the previous studies (Oluwaniyi et al. 2010; Zhao et al. 2010; Peng et al. 2013), lower essential amino acid values were detected in our study. It is thought that this may be due to seasonal, biological differences, nutritional, and environmental conditions as well as methods used for amino acid determination.

Conclusions

Our study showed that the basic food components, fatty acids, and amino acid values changes by monthly and the amounts of these changes were found generally significant ($P < 0.05$). We thought that these changes were influenced by many factors such as season, biological difference (species, height, age, sex, and sexual maturity), nutrition, and environmental conditions (water chemistry, temperature, and salinity). We also concluded that *U. moluccensis* is an important food source of the basic nutrients, fatty acids, and amino acid values in all fishing seasons.

Acknowledgements We acknowledge to Suleyman Demirel University Scientific Research Projects Coordination Unit for contributions to our Project of which No. is 4451-1-15.

References

Adeyeye EI (2009) Amino acid composition of three species of Nigerian fish: *Clarias anguillaris, Oreochromis niloticus* and *Cynoglossus senegalensis*. Food Chem 113:43–46

AOAC (2002a) Ashes content 920.153 official methods of analyses of AOAC International 17th edn. Gaithersburg, Maryland

AOAC (2002b) Fat (total, saturated and unsaturated) in foods 996.06 official methods of analyses of AOAC International 17th edn. Gaithersburg, Maryland

AOAC (2010) Microchemical determination of nitrogen-micro kjeldahl method 960.52 official methods of analyses of AOAC International 18th edn. Gaithersburg, MD

Baki B, Gonener S, Kaya D (2015) Comparison of food, amino acid and fatty acid compositions of wild and cultivated sea bass (*Dicentrarchus labrax* L., 1758). Turk J Fish Aquat Sci 15:175–179. doi:10.4194/1303-2712-v15_1_19 **(ISSN 1303-2712)**

Bligh EG, Dyer WJ (1959) A rapid method for total lipid extraction and purification. Can J Biochem Physiol 37:911–917

Broughton KS, Johnson CS, Pace BK, Liebman M, Kleppinger KM (1997) Reduced asthma symptoms with n-3 fatty acid ingestion are related to 5-series leukotriene production. Am Soc Clin Nutr 65:1011–1017

Cakli S (2007) Su urünleri isleme teknolojisi. Ege University Press, Bornova, p 696 **(ISBN: 978-975-483-761-2 Izmir, Turkey)**

Canbulat Z, Ozcan T (2008) Sut urunlerinin eikosapentaenoik asit (EPA) ve dokosahekzaenoik asit (DHA) ile zenginlestirilmesi. In: Turkey 10th Food Congress 21–23 May Erzurum, Turkey, pp 713–716

Celik M, Yanar Y, Gerek A (1999) Akdeniz'de avlanan uç barbun turunun (*Mullus barbatus, Mullus surmuletus, Upeneus moluccensis*) besin bilesenleri yonünden karsılastırılması. X. In: national water products symposium 22–24 September Adana, Turkey, pp 357–362

Conquer JA, Tierney MC, Zecevic J, Bettger WJ, Fisher RH (2000) Fatty acid analysis of blood plasma of patients with Alzheimer's disease, other types of dementia, and cognitive impairment. Lipids 35:1305–1312

Dimova N (2003) RP-HPLC analysis of amino acids with UV-detection. Bulg Acad Sci 56:75–78

Ebbesson SOE, Roman MJ, Devereux RB, Kaufman D, Fabsitz RR, MacCluer JW, Dyke B, Laston S, Wenger CR, Comuzzie AG, Romanesko T, Ebbesson LOE, Nobmann ED, Howard BV (2008) Consumption of omega-3 fatty acids is not associated with a reduction in carotid atherosclerosis: the genetics of coronary artery disease in Alaska native study. Atherosclerosis 199:346–353

Erkan N (2013) Evaluation of omega 3 (ω-3) fatty acid profile of seafood products consumed in Turkey. J Fish Sci 7:194–208. doi:10.3153/jfscom.2013020

Erkan N, Ozden Ö (2007) The changes of fatty acid and amino acid composition in sea bream (*Sparus aurata*) during irradiation process. Radiat Phys Chem 76:1636–1641

Erkan N, Selcuk A, Ozden O (2010a) Amino acid and vitamin composition of raw and cooked horse mackerel. Food Anal Methods 3:269–275

Erkan N, Ozden O, Selcuk A (2010b) Effect of frying, grilling and steaming on amino acid composition of marine fishes. J Med Food 13:1524–1531. doi:10.1089/jmf.2009.0203

Ersoy B (2006) Food composition and heavy metal contents of fishes consumed in fishing season in Northeastern Mediterranean (Adana/Karatas) Region. PhD Thesis p142 Cukurova University Institute of Applied and Natural Sciences Adana, Turkey

FAO (2013) Dietary protein quality evaluation in human nutrition. Report of an FAO Expert Consultation. Fao Food and Nutrition Paper 92 Rome. ISBN 978-92-5-107417-6

FAO/WHO (1991) Protein quality evaluation. Report of joint FAO/WHO expert consultation. FAO food and nutrition paper 51. Italy: Rome

FAO/WHO/UNU (1985) Energy and protein requirement, WHO technical report series no. 724. Geneva, Switzerland: WHO

Gamez-Meza N, Higuera-Ciapara I, Calderon de la Barca AM, Vazquez-Moreno L, Noriega-Rodriguez J, Angulo-Guerrero O (1999) Seasonal variation in the fatty acid composition and quality of sardine oil from *Sardinops sagax caeruleus* of the Gulf of California. Lipids 34:639–642

Gheshlaghi R, Scharer JM, Moo-Young M, Douglas PL (2008) Corrigendum to application of statistical design for the optimization of amino acids separation by reverse-phase HPLC. Anal Biochem 383:93–102

Gulyavuz H, Unlusayın M (1999) Su Urünleri Isleme Teknolojisi. Sahin Press, Isparta, p 366 **(ISBN:975-96897-0-7 Ankara, Turkey)**

Gumus B, Ikız R, Unlusayın M (2009) Changes in the nutrient components after hot smoking process of red mullet (*Mullus barbatus*). Istanb Univ J Fish Aquat Sci 24:24–45

Guner S, Dincer B, Alemdag N, Colak A, Tufekci M (1998) Proximate composition and selected mineral content of commercially important fish species from the Black Sea. J Sci Food Agric 78:337–342

Haris WS (1997) n-3 fatty acids and serum lipoproteins: human studies. Am J Clin Nutr 65:16455–16545

Kalogeropoulos N, Andrikopoulos NK, Hassapidou M (2004) Dietary evaluation of Mediterranean fish and molluscs pan-fried in virgin olive oil. J Sci Food Agric 84:1750–1758

Kandemir S, Polat N (2007) Seasonal variation of total lipid and total fatty acid in muscle and liver of rainbow trout (*Oncorhynchus mykiss* W., 1792) reared in Derbent Dam Lake. Turk J Fish Aquat Sci 7:27–31

Kaya Y, Erdem ME, Turan H (2014) Monthly differentiation in meat yield, chemical and amino acid composition of wild and cultured brown trout (*Salmo trutta forma fario* Linneaus, 1758). Turk J Fish Aquat Sci 14:479–486. doi:10.4194/1303-2712-v14_2_19

Kim JD, Lall SP (2000) Amino acid composition of whole body tissue of Atlantic halibut (*Hippoglossus hippoglossus*), yellowtail flounder (*Pleuronectes ferruginea*) and Japanese flounder (*Paralichthys olivaceus*). Aquaculture 187:367–373

Kuzu S (2005) The effects of different fishing seasons on the fatty acids composition of red mullet (*Mullus barbatus* L., 1758) captured from Iskenderun Bay. MSc Thesis p 39 Cukurova University Institute of Applied and Natural Sciences Adana, Turkey

Lovell RT (1981) Laboratory Manuel For Fish Feed Analysis and Fish Nutrition Studies. Department of Fisheries and Allied Aquacultures International Center for Aquaculture Auburn University, pp 65, Alabama

Luzia LA, Sampario GR, Castellucci CMN, Torres EAFS (2003) The influence of season on the lipid profiles of five commercially important species of Brazilian fish. Food Chem 83:93–97

Norrish AE, Jackson RT, Sharpe SJ, Skeaff CM (2000) Men who consume vegetable oils rich in monounsaturated fat: their dietary patterns and risk of prostate cancer (New Zealand). Cancer Causes Control 11:609–615

Oksuz A, Ozyılmaz A, Kuver S (2011) Fatty acid composition and mineral content of *Upeneus moluccensis* and *Mullus surmuletus*. Turk J Fish Aquat Sci 11:69–75

Oluwaniyi OO, Dosumu OO, Awolola GV (2010) Effect of processing methods (boiling, frying and roasting) on the amino acid composition of four marine fishes commonly consumed in Nigeria. Food Chem 123:1000–1006

Ozden O (2005) Changes in amino acid and fatty acid composition during shelf-life of marinated fish. J Sci Food Agric 85:2015–2020

Ozden O, Erkan N (2008) Comparison of biochemical composition of three aqua cultured fishes (*Dicentrarchus labrax, Sparus aurata, Dentex dentex*). Int J Food Sci Nutr 59:545–557

Ozden O, Erkan N (2011) A preliminary study of amino acid and mineral profiles of important and estimable 21 seafood species. Br Food J 113:457–469

Ozkan Y, Koca SS (2006) The efficiency of omega-3 fatty acid (fish oil) in hyperlipidemia. Fırat Tıp Dergisi 11:40–44

Ozogul Y, Polat A, Ucak I, Ozogul F (2011) Seasonal fat and fatty acids variations of seven marine fish species from the Mediterranean Sea. Eur J Lipid Sci Technol 113:1491–1498

Peng S, Chen C, Shi Z, Wang L (2013) Amino acid and fatty acid composition of the muscle tissue of yellowfin tuna (*Thunnus albacares*) and bigeye tuna (*Thunnus obesus*). J Food Nutr Res 1:42–45

Polat A, Kuzu S, Ozyurt G, Tokur B (2009) Fatty acid composition of red mullet (*Mullus barbatus*): a seasonal differentiation. J Muscle Foods 20:78–79

Rosa R, Nunes ML (2003) Nutritional quality of red shrimp, *Aristeus antennatus* (Risso), pink shrimp, *Parapenaeus longirostris* (Lucas), and Norway lobster, *Nephrops norvegicus* (Linnaeus). J Sci Food Agric 84:89–94. doi:10.1002/jsfa.1619

Ruiz-Capillas C, Moral A (2001) Changes in free amino acid during chilled storage of hake (*Merluccius merluccius* L.) in controlled atmospheres and their use as a quality control index. Eur Food Res Technol 212:302–307

Ruiz-Capillas C, Moral A (2004) Free amino acids in muscle of Norway lobster (*Nephrops novergicus* (L.) in controlled and modified atmospheres during chilled storage. Food Chem 86:85–91

Salma EO, Cyrine D, Nizar M (2016) Fatty acids and amino acids contents in *Scomber scombrus* fillets from the South East of Tunisia. Afr J Biotechnol 15:1246–1252. doi:10.5897/AJB2016.15344

Simsek A, Kırmızı S, Manasırlı M, Ozyurt G, (2009) Keserbas (*Mullus Barbatus*) ve cizgili barbun (*Upeneus moluccensis*)'un mineral ve vitamin icerikleri. XV. National Water Resources Symposium 01–04 July Rize, Turkey

Suseno SH (2015) Proximate, fatty acid and mineral composition of Tuna (*Thunnus* sp.) by-product from West Sumatra Province, Indonesia. Pak J Nutr 14:62–66

Tulgar A, Berik N (2012) Effect of seasonal changes on proximate composition of red mullet (*Mullus barbatus*) and hake (*Merluccius merluccius*) were catched from Saroz Bay. Res J Biol 2:45–50

Zhao F, Zhuang P, Song C, Shi ZH, Zhang LZ (2010) Amino acid and fatty acid compositions and nutritional quality of muscle in the pomfret, *Pampus punctatissimus*. Food Chem 118:224–227

3

Bioactivity of aqueous and organic extracts of sea cucumber *Holothuria leucospilota* (Brandt 1835) on pathogenic Candida and Streptococci

Motahereh Ghadiri · Soodabeh Kazemi · Behrooz Heidari · Mehdi Rassa

Abstract Holothurians (sea cucumbers) exist almost in all benthic marine environments and they are an important species for the commercial fisheries. In this study, the antimicrobial activity of different tissues of the sea cucumber, as a marine invertebrate, was investigated. The bacteria, *Streptococcus mutans* and *Streptococcus sobrinus*, the fungi, *Candida parapsilosis*, *Candida albicans* and *Candida glabrata*, were chosen for antimicrobial test, because they are opportunistic pathogens. The body wall, respiratory tree, and gastrointestinal tract of the sea cucumbers were utilized to make the extracts. Phosphate-buffered saline (PBS), ethanol (Et), and acetonitrile (ACN) were used as solvents for aqueous and organic extractions. The results showed that there were antimicrobial active compounds against infective streptococci and candid fungi. *Streptococcus mutans* was generally more tolerant than *Streptococcus sobrinus* to the sea cucumber extracts. The extracts of the respiratory tree with ACN and PBS, as solvents, showed the highest effect against *Candida albicans*, *Candida parapsilosis* and *Candida glabrata*. The extracts of gastrointestinal tract and the body wall with Et and PBS, respectively, had not any antifungal activity. The antimicrobial activities had been found in different extracts of the sea cucumber tissues, which could be related to the protein and non-protein compounds.

Keywords Antimicrobial activity · Candida · Sea cucumber · *Streptococcus*

Introduction

During the last decades, the marine natural products have been attracted much attentions of biologists and chemists of all around the world. Marine organisms produce natural products which can be pharmaceutically useful or being a toxic compound. During the last decade, researches on marine crustaceans, sponges, jellyfish, sea anemones, corals, bryozoans, mollusks, tunicates and echinoderms have been increased, and these investigations have been specially focused on desirable antimicrobial properties (Bhakuni and Rawat 2005; Casas et al. 2011; Haug et al. 2002). Echinoderms are a group of benthic organisms, which are exposed to relatively high concentrations of pathogens which can be harmful to the animal (Haug et al. 2002). Sea cucumbers are a class of Holothuroidea with body shape of a worm and without any spines.

M. Ghadiri · S. Kazemi · B. Heidari (✉) · M. Rassa
Department of Biology, Faculty of Science, University of Guilan, Namjoo St., Rasht 4193833697, Iran
e-mail: Bheidari@guilan.ac.ir; Behrooz1072@yahoo.com

B. Heidari
Department of Marine Sciences, Caspian Sea Basin Research Center, University of Guilan, Rasht, Iran

Many sea cucumbers have novel defensive mechanisms which compensates their lack of spines (Lewbart 2006). The sea cucumbers have multiple biological activities such as wound healing (Fredalina et al. 1999), antibacterial (Gowda et al. 2008; Haug et al. 2002; Kiani et al. 2014; Ridzwan et al. 1995), antifungal (Wang et al. 2012), agglutinating activities (De Melo et al. 2014; Gowda et al. 2008) and antioxidant properties (Osama et al. 2009).

Different habitats of human oral cavity such as teeth, gingival sulcus, tongue, cheeks, hard and soft palates, and tonsils can be colonized easily by bacteria (Dewhirst et al. 2010). The oral diseases including dental caries created by aerobic and anaerobic bacteria are of the most major health problems in the world (Aas et al. 2008; Anusavice 2002; Becker et al. 2002; Beighton 2005; Chhour et al. 2005). The streptococci are the major dental caries aetiologic agents and they are the most common pathogens isolated from the human dental plaque and play an important role in the development and progression of dental caries (Nishikawara et al. 2006; Okada et al. 2005). *Streptococcus mutans* (Clark 1924) and *Streptococcus sobrinus* (Coykendall 1974) are the two species which are responsible for the dental caries (Ghasempour et al. 2013; Nascimento et al. 2004; Okada et al. 2005). These groups of bacterial species are characterized by their ability of producing extracellular glucans from sucrose and via their acidic productions in animal and human studies (Ghasempour et al. 2013; Nishikawara et al. 2006).

The Candida strains are the main factor for serious fungal infections such as vulvovaginal candidiasis (VVC) which can infect 75% of adult women during their life (Sharanappa and Vidyasagar 2013), and it also increases the chance of cutaneous infections, vaginitis, intestinal and systemic infections (Molero et al. 1998). The Candida strains are considered to be the fourth effective factor in hematological infections for patients hospitalized and this kind of infection is responsible for about 40% of death (Anaissie et al. 2003). The opportunistic fungal strains such as *C. albicans* are more frequent than other species and they can be isolated from oral infections (Kantarcioglu and Yucel 2002). The other Candida species with direct effects can cause systemic infections in immunosuppressant patients (such as AIDS, cancer patients and solid-organ transplant recipients). Because of the pathogenic microbe resistances to the conventional antibiotics, the calls for the investigations on new antimicrobials metabolites have been increased. The marine biological reserves, especially 17 species of Holothurians (Dabbagh et al. 2011) of the Persian Gulf, have high potential for antimicrobial investigations, but their biological benefits are unknown due to their unfamiliarity. Therefore, in this work, the antimicrobial effects of various extracts of different tissues of the Persian Gulf sea cucumber on the infectious oral streptococci and Candida strains were investigated.

Materials and methods

Sampling and preparation of specimens

Live specimens of the sea cucumbers ($N = 10$) were collected via scuba diving off (4–5 m depth) from Qeshm Island of the Persian Gulf in October (Fig. 1). The samples were transferred on ice to the Marine Biology Laboratory of the University of Guilan. After dissection, all tissues of animals including body wall, respiratory tree and gastrointestinal tract were separately frozen in liquid nitrogen for 5 min and stored at − 70 °C until the extraction process. The specimens were identified using ossicles obtained from the epidermal tissues of dorsal and ventral surfaces of the body wall (Kiani et al. 2014).

Preparation of extracts

PBS extract

The tissues (body wall, respiratory tree and gastrointestinal tract) were separately homogenized (Misonix Sonicator 3000, USA) with PBS ($NaH_2PO_4{\cdot}2H_2O$, 0.1 M; NaOH, 0.2 M) (1:5 w/v) for 5 min. The obtained samples were shaken (Infros, RFI-150) on ice for 2 h at 120 rpm and then they were centrifuged at 4000 rpm for 10 min at 4 °C. The supernatants were collected and stored at 4 °C, and the residues were centrifuged in two volumes of PBS. Then, the combined supernatants were centrifuged and lyophilized by freeze–dryer

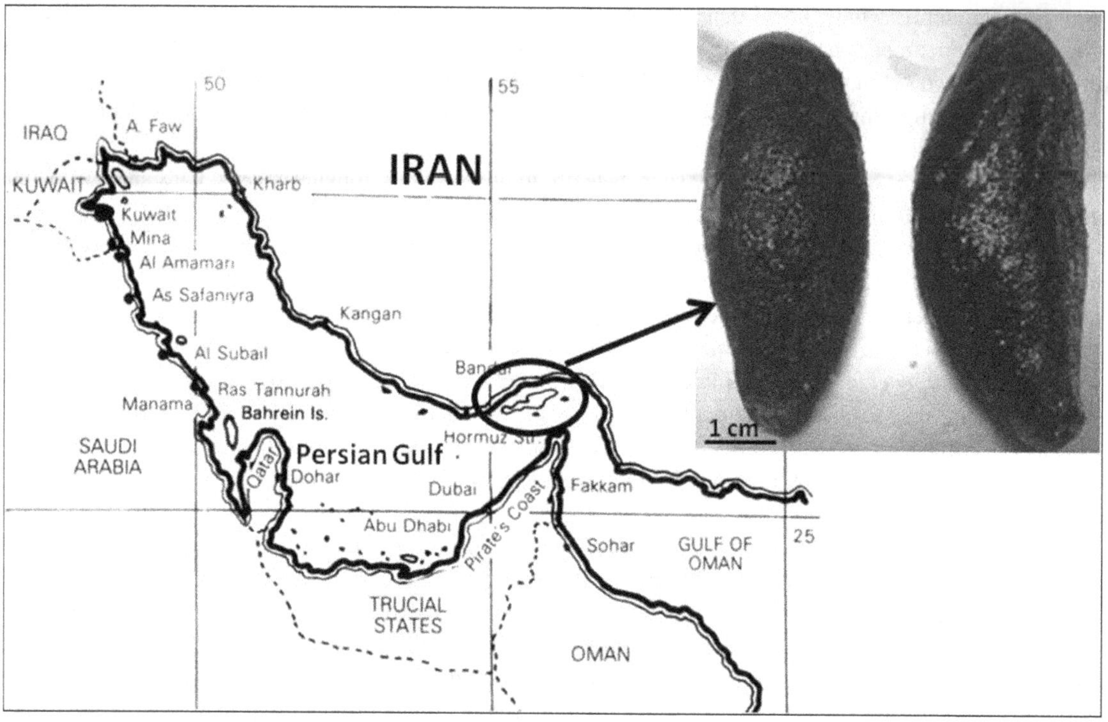

Fig. 1 Map of the Persian Gulf indicating the sampling area and the sea cucumber collected for the study

(Christ Alpha 1–2/LD, Osterode am Harz, Germany) for 9 h (− 55 °C and 0.02 mbar) and they were stored at − 70 °C until antibacterial test.

Et and ACN extracts

The organic extracts have been prepared with Et 96% and ACN (Merck Schuchardt, Germany). As above, all tissues were separately homogenized in Et and ACN with combination of (1:5 w/v) for 5 min. Subsequently, the solutions were shaken at 40 °C for 24 h (120 rpm) and centrifuged at 4000 rpm for 10 min at 4 °C. The supernatants were sampled and the residues were centrifuged in two volumes of the solvents. Finally, the collected supernatants were lyophilized by freeze-dryer for 9 h (− 55 °C and 0.02 mbar) and stored at − 70 °C until antibacterial test.

Bacterial and yeast strains

The Gram-positive bacteria *Streptococcus mutans* (ATCC 35668) and *Streptococcus sobrinus* (ATCC 27607) were purchased from the Persian Type Culture Collection (PTCC, Tehran, Iran) and were used as test microorganisms. The strains were grown on blood agar (Biolife, Italy) containing 7% fresh horse blood and maintained at 37 °C for 48 h. The candida strains *Candida albicans* (5027 ATCC 10231), *C. parapsilosis* (5297 DSM 11226) were obtained from the Persian Type Culture Collection (PTCC, Tehran, Iran) as well as *C. glabrata* (30005) was purchased from Iranian Biological Resource Center (IBRC-M, Tehran, Iran).

Antibacterial test

The antibacterial activities of the aqueous (PBS) and organic extractions were evaluated by the well diffusion method (Kiani et al. 2014). *S. mutans* and *S. sobrinus* were cultured in blood agar for 24 h. The crude aqueous and organic extracts were tested separately with two concentrations of 600 and 1500 μg were dissolved in 40 μL of double distilled water (d.d.H_2O) and dimethyl sulfoxide (DMSO) (Merck, Germany) as solvents, respectively. After that, the dissolved extracts were transferred into the wells (diameter 6 mm), punched out on plates, containing the bacteria and incubated at 37 °C for 24 h. The antibacterial activities were determined by measuring the diameter of zone of inhibition (mm).

Controls

Vancomycin (30 μg disc^{-1}), Tetracycline (30 μg disc^{-1}), Chloramphenicol (30 μg disc^{-1}), Nalidixic acid (30 μg disc^{-1}), Sulfadiazine (1.0 μg disc^{-1}), Oxacillin (1 μg disc^{-1}), Bacitracin (15 μg disc^{-1}), and Cefepime (30 μg disc^{-1}) were utilized as positive controls. DMSO (40 μL well^{-1}) and dd H_2O (40 μL well^{-1}) were also tested as negative controls to ensure that they did not interfere with the tests.

Antifungal test

The antifungal activities of the sea cucumber extracts were evaluated using Sabouraud Dextrose Agar (SDA) well diffusion method (Periyasamy et al. 2012; Stepanovic et al. 2003) against *Candida albicans* (Persian Type Culture Collection (PTCC): 5027 ATCC 10231), *C. parapsilosis* (PTCC: 5297 DSM 11226) and *C. glabrata* (Iranian Biological Resource Center (IBRC-M): 30005). All Candida strains were unloaded from lyophilized ampoules to SDA media after dissolving in 0.5 mL of the distilled water and cultivated by linear method. The organic and aqueous extracts with two concentrations of 600 and 1500 μg were dissolved in 30 μL of DMSO and 30 μL of distilled water, respectively. All extractions were added to the labeled wells with diameter 6 mm on the plates containing the fungi, and then, the loaded plates were incubated at 37 °C for 24 h.

Controls

Three antibiotics drugs including Ketoconazole, Fluconazole and Nystatin (with concentration 30 μg mL^{-1}) were prepared to be used as positive controls, DMSO (40 μL well^{-1}) and distilled water (40 μL well^{-1}) were also utilized as negative controls.

Hemolytic assay

To test whether the sea cucumber contain factors that display toxic effect on eukaryotic cells, the hemolytic activity of aqueous and organic extracts obtained from body wall, respiratory tree and gastrointestinal tract was determined. The assay was performed using the method described by Haug et al. (2002). Briefly, heparinized horse red blood cells (RBCs) washed three times in PBS (pH 7) and they were suspended in PBS to reach to a hematocrit value of 10%. The samples were diluted to a protein concentration of 500 μg mL^{-1}. The test mixture consisted of 40 μL PBS, 50 μL extract, and 10 μL of the RBCs suspension which was incubated at 37 °C for 1 h and then it was centrifuged at 3000 rpm for 10 min. Finally, the absorbance of the supernatants (60 μL) was measured at wavelength of 550 nm. Baseline hemolysis and 100% hemolysis were defined as the amount of hemoglobin released in the presence of PBS and 0.1% Triton X-100 (Sigma), respectively.

Specimen determination

The specimens were identified using the ossicle of the sea cucumber. After taking digital photographs of ossicle, the identification of species was performed via the diagnostic keys (Price 1981).

Statistical analysis

The raw data were initially tested versus the assumptions of normality and homogeneity of variance. Levene's test and Kolmogorov–Smirnov test were used to assess the homogeneity and normality of variance, respectively. The variances of the populations from which different samples were drawn were equal. The one-way ANOVA followed by Duncan test with a confidence level of 95% were used to determine the differences between various extracts using IBM SPSS Statistics 19. The significant differences between the two concentrations of 600 and 1500 μg $well^{-1}$ were assessed by independent *t* test. All experiments (treatments and controls) were carried out for three times and the obtained data were reported as the mean ± SD.

Results

Due to the dorsal and ventral ossicles obtained from epidermal tissues of the body wall, it was concluded that the studied species was *Holothuria leucospilota* (Brandt 1835) (Fig. 2).

Due to the results, all tissues of the echinoderm species showed antimicrobial activities against oral streptococci and *Candida* sp.

Extract activities against *S. Sobrinus* and *S. mutans*

At concentration of 1500 μg $well^{-1}$ of the various extracts of different tissues, antibacterial effects against *S. sobrinus* and *S. mutans* were observed (Fig. 3). The maximum diameter of inhibition zones (statistically significant, at $p < 0.05$ level) was belonged to the ACN extract of the respiratory tree and Et extract of the body wall, while the other extracts showed similar inhibitory effects on the growth of *S. sobrinus* (statistically insignificant, at $p > 0.05$ level) (Fig. 3).

At concentration of 1500 μg.$well^{-1}$, the maximum diameter of inhibition zone against *S. mutans* was obtained by the PBS extract of gastrointestinal tract (9.67 mm) while the ACN extracts of gastrointestinal tract

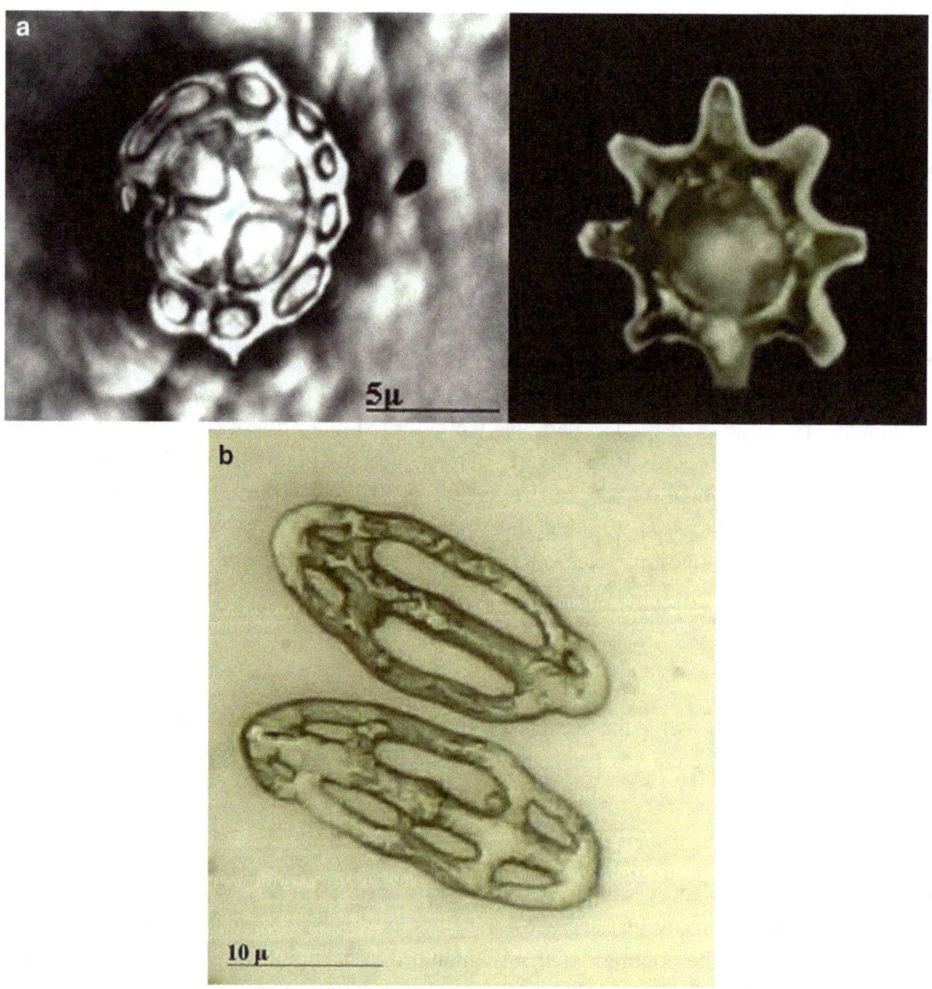

Fig. 2 The dorsal (**a**) and ventral (**b**) ossicles of sea cucumber that were used to determine the species

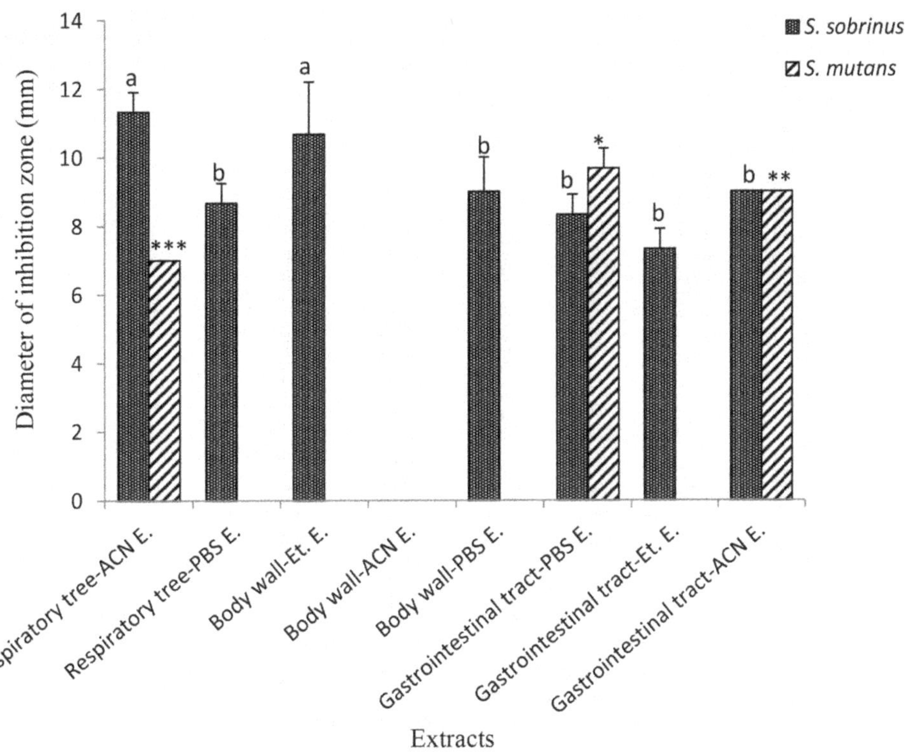

Fig. 3 The mean diameter of inhibition zone for *S. sobrinus* and *S. mutans* under the effect of different extracts of *H. leucospilota* with concentrations of 1500 μg $well^{-1}$. Mean values bearing different superscripts are significantly different ($p < 0.05$)

and respiratory tree had less significant inhibitory effects, respectively ($p < 0.05$) (Fig. 3). The other extracts had no effects on *S. mutans*.

The ACN and Et extracts of gastrointestinal tract, at concentration of 600 μg $well^{-1}$, had no antibacterial effects against *S. sobrinus*. The highest antibacterial effect on *S. sobrinus* was observed for the ACN extract of the respiratory tree ($p < 0.05$) (Fig. 4). There was no significant difference ($p > 0.05$) between the antibacterial effects of the various extracts of the body and the PBS extract of gastrointestinal tract (Fig. 4). The Et extract of the respiratory tree had no effect on *S. sobrinus* at both concentrations.

In concentration of 600 μg $well^{-1}$, only the PBS extract of gastrointestinal tract had inhibitory effect on the growth of *S. mutans* (Fig. 4).

The comparison of antibacterial effects of two concentrations using statistical independent t test showed that the extracts with concentration of 1500 μg $well^{-1}$ had overall more significant effect ($p < 0.05$) against *S. sobrinus* and *S. mutans* (Fig. 5a, b).

Activity of extracts with a concentration of 600 μg.$well^{-1}$ against *Candida sp.*

The ACN extract of the respiratory tree and the body wall had significantly the maximum and minimum inhibitory effects on the growth of *C. parapsilosis*, respectively ($p < 0.05$) (Fig. 6). The other extracts had no significant antifungal activity ($p > 0.05$).

The ACN extracts of the body wall, respiratory tree and gastrointestinal tract showed the highest antifungal activity on *C. albicans* and followed by the Et extracts of the respiratory tree and body wall ($p < 0.05$) (Fig. 6).

The maximum inhibitory activity against *C. glabrata* belonged to the PBS extract of the respiratory tree and followed by the ACN extract of the body wall ($p < 0.05$). In addition, there were no significant differences between antifungal effects of the ACN extract of the respiratory tree and the PBS extract of gastrointestinal tract ($p > 0.05$) (Fig. 6).

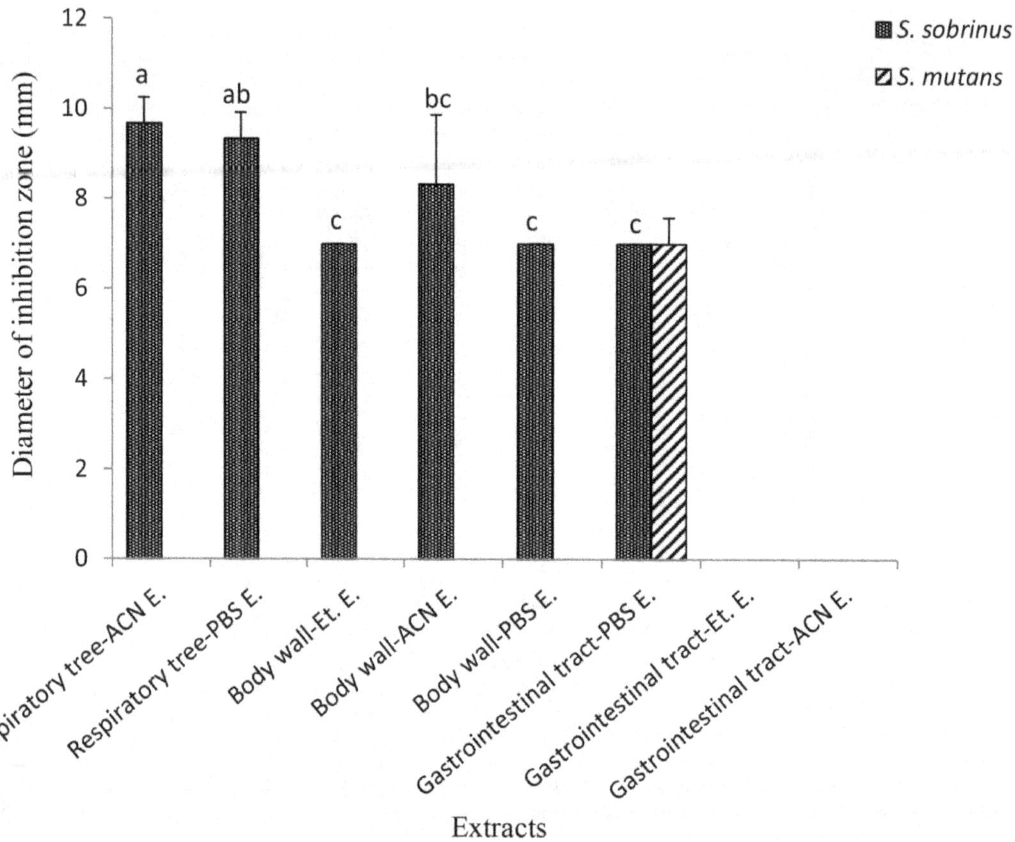

Fig. 4 The mean diameter of inhibition zone for *S. sobrinus* and *S. mutans* under the effect of different extracts of *H. leucospilota* with concentrations of 600 μg well^{-1}. Mean values bearing different superscripts are significantly different ($p < 0.05$)

Activity of extracts with a concentration of 1500 μg.well^{-1} against *Candida sp.*

The Et. extracts of all tissues of the sea cucumber had significantly higher antifungal activities than the other extracts against *C. parapsilosis* ($p < 0.05$) (Fig. 6). The ACN extracts of all tissues had the same antifungal effects ($p > 0.05$).

The maximum inhibitory activity against *C. albicans* remarkably belonged to the ACN extract of the respiratory tree ($p < 0.05$) (Fig. 7). The ACN and Et extracts of the gastrointestinal tract as well as the Et extract of the body wall had the same inhibitory effects ($p > 0.05$).

The extracts of the different tissues of the sea cucumber showed the diverse significant antifungal effects on *C. glabrata* (Fig. 7). The ACN extract of the respiratory tree had significantly the maximum inhibitory effect ($p < 0.05$). The Et and PBS extracts like ACN extracts of the respiratory tree and gastrointestinal tract had the minimum antifungal effects ($p < 0.05$). In addition, the ACN extract of the body wall and PBS extract of the gastrointestinal tract had higher effect than Et extract of the body wall ($p < 0.05$) (Fig. 7).

Hemolytic activity of the sea cucumber extracts

The hemolytic activity of the PBS, Et and ACN extracts from the body wall, respiratory tree and gastrointestinal tract of *H. leucospilota* was evaluated using horse red blood cells. Several extracts showed high hemolytic activity (Table 1). In general, the highest and lowest hemolytic activities were obtained from the ACN and PBS extracts of all tissues, respectively. In addition, the antimicrobial activity of the positive and negative controls against the bacteria and fungi was summarized in Figs. 8 and 9.

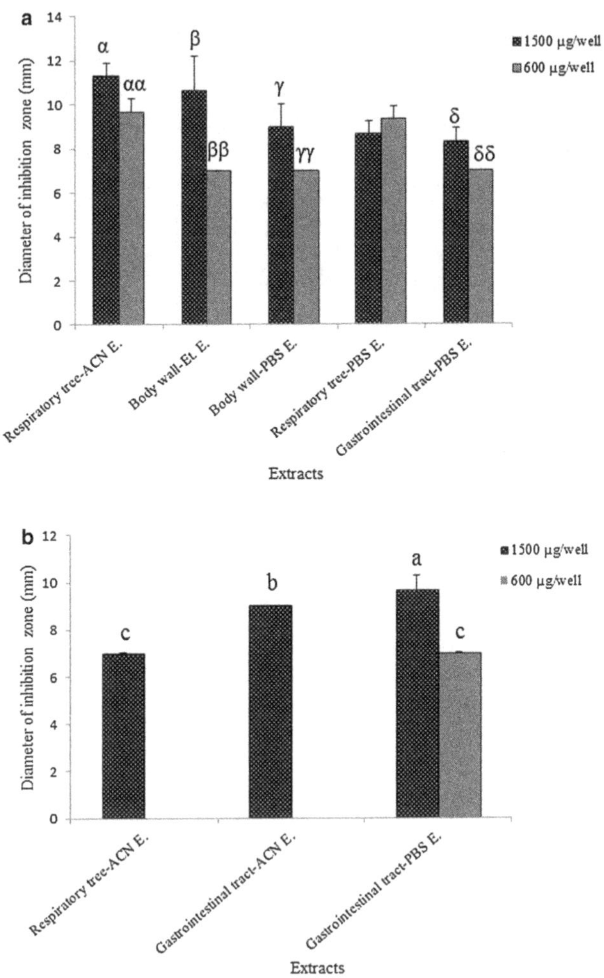

Fig. 5 The comparison between two concentrations of the extracts against *S. sobrinus* (**a**), *S. mutans* (**b**). Mean values bearing different superscripts are significantly different ($p < 0.05$)

Discussion

Marine organisms have high positions in traditional medical sciences and they also utilize as the main sources for pharmacological investigations (Fenical 1996). It has been discovered that echinoderms have stronger antibacterial effects than Porifera, Mollusca, Bryozoa or Annelida (Ridzwan et al. 1995). The sea cucumbers are benthic animals with no shell and they do not have the ability to escape from predators but they have some effective chemical responses. In this study, it had been tried to evaluate the antimicrobial effects of the various extracts of the different parts of the sea cucumber *Holothuria leucospilota* on oral streptococci and Candida strains. These extracts with different solvents (Et, ACN and PBS) allowed us to investigate the presence of antimicrobial activity of the substances potentially presented in the lipid- or water-soluble fraction. Since the antimicrobial effects in the different tissues of *H. leucospilota* with various solvents were found, therefore it was concluded that the protein and non–protein factors could be responsible for such effects.

Antibacterial effects of the extracts

The antibacterial activity has previously been described in some species of echinoderms (Gowda et al. 2008; Haug et al. 2002; Kazemi et al. 2016; Kiani et al. 2014; Ridzwan et al. 1995). Haug et al. (2002) studied the antibacterial activity of different parts of the green sea urchin *Strongylocentrotus droebachiensis* (Müller, 1776), the common starfish *Asterias rubens* (Linnaeus, 1758) and the sea cucumber *Cucumaria frondosa*

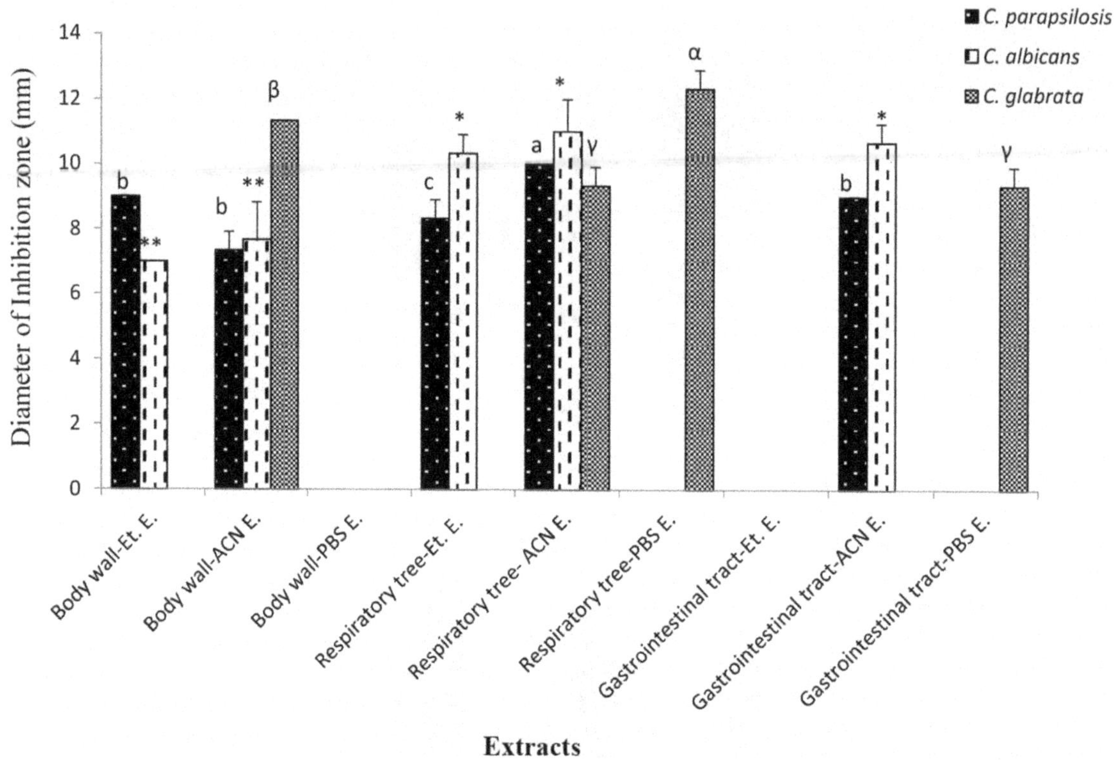

Fig. 6 The mean diameter of inhibition zone for *Candida* sp. under the effect of different extracts of *H. leucospilota* with concentrations of 600 μg well^{-1}. Mean values bearing different superscripts are significantly different ($p < 0.05$)

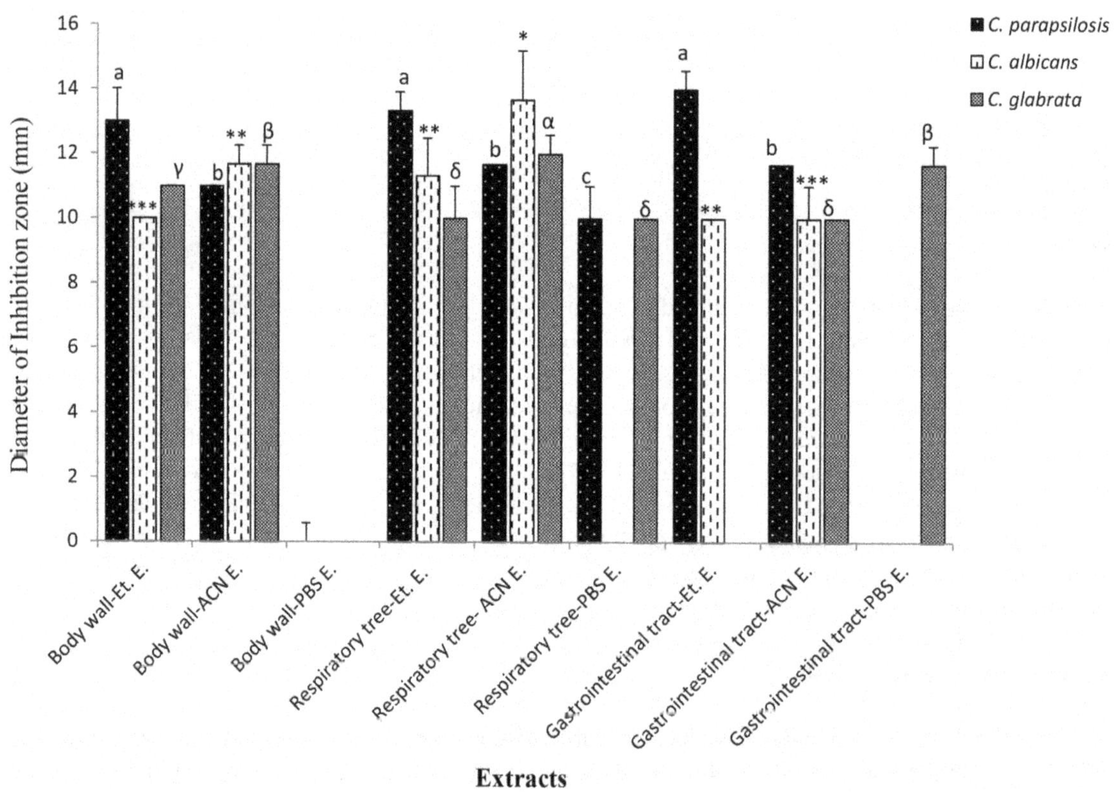

Fig. 7 The mean diameter of inhibition zone for *Candida* sp. under the effect of different extracts of *H. leucospilota* with concentrations of 1500 μg well^{-1}. Mean values bearing different superscripts are significantly different ($p < 0.05$)

Table 1 Hemolytic activity in the sea cucumber (*H. leucospilota*) extracts against the red blood cells

Tissues	Extracts	Hemolytic activity (%)[a]
Body wall	PBS	20
	ACN	80
	Ethanol (96%)	60
Respiratory tree	PBS	30
	ACN	80
	Ethanol (96%)	80
Gastrointestinal tract	PBS	20
	ACN	80
	Ethanol (96%)	60

[a]The data is presented as % hemolysis compared to control (Triton X-100)

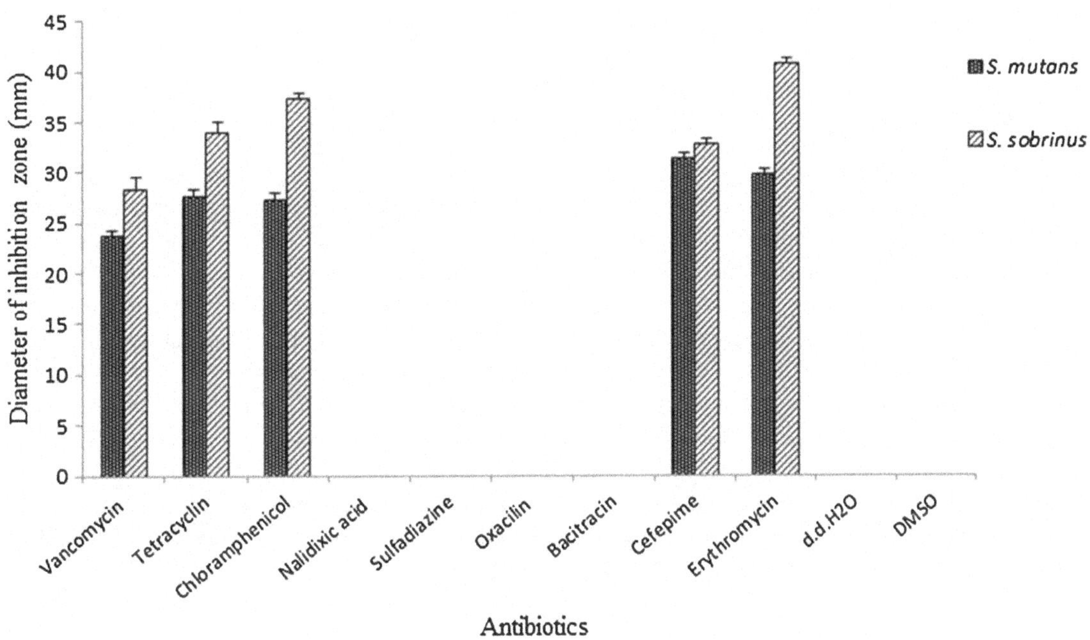

Fig. 8 Antibiogram and solvents (d.d.H_2O) tests against *S. mutans* and *S. sobrinus*. The concentration of antibiotics and solvents was 30 μg mL^{-1} and μL $well^{-1}$, respectively

(Gunnerus, 1767) against Gram-positive and -negative bacteria. They showed antibacterial activities in the extracts of several tissues including the body wall, coelomocyte and especially the gastrointestinal organ and the eggs from *A. rubens* and *C. frondosa*. Ridzwan et al. 1995 tested the extracts of the sea cucumbers *Holothuria atra* (Jaeger, 1933), *H. scabra* (Jaeger, 1833) and *Bohadschia argus* (Jaeger, 1833) against seven species of bacteria and they found that the lipid and methanol extracts had no inhibitory activities, whereas the PBS extract had severe inhibitory activity. The methanol–acetone extract of the body wall of the sea cucumber *Parastichopus parvimensis* (Clark, 1913) had antibacterial properties against *Bacillus subtilis* (Ehrenberg 1835) and *Escherichia coli* (Migula, 1895) (Villasin and Pomory 2000). The present study demonstrates the presence of antibacterial substances in the body wall, respiratory tree and gastrointestinal tract of the sea cucumber, *H. leucospilota*. As the most tropical sea cucumbers feed on live microorganisms and the organic contents found in sand or slime (Rowe 1969), there is the possibility that some pathogenic microorganisms are taken together with the food substances. Therefore, the presence of some active antibacterial substance(s) is predictable for the body defense in different parts of the sea cucumber.

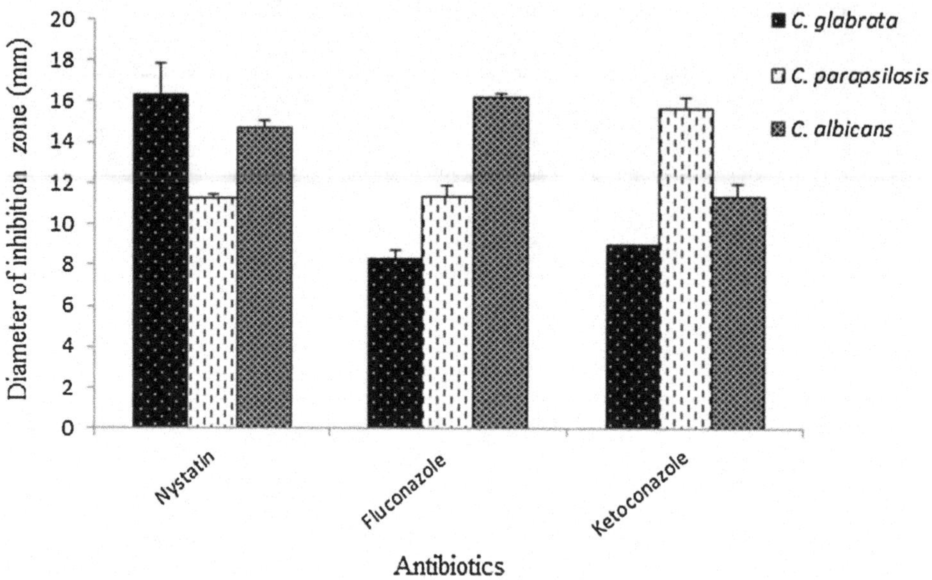

Fig. 9 Antibiogram tests against Candid strains. The concentration of antibiotics was 30 μg mL^{-1}

In the present study, among the tissues of sea cucumber, almost all extracts of the gastrointestinal tract had effect on both species of *Streptococcus*, whereas the effects of the extracts of body wall and respiratory tree were different. All of the extracts prepared from the body wall did not show any inhibitory effect on *S. mutans*, whereas its antibacterial activity against *S. sobrinus* indicated the presence of the effective active compounds in this tissue. The concentrated extracts of all tissues (1500 μg $well^{-1}$) had more significant inhibitory effects than other ones (600 μg $well^{-1}$). The different effects of the sea cucumber extracts can be related to the animal's habitat and its feeding on microorganisms presented in the sea bed. In addition, the antibacterial activity might be both due to the factors of the innate immune system or bacterial symbionts (Strahl et al. 2002) living on the organisms.

According to the results, as some extracts had shown antibacterial activities against these two oral streptococci, they consist of compounds which can be used for further and more in depth studies due to inhibition of oral infections caused by these bacteria.

Antifungal effects of the extracts

The antifungal activity has previously been described in some species of echinoderms (Batrakov et al. 1980; Kumar et al. 2007; Mohammadizadeh et al. 2013; Mokhlesi et al. 2011). The antifungal activity of the extracts of the respiratory tree was more than the extracts of the body wall and intestine of the sea cucumber, *Holothuria scabra,* against *Candida albicans* and *Aspergillus niger* (Mohammadizadeh et al. 2013). The different organs of the sea cucumber, *Bohadschia marmorata*, including Cuvier organ, body wall and coelomic fluid showed antifungal activities in which the Et extract of the body wall and water–alcohol extract of the Cuvier organ had the highest effects (Mokhlesi et al. 2011). In the present study, the organic extracts (ACN and Et) of the different tissues of the sea cucumber, *Holothuria leucospilota,* had generally more antifungal effects than the PBS extracts. The observed differences between the results of this study and the other ones could be attributed to the animal habitat and the sea bed that the cucumber feed on it. In addition, the extracts of the respiratory tree had generally more antifungal activity against Candida strains which was followed by the body wall and gastrointestinal tract extracts, respectively. The antifungal activity of the respiratory tree in the sea cucumber could be referred to the defensive role of it which was demonstrated by Purcell et al. (2012). The antimicrobial activity of the body wall is evident as the first defensive line.

The studied species possesses hemolytic activity in their tissues. Several hemolytic factors had been isolated from echinoderms, including proteins (Canicatti and D'Ancona 1990), lectins (Hatakeyama et al. 1994), complement-like factors (Bertheussen 1983), and the variety of saponins (Iorizzi et al. 2001). In the present study, it seems that the hemolytic activities of the extracts generally are higher than their antimicrobial activities. Some extracts had highly hemolytic activity without any effects on streptococci (e.g., Et extract of respiratory tree) while some extracts showed more antibacterial activity than the hemolytic one (e.g., the PBS extract of respiratory tree). From the pharmaceutical point of view, it is a requirement that the antibacterial drugs have no side effects such as hemolytic activity (Haug et al. 2002).

Conclusion

The present study shows that the sea cucumber, *Holothuria leucospilota*, as an echinoderm, can be a source of antibiotics. *Streptococcus sobrinus* had generally been less tolerant than *S. mutans* against the sea cucumber extracts, therefore almost all extracts except Et extract of respiratory tree can inhibit the growth of *S. sobrinus*. The growth inhibition of *Streptococcus sobrinus* and *S. mutans* was generally controlled better by PBS and organic extracts, respectively. The extract of the respiratory tree had the maximum antifungal effect against Candida in comparison to the other tissues. In addition, it seemed that the sea cucumber extracts controlled the growth of *Candida parapsilosis* better than the other candida fungi. Finally, the extracts of the sea cucumber should be further analyzed to isolate, purify and determine the chemical structure of the antimicrobial compounds for their applications in the medicine and dentistry industries.

References

Aas JA, Griffen AL, Dardis SR, Lee AM, Olsen I, Dewhirst FE, Leys EJ, Paster BJ (2008) Bacteria of dental caries in primary and permanent teeth in children and young adults. J Clin Microbiol 46:1407–1417

Anaissie EJ, Ginnis M, Pfaller MA (2003) Clinical mycology, 2nd edn. Churchill Livingstone, Oxford

Anusavice K (2002) Dental caries: risk assessment and treatment solutions for an elderly population. Compend Contin Educ Dent 23:12–20

Batrakov SG, Girshovich ES, Drozhzhin NS (1980) Triterpene glycosides with Antifungal activity isolated from the sea cucumber, *Cucumaria japonica*. Antibiotiki 25:408–411

Becker MR, Paster BJ, Leys EJ, Moeschberger ML, Kenyon SG, Galvin JL, Boches SK, Dewhirst FE, Griffen AL (2002) Molecular analysis of bacterial species associated with childhood caries. J Clin Microbiol 40:1001–1009

Beighton D (2005) The complex oral microflora of high-risk individuals and groups and its role in the caries process. Community Dent Oral Epidemiol 33:248–255

Bertheussen K (1983) Complement-like activity in sea urchin coelomic fluid. Dev Comp Immunol 7:21–31

Bhakuni DS, Rawat DDS (2005) Bioactive marine natural products. Springer, New York

Canicatti C, D'Ancona G (1990) Biological protective substances in *Marthasterias glacialis* (Asteroidea) epidermal secretion. J Zool 222:445–454

Casas SM, Comesaña P, Cao A, Villalba A (2011) Comparison of antibacterial activity in the hemolymph of marine bivalves from Galicia (NW Spain). J Invertebr Pathol 106:343–345

Chhour KL, Nadkarni MA, Byun R, Martin FE, Jacques NA, Hunter N (2005) Molecular analysis of microbial diversity in advanced caries. J Clin Microbiol 43:843–849

Dabbagh AR, Sedaghat MR, Rameshi H, Kamrani E (2011) Breeding and larval rearing of the sea cucumber Holothuria leucospilota Brandt (Holothuria vegabunda Selenka) from the northern Persian Gulf, Iran. SPC Beche-De-Mer Information Bullet 31:35–38

De Melo AA, Carneiro RF, De Melo Silva W, Moura Rda M, Silva GC, De Sousa OV, Nascimento KS, Saker-Sampaio S, Nagano CS, Cavada BS, Sampaio AH (2014) HGA-2, a novel galactoside-binding lectin from the sea cucumber *Holothuria grisea* binds to bacterial cells. Int J Biol Macromol 64:435–442

Dewhirst FE, Chen T, Izard J, Paster BJ, Tanner ACR, Yu WH, Lakshmanan A, Wade GW (2010) The human oral microbiome. J Bacteriol 192:5002–5017

Fenical W (1996) Marine biodiversity and the medicine cabinet the status of new drugs from marine organisms. Oceanography 8:23–27

Fredalina BD, Ridzwan BH, Zainalabdin AA, Kaswandi MA, Zaiton H, Zali I, Kittakoop P, Jais AM (1999) Fatty acid composition in local sea cucumber, *Stichopus chloronotus* for wound healing. Gen Pharmacol 33:337–340

Ghasempour M, Rajabnia R, Irannejad A, Hamzeh M, Ferdosi E, Bagheri M (2013) Frequency, biofilm formation and acid susceptibility of *Streptococcus mutans* and *Streptococcus sobrinus* in saliva of preschool children with different levels of caries activity. Dent Res J 10:440–445

Gowda NM, Goswami U, Islam KM (2008) T-antigen binding lectin with antibacterial activity from marine invertebrate, sea cucumber (*Holothuria scabra*): Possible involvement in differential recognition of bacteria. J Invertebr Pathol 99:141–145

Hatakeyama T, Kohzaki H, Nagatomo H, Yamasaki N (1994) Purification and characterization of four Ca2þ-dependent lectins from the marine invertebrate, *Cucumaria echinata*. J Biochem 116:209–214

Haug T, Kjuu AK, Styrvold OB, Sandsdalen E, Olsen OM, Stensvag K (2002) Antibacterial activity in *Strongylocentrotus droebachiensis* (Echinoidea), *Cucumaria frondosa* (Holothuroidea), and *Asterias rubens* (Asteroidea). J Invertebr Pathol 1:94–102

Iorizzi M, DeMarino S, Zollo F (2001) Steroidal oligoglycosides from the asteroidea. Curr Org Chem 5:951–973

Kantarcioglu AS, Yucel A (2002) The presence of fluconazole-resistant *Candida dubliniensis* strain among *Candida albicans* isolated from immunocompromised or otherwise debilitated HIV negative Turkish patients. Revista Iberoamericana de Micología 19:44–48

Kazemi S, Heidari B, Rassa M (2016) Antibacterial and hemolytic effects of aqueous and organic extracts from different tissues of sea urchin *Echinometra mathaei* on pathogenic streptococci. Inter Aqua Res 8:299–308

Kiani N, Heidari B, Rassa M, Kadkhodazadeh M, Heidari B (2014) Antibacterial activity of the body wall extracts of sea cucumber (Invertebrata; Echinodermata) on infectious oral streptococci. J Basic Clin Physiol Pharmacol 25:367–373

Kumar R, Chaturvedi AK, Shukla PK, Lakshmi V (2007) Antifungal activity in triterpene glycosides from the sea cucumber *Actinopyga lecanora*. Bioorg Med Chem Lett 17:4387–4391

Lewbart GA (2006) Invertebrate medicine, 1st edn. Blackwell Pub, Iowa

Mohammadizadeh F, Ehsanpor M, Afkhami M, Mokhlesi A, Khazaali A, Montazeri S (2013) Evaluation of antibacterial, antifungal and cytotoxic effects of *Holothuria scabra* from the North Coast of the Persian Gulf. J Mycol Med 23:225–229

Mokhlesi A, Saeidnia S, Gohari AR, Shahverdi AR, Moghaddam KM, ES'haghi N (2011) Antibacterial, antifungal and cytotoxic activities of *Bohadschia marmorata* a sea cucumber from North coastal of Persian. Pharmacologyonline 3:1029–1038

Molero G, Diez-Orejas R, Navarro-Garcia F, Monteoliva L, Pla J, Gil C, Sánchez-Pérez M, Nombela C (1998) *Candida albicans*: genetics, dimorphism and pathogenicity. Int Microbiol 1:95–106

Nascimento MM, Lemos JA, Abranches J, Gonçalves RB, Burne RA (2004) Adaptive acid tolerance response of *Streptococcus sobrinus*. J Bacteriol 186:6383–6390

Nishikawara F, Katsumura S, Ando A, Tamaki Y, Nakamura Y, Sato K, Nomura Y, Hanada N (2006) Correlation of cariogenic bacteria and dental caries in adults. J Oral Sci 48:245–251

Okada M, Soda Y, Hayashi F, Doi T, Suzuki J, Miura K, Kozai K (2005) Longitudinal study of dental caries incidence associated with *Streptococcus mutans* and *Streptococcus sobrinus* in pre-school children. J Med Microbiol 54:661–665

Osama YA, Ridzwan BH, Muhammad T, Jamaludin MD, Masa-Aki I, Zali BI (2009) In vitro antioxidant and antiproliferative activities of three Malaysian sea cucumber species. Eur J Sci Res 37:376–387

Periyasamy N, Srinivasan M, Balakrishnan S (2012) Antimicrobial activities of the tissue extracts of *Babylonia spirata* Linnaeus, 1758 (Mollusca: Gastropoda) from Thanzhanguda, southeast coast of India. Asian Pac J Trop Biomed 2:36–40

Price ARG (1981) Studies on the echinoderm fauna of the western Gulf. J Nat Hist 15:1–15

Purcell SW, Samyn Y, Conand C (2012) Commercially important sea cucumbers of the world. FAO Species Catalogue for Fishery Purposes 6:1–149

Ridzwan BH, Kaswandi MA, Azman Y, Fuad M (1995) Screening for antibacterial agents in three species of sea cucumbers from coastal areas of Sabah. Gen Pharmacol 26:1539–1543

Rowe FW (1969) A review of the family Holothuriidae (Holothuroidea: Aspidochirotida). Bull Br Mus Nat Hist (Zool) 18:119–170

Sharanappa R, Vidyasagar GM (2013) Anti-Candida activity of medicinal plants. Int J Pharm Pharm Sci 5:9–16

Stepanovic S, Antic N, Dakic I, Svabicvlahovic M (2003) In vitro antimicrobial activity of propolis and synergism between propolis and antimicrobial drugs. Microbiol Res 158:353–357

Strahl E, Dobson W, Lundie JL (2002) Isolation and screening of brittlestar-associated bacteria for antibacterial activity. Curr Microbiol 44:450–459

Villasin J, Pomory CM (2000) Antibacterial activity of extracts from the body wall of *Parastichopus parvimensis* (Echinodermata: Holothuroidea). Fish Shellfish Immunol 10:465–467

Wang Z, Zhang H, Yuan W, Gong W, Tang H, Liu B, Li L, Yia Y, Zhanga W (2012) Antifungal nortriterpene and triterpene glycosides from the sea cucumber *Apostichopus japonicus* Selenka. Food Chem 132:295–300

4

α-Melanocyte-stimulating hormone directly increases the plasma calcitonin level and involves calcium metabolism in goldfish

Takahiro Ikari · Yuki Kobayashi · Yoichiro Kitani · Toshio Sekiguchi · Masato Endo · Akira Kambegawa · Kiyoshi Asahina · Atsuhiko Hattori · Yoshiaki Tabuchi · Thumronk Amornsakun · Kanta Mizusawa · Akiyoshi Takahashi · Nobuo Suzuki

Abstract The effects of α-melanocyte-stimulating hormone (α-MSH) on calcium metabolism were examined with goldfish. The scales on the left side of goldfish bodies were removed to allow the regeneration of scales under anesthesia. Thereafter, the influences of α-MSH injection (low dose: 0.1 μg/g body weight; high dose: 1 μg/g body weight) on plasma calcitonin (calcium-regulating hormone) and the calcium content of the scales were investigated. Ten days after removing the scales, we measured the plasma calcitonin and calcium content of both regenerating scales on the left side and ontogenic scales on the right side. At both doses of α-MSH injection, plasma calcitonin concentrations in the α-MSH-treated group were significantly higher than those in the control group. The mRNA expressions of α-MSH-receptors were detected in the ultimobranchial glands (secretory organ of calcitonin), indicating that α-MSH directly functions in ultimobranchial glands and promotes calcitonin secretion. Furthermore, we found that the calcium content of regenerating scales in α-MSH-treated goldfish was higher than that in control goldfish, while the calcium content of ontogenic scales on the right side was significantly decreased by α-MSH injection. There was a significant co-relationship between plasma calcitonin and the calcium content of regenerating scales. The mRNA expression of calcitonin receptors in regenerating scales was remarkably higher than that in ontogenic scales. These results imply that calcitonin functions to promote scale regeneration resulting from the inhibition of bone resorption because

---

T. Ikari · Y. Kitani · T. Sekiguchi · N. Suzuki (✉)
Noto Marine Laboratory, Division of Marine Environmental Studies, Institute of Nature and Environmental Technology, Kanazawa University, Noto-Cho, Ishikawa 927-0553, Japan
e-mail: nobuos@staff.kanazawa-u.ac.jp

Y. Kobayashi
Graduate School of Integrated Arts and Sciences, Hiroshima University, Higashi-Hiroshima 739-8521, Hiroshima, Japan

M. Endo
Department of Marine Biosciences, Division of Marine Science, Tokyo University of Marine Science and Technology, Minato-Ku, Tokyo 108-8477, Japan

A. Kambegawa
The Kambegawa Laboratory, Komae, Tokyo 201-0013, Japan

K. Asahina
Department of Marine Science and Resources, College of Bioresource Sciences, Nihon University, Fujisawa 252-0813, Kanagawa, Japan

A. Hattori
Department of Biology, College of Liberal Arts and Sciences, Tokyo Medical and Dental University, Ichikawa, Chiba 272-0827, Japan

calcitonin suppresses osteoclastic activity. Thus, we are the first to demonstrate the interaction between α-MSH and calcitonin in teleosts.

Keywords α-Melanocyte-stimulating hormone · Calcitonin · Calcium · Fish scales · Ultimobranchial glands · Scale regeneration

Introduction

Alpha-melanocyte-stimulating hormone (α-MSH), which is composed of 13 amino acids, is a proteolytic cleavage product generated from adrenocorticotropic hormone (Takahashi and Kawauchi 2006; Brzoska et al. 2008; Takahashi and Mizusawa 2013). The amino acid sequence of α-MSH is well conserved among vertebrates and functions in several tissues via melanocortin receptors (Brzoska et al. 2008; Takahashi and Mizusawa 2013; Dores et al. 2016). Recently, a new function of α-MSH has been determined. In mammals, namely, it has been reported that α-MSH directly functions in bone metabolism (Farooqi et al. 2000) and promotes bone resorption (Cornish et al. 2003). Furthermore, Cornish et al. (2003) reported that trabecular bone volume was reduced by 22% with the administration of α-MSH to mice (20 injections of 4.5 μg/day). Therefore, α-MSH is one hormone that regulates bone metabolism in mammals.

In teleosts as well as mammals, we previously demonstrated that α-MSH functions in goldfish scales and induces hypercalcemia resulting from the promotion of bone resorption (Ishizu et al. 2018). In vertebrates, including teleosts, the plasma calcium concentration is maintained at a constant level by hormonal regulation. This implies that hypocalcemic hormones such as calcitonin respond to hypercalcemia in goldfish (*Carassius auratus*). In eels, *Anguilla japonica* (freshwater teleosts), plasma calcitonin levels actually increased with the rise of plasma calcium caused by the dietary uptake of calcium (Suzuki et al. 1999). In stonefish, *Inimicus japonicus* (marine teleosts) also, both plasma calcium and calcitonin levels increased after the administration of a high-calcium solution into the stomach (Kaida and Sasayama 2003). Thus, we focused on the relationship between α-MSH and calcitonin.

On the other hand, teleost scales are functional internal calcium reservoirs involved in calcium metabolism, particularly in freshwater teleosts such as goldfish (Mugiya and Watabe 1977; Suzuki et al. 2008, 2016). Additionally, it is known that teleost scales regenerate after being removed (Bereiter-Hahn and Zylberberg 1993; Suzuki et al. 2009; Yoshikubo et al. 2005). During scale regeneration, both osteoblastic and osteoclastic activities in regenerating scales were higher than those in ontogenic scales (Yoshikubo et al. 2005). As α-MSH induced hypercalcemia (Ishizu et al. 2018), we strongly believe that α-MSH involves calcium metabolism related to scale regeneration.

Thus, in the present study, the influence of α-MSH injection on calcium regulation during scale regeneration was examined in goldfish. In brief, after α-MSH administration, plasma calcitonin (calcium-regulating hormone) and the calcium content of both regenerating and ontogenic scales were investigated in the present study. Furthermore, we analyzed the expression of α-MSH and calcitonin receptors. Our investigation is the first to demonstrate the interaction between α-MSH and calcitonin in teleosts.

Y. Tabuchi
Division of Molecular Genetics Research, Life Science Research Center, University of Toyama, Toyama 930-0194, Japan

T. Amornsakun
Fisheries Technology Program, Faculty of Science and Technology, Prince of Songkla University, Pattani 94000, Pattani, Thailand

K. Mizusawa · A. Takahashi
School of Marine Biosciences, Kitasato University, Sagamihara, Kanagawa 252-0373, Japan

Materials and methods

Animals

One pair of female and male goldfish (*Carassius auratus*) was artificially fertilized at the Graduate School of Marine Science and Technology, Tokyo University of Marine Science and Technology (Suzuki et al. 2009). After the hatched goldfish grew to a body length of about 12 cm, these fish (n = 40) were transferred to Noto Marine Laboratory at Kanazawa University, and used for an in vivo experiment. To avoid the effects of sex hormones, non-breeding goldfish were used in the present study. All experimental procedures were conducted in accordance with the Guide for the Care and Use of Laboratory Animals prepared by Kanazawa University.

Effects of α-MSH on plasma calcitonin and the calcium content of both regenerating and ontogenic scales in goldfish

In both control and experimental groups, normally developed scales on the left sides of goldfish were removed to allow the regeneration of scales under anesthesia with 0.03% ethyl 3-aminobenzoate and methanesulfonic acid salt (Sigma-Aldrich, Inc., St. Louis, MO, USA) neutralized with 0.03% sodium bicarbonate. Thereafter, the influence of α-MSH on the calcium metabolism regarding scale regeneration was studied. Namely, α-MSH was administered at a low dose (0.1 μg/g body weight) or a high dose (1 μg/g body weight) to goldfish every other day in the morning. During all experimental periods, goldfish were kept at 26 °C under a daily photoperiod cycle of 12-h light: 12-h darkness (Suzuki et al. 2009). Saline (0.9% NaCl) was injected into the goldfish in the control group in the same manner as in the experimental group (each n = 8). During the experimental periods, both experimental and control goldfish were fed every morning. Ten days after scale removal, we collected both the regenerating scales on the left side, and ontogenic scales on the right side. The scale-calcium content [mg/dry weight (mg) of scale] was determined using the Calcium C kit (Wako Pure Chemical Industries, Ltd., Osaka, Japan) with a microplate reader after the dried scale (60 °C, 12 h) had been dissolved in nitric acid and then neutralized by NaOH (Suzuki et al. 2004, 2011). After removing the regenerating scales and ontogenic scales, blood samples were collected from the caudal vessel using a heparinized capillary from anesthetized goldfish. After centrifugation at 15,000 rpm for 3 min, the plasma was immediately frozen and kept at − 80 °C.

Plasma calcitonin levels were measured by enzyme-linked immunosorbent assay (ELISA). The procedures of ELISA have been described in detail by Suzuki (2001). The detection limit was 25 pg/ml (Suzuki 2001). The specificity of anti-salmon calcitonin serum (No. 626, Cosmo Bio Co., Ltd., Tokyo, Japan) was checked using peptide hormones (N-terminal fragment of 1–34 bovine parathyroid hormone and human calcitonin gene-related peptide). This anti-serum did not cross-react to these peptide hormones.

The mRNA expressions of melanocortin receptors (MCRs) in the ultimobranchial glands of goldfish

We examined the mRNA expression of melanocortin receptors (MCRs) in the ultimobranchial glands, which are the secretory organ of calcitonin. The ultimobranchial glands were dissected from goldfish (three males and three females) under anesthesia with 0.03% ethyl 3-aminobenzoate and methanesulfonic acid salt (Sigma-Aldrich) neutralized with 0.03% sodium bicarbonate. Also, the brains were removed from goldfish (one male and one female) as a positive control (Mizusawa et al. 2018). Total RNAs were prepared from dissected ultimobranchial glands and brains using a total RNA isolation kit (NucleoSpin RNA II, Takara Bio, Inc., Shiga, Japan) and treated with DNase I (RNase-Free DNase Kit, Takara Bio, Inc.) for 15 min at room temperature to remove residual genomic DNA. Complementary DNA synthesis was also performed using a kit (PrimeScript™ II 1st strand cDNA Synthesis Kit, Takara Bio, Inc.).

The specific primers for the MCRs (melanocortin 1 receptor: MC1R; melanocortin 2 receptor: MC2R; melanocortin 3 receptor: MC3R; melanocortin 4 receptor: MC4R; melanocortin 5 receptor: MC5R) (Table 1) were used (Kobayashi et al. 2011). The sample on which PCR was performed with the addition of reverse transcriptase was compared with a sample on which PCR was performed without reverse transcriptase as a negative control. β-actin was used as a housekeeping gene (Kobayashi et al. 2011). The primer sequences of β-actin are described in Table 1. The conditions for PCR amplification were 40 for MCRs or 35 for β-actin

Table 1 Custom oligonucleotide primers used for the expression analysis of MCRs in the ultimobranchial glands of goldfish

Target	Nucleotide sequence
MC1R	FW 5′-GCT TGT CAC GGC AAA GAT GT-3′
MC1R	RV 5′-TGG CTT GTC GGC GAC TCT TA-3′
MC2R	FW 5′-ACA CCT GAA CGG TCG TTT CG-3′
MC2R	RV 5′-CTC AAG CCA CTT TGT CTC TG-3′
MC3R	FW 5′-TGT CTG TTC TTCCCC ATC TC-3′
MC3R	RV 5′-GGC GAT TGT TTA GTA CAG CA-3′
MC4R	FW 5′-TGC CTC CGA AAC GGT AGT GA-3′
MC4R	RV 5′-GCT GAT AAG GCA GAT GAG AA-3′
MC5R	FW 5′-CTG TCA CTT TGG GCC ATC AG-3′
MC5R	RV 5′-TCT GAT GAA ATG GTC CTC CA-3′
β-actin	FW 5′-TGA AGT ACC CCA TCG AGG CA-3′
β-actin	RV 5′-AGG ATC TTC ATG AGG TAG TC-3′

FW forward primer, *RV* reverse primer

cycles of denaturation for 15 s at 94 °C, annealing for 30 s at 60 °C, and extension for 40 s at 72 °C. PCR products were electrophoresed on 3% agarose gel (Agarose S, Nippon Gene Co., Ltd., Toyama, Japan) and visualized with 0.025% ethidium bromide. Photographs were taken using a Densitograph (ATTO Corp., Tokyo, Japan).

Comparison of calcitonin receptor mRNA expression in regenerating scales and ontogenic scales of goldfish

To analyze calcitonin receptor mRNA expression, both regenerating and ontogenic scales from goldfish ($n = 10$) were prepared again.

Total RNAs were prepared from goldfish scales using a total RNA isolation kit (NucleoSpin RNA II, Takara Bio, Inc.), as described above. The PCR amplification was analyzed with a real-time PCR apparatus (Mx3000pTM, Stratagene, La Jolla, CA, USA) using SYBR Premix Ex Taq (Takara Bio, Inc.) (Suzuki et al. 2011; Sato et al. 2017). Real-time qPCR was performed using the specific primer set (forward primer: 5′-AAAGCAGAGCCCACCACTGA-3′; reverse primer: 5′-CTGCTGCAGAACGAACCTGT-3′) for calcitonin receptors (Ikari et al. 2018). The annealing temperature for calcitonin receptors was 55 °C (Ikari et al. 2018). The mRNA expression level of calcitonin receptors was normalized to the mRNA expression level of β-actin (forward primer: 5′-CGAGCGTGGCTACAGCTTCA-3′; reverse primer: 5′-GCCCGTCAGGGAGCTCA-TAG-3′) as a housekeeping gene (Azuma et al. 2007). The annealing temperature for β-actin was 60 °C (Azuma et al. 2007). The results are shown as the mean ± SEM ($n = 10$ individual goldfish).

Statistical analysis

All results are expressed as the mean ± SE. The statistical significance between the control and experimental groups was assessed by Student's *t* test or by one-way ANOVA followed by Dunnett's test. In all cases, the selected significance level was $p < 0.05$.

Results

Effects of α-MSH injection on plasma calcitonin in goldfish

The plasma calcitonin levels are indicated in Fig. 1. α-MSH was injected at a low dose (0.1 μg/g body weight) or a high dose (1 μg/g body weight) into goldfish every other day. At both doses, plasma calcitonin levels in α-MSH-injected goldfish increased significantly as compared with those in control goldfish.

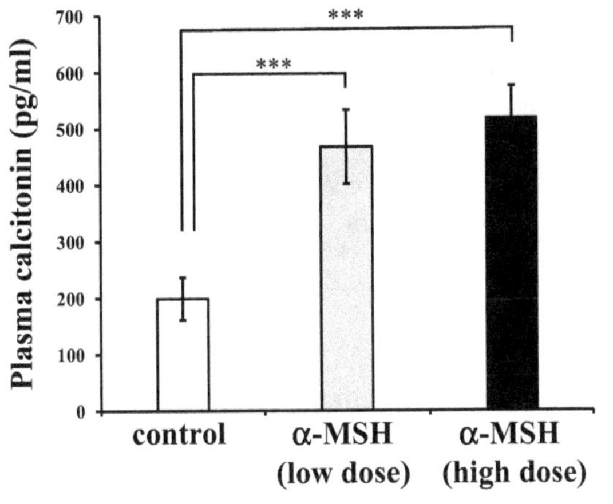

Fig. 1 Effects of α-MSH on plasma calcitonin levels in goldfish. α-MSH was injected at a low dose (0.1 μg/g body weight) or a high dose (1 μg/g body weight) into goldfish every other day. Ten days after injection, we collected plasma samples and analyzed their calcitonin levels. ***Indicates a statistically significant difference at $p < 0.001$ from the values in control scales. $n = 8$ samples; one sample from one fish

The mRNA expressions of MCRs in ultimobranchial glands of goldfish

To examine the interaction between α-MSH and calcitonin, the mRNA expressions of MCRs were analyzed in the ultimobranchial glands. In males (Fig. 2b) and females (Fig. 2d), MCR1, 2, 3, 4, and 5 were detected in the ultimobranchial glands of goldfish, although variation in the expression level of each individual was observed. β-actin, as a housekeeping gene, was expressed in each ultimobranchial gland of goldfish (Fig. 2e). On the other hand, the brains of males (Fig. 2a) and females (Fig. 2c), used as a positive control expressed all MCRs. In the sample on which PCR was performed without reverse transcriptase, specific bands were not detected (data not shown).

Effects of α-MSH injection on the calcium content of both regenerating and ontogenic scales in goldfish

The calcium content of both regenerating and ontogenic scales in goldfish is shown in Fig. 3. Ten days after the removal of scales, the calcium content of regenerating scales in goldfish treated with high doses of α-MSH was significantly higher than that in control goldfish. On the other hand, the calcium content of ontogenic scales on the right side was significantly decreased by high-dose injections of α-MSH. There was a significant co-relationship between the plasma calcitonin and calcium content of regenerating scales in both the control and experimental groups (Fig. 4a), although the plasma calcitonin level did not have a co-relationship with the calcium content of the ontogenic scales (Fig. 4b).

Comparison of calcitonin receptor mRNA expression of regenerating and ontogenic scales of goldfish

The mRNA expression of calcitonin receptors was detected in both regenerating and ontogenic scales (Fig. 5). The calcitonin receptor mRNA expression in regenerating scales was remarkably higher than that in ontogenic scales (Fig. 5). There was a significant difference between regenerating scales and ontogenic scales.

Discussion

Using α-MSH-injected goldfish, we are the first to demonstrate the interaction between α-MSH and calcitonin in teleosts. In brief, after α-MSH administration (low dose: 0.1 μg/g body weight; high dose: 1 μg/g body weight), the plasma calcitonin level increased significantly at both doses (Fig. 1). To examine the direct influence of α-MSH on calcitonin production, MCR mRNA expression was measured in the ultimobranchial glands of goldfish. As a result, MCR1, 2, 3, 4, and 5 were detected in goldfish ultimobranchial glands (Fig. 2).

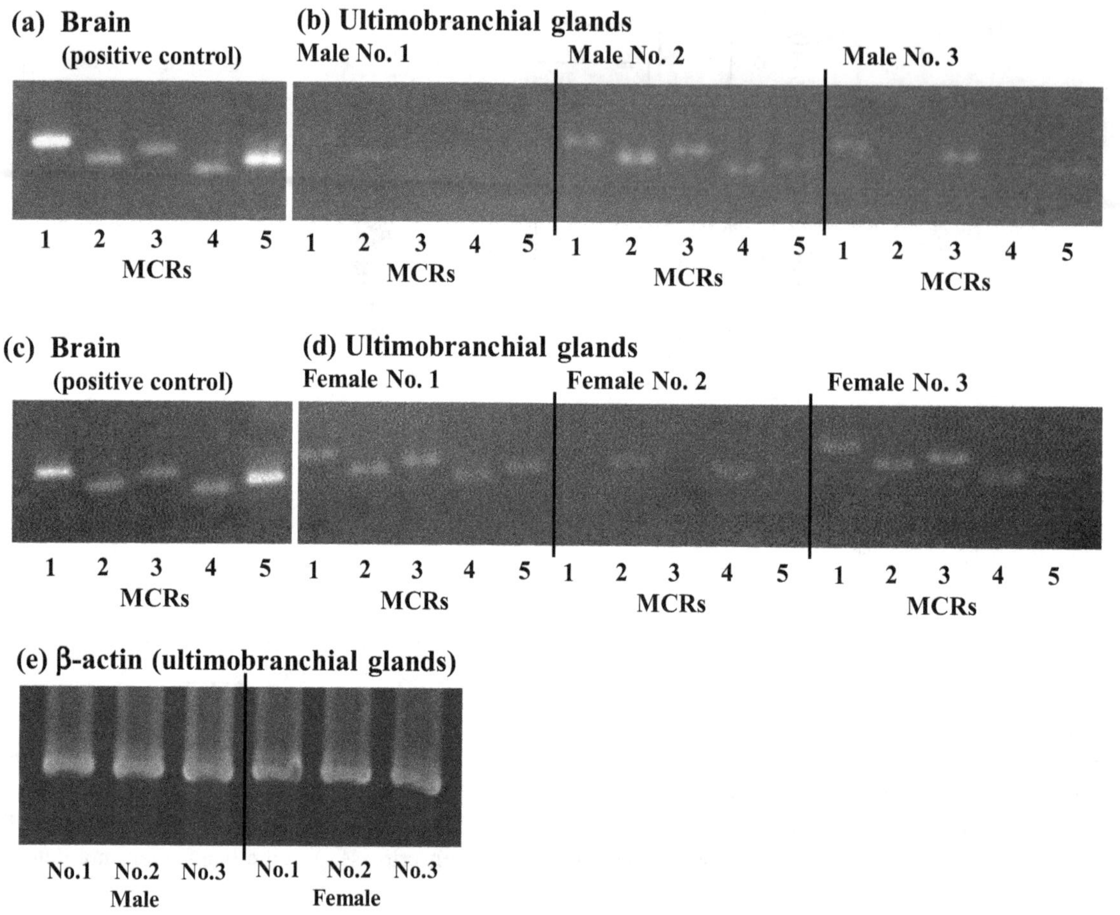

Fig. 2 The expressions of melanocortin receptors' (MCRs: MC1R, MC2R, MC3R, MC4R, and MC5R) mRNA in the brains (**a** male; **c** female) and ultimobranchial glands (**b** male; **d** female) of goldfish. The mRNA was prepared from the brains or ultimobranchial glands of male and female goldfish. The mRNA expressions of MCRs were analyzed in brains or ultimobranchial glands. β-actin was used as a housekeeping gene (**e**) (Kobayashi et al. 2011)

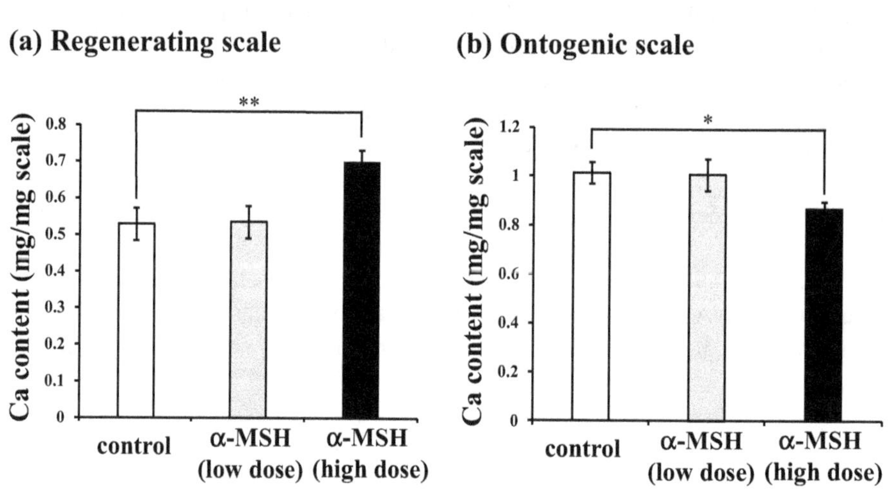

Fig. 3 Effects of α-MSH on calcium content in regenerating and ontogenic goldfish scales. α-MSH was injected at a low dose (0.1 μg/g body weight) or a high dose (1 μg/g body weight) into goldfish every other day. Ten days later, we analyzed the scale-calcium content in goldfish. * and **indicate statistically significant differences at $p < 0.05$ and $p < 0.01$, respectively, from the values in the control scales. $n = 8$ samples; one sample from one fish

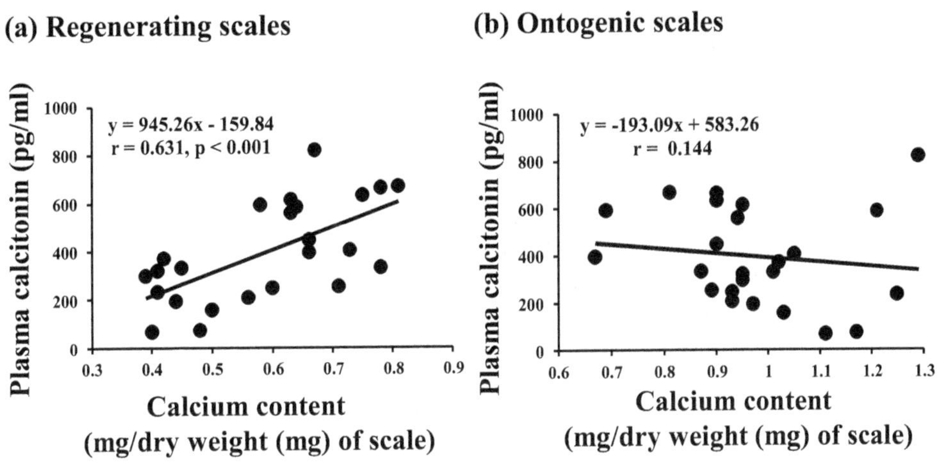

Fig. 4 Relationship between plasma calcitonin levels and scale-calcium content (**a** regenerating scales; **b** ontogenic scales). On the basis of data from Figs. 1 and 3, the relationship between plasma calcitonin level and scale-calcium content was examined. The data of control, low-dose-treated, and high-dose-treated goldfish were combined. There was a significant co-relationship ($r = 0.631$, $p < 0.001$) between the plasma calcitonin and calcium content of regenerating scales

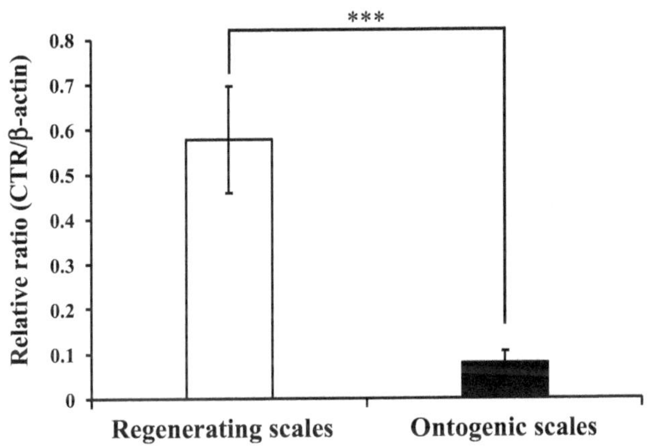

Fig. 5 Expression analysis of calcitonin receptor mRNA in regenerating and ontogenic scales. Ten days after removal of the scales, mRNA was prepared from regenerating and ontogenic scales. The mRNA expressions of calcitonin receptors were analyzed in scales. ***indicates a statistically significant difference at $p < 0.001$ from the values in the control scales. $n = 10$ samples; one sample from one fish. β-actin was used for normalization to each mRNA expression level

This implies that α-MSH directly acts on ultimobranchial glands and promotes the secretion of calcitonin. We believe that the secreted calcitonin functions in regenerating scales and facilitates scale bone formation (Fig. 3a) because the plasma calcitonin level has a co-relationship with the calcium content of the regenerating scales (Fig. 4a) but not with the content of ontogenic scales (Fig. 4b). In addition, the calcitonin receptor mRNA expression in regenerating scales was significantly higher than that in ontogenic scales (Fig. 5), supporting the conclusion that calcitonin functions in regenerating scales.

In the brain of goldfish (Mizusawa et al. 2018) and the barfin flounder *Verasper moseri* (Takahashi et al. 2014), MCR1, 2, 3, 4, and 5 were expressed. In the case of goldfish ultimobranchial glands, MCR1, 2, 3, 4, and 5 were detected, although variation in the expression level of each individual was observed (Fig. 2). MCR mRNA expression in the ultimobranchial glands was different among individuals (Fig. 2). In male No. 2 and female Nos. 1 and 3, all MCRs were expressed in the ultimobranchial glands of goldfish. In other males (Nos. 1 and 3) and a female (No. 2), however, only two or three types of MCRs were detected in their ultimobranchial glands, at least in the present conditions. This MCR expression in ultimobranchial glands may be related to physiological conditions within individuals. In feeding, calcitonin has some functions in teleosts, as described in the "Introduction". In eels, *Anguilla japonica* (freshwater teleosts), and stonefish, *Inimicus japonicus* (marine teleosts), plasma calcitonin levels increased with the rise of plasma calcium caused by the dietary uptake of calcium (Suzuki et al. 1999; Kaida and Sasayama 2003). It is known that α-MSH functions in feeding behavior (for a review, see Metz et al. 2006) and induces anorexigenic actions in goldfish (Kojima et al. 2010). During the feeding period, therefore, α-MSH may have some relationship with calcitonin. In our next study, we would like to investigate the interaction between α-MSH and calcitonin during feeding time.

We previously indicated that α-MSH functions in scales and promotes bone resorption in goldfish (Ishizu et al. 2018). In ontogenic scales, the calcium content decreased (Fig. 3b), and might accelerate scale-bone resorption by α-MSH. In regenerating scales, however, the calcium content increased (Fig. 3a), suggesting that calcitonin has some role in this phenomenon. Calcitonin is a hypocalcemic hormone resulting from the inhibition of osteoclastic activities in mammals (Azria 1989). Furthermore, in teleosts as well as mammals, it has been demonstrated that calcitonin suppresses osteoclastic activity (Suzuki et al. 2000; Sekiguchi et al. 2009, 2017). As goldfish-calcitonin suppressed osteoclastic activity in the scales of goldfish (Suzuki et al. 2000), calcitonin induced by α-MSH-injection seems to function in regenerating scales and promoting scale regeneration.

On the other hand, we recently discovered a new function of calcitonin (Kase et al. 2017). We found that sardine procalcitonin was composed of procalcitonin amino-terminal cleavage peptide (N-proCT), calcitonin, and procalcitonin carboxyl terminal cleavage peptide (C-proCT). As compared with C-proCT, N-proCT has been highly conserved among teleosts, reptiles, and birds, which suggests that N-proCT has some bioactivities. To compare the bioactivities of calcitonin and N-proCT, we examined their bioactivities for osteoblasts and osteoclasts using our assay system with goldfish scales that consisted of osteoblasts and osteoclasts. As a result, sardine N-proCT (10^{-7} M) activated osteoblastic marker enzyme activity, while sardine calcitonin did not change. On the other hand, sardine calcitonin (10^{-9} to 10^{-7} M) suppressed osteoclastic marker enzyme activity, although sardine N-proCT did not influence enzyme activity. In α-MSH-injected goldfish, N-proCT might function in osteoblasts to regenerate scales and promote bone formation because osteoblastic activity and hormonal responses in regenerating scales were considerably higher than those in ontogenic scales (Yoshikubo et al. 2005). Rat-thyroid levels of calcitonin and N-proCT increase in parallel in vivo (Burns et al. 1989), suggesting that both calcitonin and N-proCT function in osteoclasts and osteoblasts, respectively, to regenerate scales.

Teleost scales are functional internal calcium reservoirs during periods of increased calcium demand (Mugiya and Watabe 1977; Bereiter-Hahn and Zylberberg 1993) and are a good model for analyzing bone metabolism (Vieira et al. 2011; de Vrieze et al. 2014a, b; Suzuki et al. 2016; Carnovali et al. 2016; Witten et al. 2017; Pinto et al. 2017). The osteogenesis in regenerating scales was very similar to that in calvarial bone (Yoshikubo et al. 2005; Thamamongood et al. 2012). In the present study, we can easily analyze α-MSH function regarding bone metabolism. Thus, we conclude that regenerating scales can be utilized as a model for in vivo osteogenesis.

Acknowledgements This study was supported in part by grants to N.S. (Grant-in-Aid for Scientific Research [C] No. 16K07871 by JSPS, to T.S. (Grant-in-Aid for Scientific Research [C] No. 18K06312 by JSPS), and to A.H. (Grant-in-Aid for Scientific Research [C] No. 18K11016 by JSPS). This work was partly supported by the cooperative research program of the Institute of Nature and Environmental Technology, Kanazawa University, Accept Number 18014.

Conflict of interest The authors have no competing interest to declare.

References

Azria M (1989) The calcitonins: physiology and pharmacology. Karger, Basel

Azuma K, Kobayashi M, Nakamura M, Suzuki N, Yashima S, Iwamuro S, Ikegame M, Yamamoto T, Hattori A (2007) Two osteoclastic markers expressed in multinucleate osteoclasts of goldfish scales. Biochem Biophys Res Commun 362:594–600

Bereiter-Hahn J, Zylberberg L (1993) Regeneration of teleost fish scale. Comp Biochem Physiol A 105:625–641

Brzoska T, Luger TA, Maaser C, Abels C, Böhm M (2008) α-melanocyte-stimulating hormone and related tripeptides: biochemistry, antiinflammatory and protective effects in vitro and in vivo, and future perspectives for the treatment of immune-mediated inflammatory diseases. Endocr Rev 29:581–602

Burns DM, Birnbaum RS, Roos BA (1989) A neuroendocrine peptide derived from the amino-terminal half of rat procalcitonin. Mol Endocrinol 3:140–147

Carnovali M, Luzi L, Banfi G, Mariotti M (2016) Chronic hyperglycemia affects bone metabolism in adult zebrafish scale model. Endocrine 54:808–817

Cornish J, Callon KE, Mountjoy KG, Bava U, Lin JM, Myers DE, Naot D, Reid IR (2003) α-melanocyte-stimulating hormone is a novel regulator of bone. Am J Physiol Endocrinol Metab 284:E1181–E1190

de Vrieze E, Moren M, Metz JR, Flik G, Lie KK (2014a) Arachidonic acid enhances turnover of the dermal skeleton: studies on zebrafish scales. PLoS One 9:e89347

de Vrieze E, van Kessel MAHJ, Peters HM, Spanings FAT, Flik G, Metz JR (2014b) Prednisolone induces osteoporosis-like phenotype in regenerating zebrafish scales. Osteoporos Int 25:567–578

Dores RM, Liang L, Davis P, Thomas AL, Petko B (2016) 60 years of POMC: melanocortin receptors: evolution of ligand selectivity for melanocortin peptides. J Mol Endocrinol 56:T119–T133

Farooqi IS, Yeo GSH, Keogh JM, Aminian S, Jebb SA, Butler G, Cheetham T, O'Rahilly S (2000) Dominant and recessive inheritance of morbid obesity associated with melanocortin 4 receptor deficiency. J Clin Invest 106:271–279

Ikari T, Sekiguchi T, Urata M, Furusawa Y, Ikegame M, Kinoshita Y, Kitamura K, Nakabayashi I, Horita M, Tabuchi Y, Hattori A, Srivastav AK, Suzuki N (2018) Sequencing and expression analysis of calcitonin receptor in the scales of goldfish, *Carassius auratus*. Int J Zool Inv 4:1–10

Ishizu H, Sekiguchi T, Ikari T, Kitamura K, Kitani Y, Endo M, Urata M, Kinoshita Y, Hattori A, Srivastav AK, Mishima H, Mizusawa K, Takahashi A, Suzuki N (2018) α-melanocyte-stimulating hormone promotes the bone resorption resulting from increased osteoblastic and osteoclastic activities in goldfish. Gen Comp Endocrinol 262:99–105

Kaida N, Sasayama Y (2003) Dynamics of plasma Ca and calcitonin levels in stonefish (*Inimicus japonicus*) administered a high-Ca solution into the stomach. Zool Sci 20:353–356

Kase Y, Ikari T, Sekiguchi T, Sato M, Ogiso S, Kawada T, Matsubara S, Satake H, Sasayama Y, Endo M, Kitamura K, Hattori A, Watanabe TX, Maruyama Y, Watanabe Y, Funahashi H, Kambegawa A, Suzuki N (2017) Sardine procalcitonin amino-terminal cleavage peptide has a different action from calcitonin and promotes osteoblastic activity in the scales of goldfish. Comp Biochem Physiol A 211:77–83

Kobayashi Y, Chiba H, Mizusawa K, Suzuki N, Cerdá-Reverter JM, Takahashi A (2011) Pigment-dispersing activities and cortisol-releasing activities of melanocortins and their receptors in xanthophores and head kidneys of the goldfish *Carassius auratus*. Gen Comp Endocrinol 173:438–446

Kojima K, Amiya N, Kamijo M, Kageyama H, Uchiyama M, Shioda S, Matsuda K (2010) Relationship between α-melanocyte-stimulating hormone- and neuropeptide Y-containing neurons in the goldfish hypothalamus. Gen Comp Endocrinol 167:366–372

Metz JR, Peters JJM, Flik G (2006) Molecular biology and physiology of the melanocortin system in fish: a review. Gen Comp Endocrinol 148:150–162

Mizusawa K, Yamamura Y, Kasagi S, Cerdá-Reverter JM, Takahashi A (2018) Expression of genes for melanotropic peptides and their receptors for morphological color change in goldfish *Carassius auratus*. Gen Comp Endocrinol 264:138–150

Mugiya Y, Watabe N (1977) Studies on fish scale formation and resorption II: effect of estradiol on calcium homeostasis and skeletal tissue resorption in the goldfish, *Carassius auratus*, and the killifish, *Fundulus heteroclitus*. Comp Biochem Physiol A 57:197–202

Pinto PIS, Estêvão MD, Santos S, Andrade A, Power DM (2017) In vitro screening for estrogenic endocrine disrupting compounds using Mozambique tilapia and sea bass scales. Comp Biochem Physiol C 199:106–113

Sato M, Yachiguchi K, Motohashi K, Yaguchi Y, Tabuchi Y, Kitani Y, Ikaria T, Ogiso S, Sekiguchi T, Hai TN, Huong DTT, Hoang NV, Urata M, Mishima H, Hattori A, Suzuki N (2017) Sodium fluoride influences calcium metabolism resulting from the suppression of osteoclasts in the scales of nibbler fish *Girella punctate*. Fish Sci 83:543–550

Sekiguchi T, Suzuki N, Fujiwara N, Aoyama M, Kawada T, Sugase K, Murata Y, Sasayama Y, Ogasawara M, Satake H (2009) Calcitonin in a protochordate, *Ciona intestinalis*: the prototype of the vertebrate calcitonin/calcitonin gene related peptide superfamily. FEBS J 276:4437–4447

Sekiguchi T, Shiraishi A, Satake H, Kuwasako K, Takahashi H, Sato M, Urata M, Wada S, Endo M, Ikari T, Hattori A, Srivastav AK, Suzuki N (2017) Calcitonin-typical suppression of osteoclastic activity by amphioxus calcitonin superfamily peptides and insights into the evolutionary conservation and diversity of their structures. Gen Comp Endocrinol 246:294–300

Suzuki N (2001) Calcitonin-like substance in the plasma of Cyclostomata and its putative role. Comp Biochem Physiol B 129:319–326

Suzuki N, Suzuki D, Sasayama Y, Srivastav AK, Kambegawa A, Asahina K (1999) Plasma calcium and calcitonin levels in eels fed a high calcium solution or transferred to seawater. Gen Comp Endocrinol 114:324–329

Suzuki N, Suzuki T, Kurokawa T (2000) Suppression of osteoclastic activities by calcitonin in the scales of goldfish (freshwater teleost) and nibbler fish (seawater teleost). Peptides 21:115–124

Suzuki N, Yamamoto K, Sasayama Y, Suzuki T, Kurokawa T, Kambegawa A, Srivastav AK, Hayashi S, Kikuyama S (2004) Possible direct induction by estrogen of calcitonin secretion from ultimobranchial cells in the goldfish. Gen Comp Endocrinol 138:121–127

Suzuki N, Somei M, Seki A, Reiter RJ, Hattori A (2008) Novel bromomelatonin derivatives as potentially effective drugs to treat bone diseases. J Pineal Res 45:229–234

Suzuki N, Kitamura K, Omori K, Nemoto T, Satoh Y, Tabata MJ, Ikegame M, Yamamoto T, Ijiri K, Furusawa Y, Kondo T, Takasaki I, Tabuchi Y, Wada S, Shimizu N, Sasayama Y, Endo M, Takeuchi T, Nara M, Somei M, Maruyama Y, Hayakawa K, Shimazu T, Shigeto Y, Yano S, Hattori A (2009) Response of osteoblasts and osteoclasts in regenerating scales to gravity loading. Biol Sci Space 23:211–217

Suzuki N, Danks JA, Maruyama Y, Ikegame M, Sasayama Y, Hattori A, Nakamura M, Tabata MJ, Yamamoto T, Furuya R, Saijoh K, Mishima H, Srivastav AK, Furusawa Y, Kondo T, Tabuchi Y, Takasaki I, Chowdhury VS, Hayakawa K, Martin TJ (2011) Parathyroid hormone 1 (1–34) acts on the scales and involves calcium metabolism in goldfish. Bone 48:1186–1193

Suzuki N, Kitamura K, Hattori A (2016) Fish scale is a suitable model for analyzing determinants of skeletal fragility in type 2 diabetes. Endocrine 54:575–577

Takahashi A, Kawauchi H (2006) Evolution of melanocortin systems in fish. Gen Comp Endocrinol 148:85–94

Takahashi A, Mizusawa K (2013) Posttranslational modifications of proopiomelanocortin in vertebrates and their biological significance. Front Endocrinol 4:143

Takahashi A, Mizusawa K, Amano M (2014) Multifunctional roles of melanocyte-stimulating hormone and melanin-concentrating hormone in fish: evolution from classical body color change. Aqua BioSci Monogr 7:1–46

Thamamongood TA, Furuya R, Fukuba S, Nakamura M, Suzuki N, Hattori A (2012) Expression of osteoblastic and osteoclastic genes during spontaneous regeneration and autotransplantation of goldfish scale: a new tool to study intramembranous bone regeneration. Bone 50:1240–1249

Vieira FA, Gregório SF, Ferraresso S, Thorne MA, Costa R, Milan M, Bargelloni L, Clark MS, Canario AVM, Power DM (2011) Skin healing and scale regeneration in fed and unfed sea bream, *Sparus auratus*. BMC Genom 12:490

Witten PE, Harris MP, Huysseune A, Winkler C (2017) Small teleost fish provide new insights into human skeletal diseases. Methods Cell Biol 138:321–346

Yoshikubo H, Suzuki N, Takemura K, Hoso M, Yashima S, Iwamuro S, Takagi Y, Tabata MJ, Hattori A (2005) Osteoblastic activity and estrogenic response in the regenerating scale of goldfish, a good model of osteogenesis. Life Sci 76:2699–2709

5

Survival time of juvenile pen shell *Atrina pectinata* (Bivalvia: Pinnidae) in hyposaline water

Takeo Kurihara · Shoji Nakano · Yukihiko Matsuyama · Kazumasa Hashimoto · Katsumasa Yamada · Atsushi Ito · Masaei Kanematsu

Abstract Survival time of juvenile pen shell *Atrina pectinata* (shell length $\approx$ 34 mm) exposed to hyposaline water without sediments was investigated in laboratory experiments. The bivalves were individually maintained in rearing cups in which salinity was decreased either gradually (over 1–11 h to 12 salinities ranging from 0 to ca. 34 ppt) or rapidly (directly to 21 salinities) at 12 and 24 °C. When the environmental salinity was gradually decreased, survival time differed markedly between the bivalves exposed to: salinities $\leq$ 18.6 ppt (survival time < 20 h in general) vs. $\geq$ 21.1 ppt (> 83 h) at 24 °C; and between salinities $\leq$ 15.1 ppt (< 34 h) vs. $\geq$ 17.2 ppt (> 83 h) at 12 °C. When the environmental salinity was rapidly decreased, survival time, again, differed markedly between the bivalves exposed to: salinities $\leq$ 15.9 ppt (< 20 h in general) vs. $\geq$ 17.7 ppt (> 140 h) at 24 °C; and salinities $\leq$ 21.3 ppt (< 38 h in general) vs. $\geq$ 22.3 ppt (> 140 h) at 12 °C. Such sharp decreases in survival time due to reduced salinity led to the selected, statistical models that assume either a sharp exponential increase or stepwise hike in hazard (instantaneous mortality rate) according to salinity reduction. These data and models indicate that juvenile *A. pectinata* should not be kept at salinities < 21 ppt for periods > 1 day in environments without sediments.

Keywords Pen shell · *Atrina pectinata* Lineage 2 · Survival analysis · Low salinity · Gradual and rapid decreases · Temperature

Introduction

Salinity is one of major environmental factors determining the physiology and behavior of marine animals (Kinne 1966). A change in environmental salinity leads to a difference in osmotic pressure between the environment and tissue cells, which is stressful for marine animals. In shallow seas salinity changes owing to various factors (e.g. tidal currents, terrestrial runoff, rain). Such changes often lead to the mortality of various

T. Kurihara (✉) · S. Nakano · Y. Matsuyama · K. Hashimoto · K. Yamada
Seikai National Fisheries Research Institute, Japan Fisheries Research and Education Agency, 1551-8 Taira-machi, Nagasaki 851-2213, Japan
e-mail: takeo@affrc.go.jp

A. Ito · M. Kanematsu
National Research Institute of Fisheries and Environment of Inland Sea, Japan Fisheries Research and Education Agency, 2-17-5 Maruishi, Hatsukaichi, Hiroshima 739-0452, Japan

animals within such areas (e.g. Kinne 1966; Gillanders and Kingsford 2002). The influence of salinity changes on survival has been studied for a diverse range of animals (Newell 1976).

These studies include hyposaline endurance of marine bivalves (e.g. Leyva-Valencia et al. 2001; Marsden 2004; Rupp and Parsons 2004; Holmes and Miller 2006; Matsuda et al. 2008; McLeod and Wing 2008; Jansen et al. 2009; Yuan et al. 2010; Yamamoto and Handa 2011). The studies on bivalves revealed that their endurance of hyposaline conditions depends on various factors including species specificity, life stage, temperature, and the rate of salinity change. That is, some species tolerate extremely low salinities such as freshwater (e.g., *Corbula gibba*, *Austrovenus stutchburyi*, *Phragmites australis*, *Mytella charruana*; Holmes and Miller 2006; McLeod and Wing 2008; Yuan et al. 2010), whereas some species cannot tolerate even 75% seawater (e.g., *Patinopecten yessoensis*, *Pecten albicans*; Funakoshi et al. 1985). Juveniles and adults often show different hyposaline endurance (e.g., *Chlamys opercularis*, *Mytella charruana*, *Perna viridis*; Paul 1980; Yuan et al. 2016). Extremely high or low water temperature leads to lowered hyposaline endurance in some species (*Chlamys opercularis*, *Mytella charruana*, *Perna viridis*; Paul 1980; Yuan et al. 2016). More rapid decreases in salinity possibly lead to lower hyposaline endurance, since some bivalves cannot change the densities of osmolites in their cells quickly enough (Shumway et al. 1977).

The pen shell *Atrina pectinata* (called "tairagi" in Japanese; see "Materials" for its taxonomic problems) is an important fishery resource in Japan. For example, its catch exceeded a value of 1 billion yen (about 9 million US dollars) for the Ariake Sea in Saga Prefecture in 1990 (Koga and Aramaki 2013). The resource has greatly decreased in recent years especially in the Ariake Sea (Koga and Aramaki 2013). For example, the catch (wet weight of soft tissue and shells) decreased from > 10,000 metric tons in the 1960s to < 10 metric tons in the 2010s in Saga Prefecture. To regain the stock level in Japan, artificial hatchery-rearing techniques have been recently developed (Fisheries Research Agency et al. 2017). In addition, the transplantation of juveniles has been examined on tidal muddy flats (Aramaki and Tsukuda 2014; Fisheries Research Agency et al. 2017), which are promising transplantation sites because transplanted wild pen shells have often survived better on tidal flats than in subtidal areas in the Ariake Sea (Yoshida et al. 2007).

Occasionally in rearing systems and frequently on tidal flats *A. pectinata* juveniles will be exposed to low salinities < 20 ppt owing to rain and terrestrial runoff (Kurihara 2017; Kurihara personal data). *A. pectinata* juveniles are sometimes reared in hatcheries without sediments (Fisheries Research Agency et al. 2017) and are often transplanted onto tidal flats without being embedded into the sediments (Aramaki and Tsukuda 2014), because such treatment using sediments is labor intensive. Therefore, *A. pectinata* juveniles can be exposed to low-salinity water without being covered by sediments, and thus its endurance against low salinity without sediments needs to be tested. In addition, as with foregoing other species (*Chlamys opercularis*, *Mytella charruana*, *Perna viridis*), *A. pectinata* juveniles may also change their hyposaline endurance in response to water temperature and rapidness of salinity decrease, and thus such potential interactions also needs to be studied. At present, information on hyposaline endurances of *A. pectinata* is only available for the adult specimens in an experiment without sediments by Yamamoto and Handa (2011) in which the hyposaline endurance was tested at one level of water temperature (28 °C) with salinity decreased rapidly.

We conducted laboratory experiments with environmental salinity being decreased either gradually or rapidly in either warm or cold water to examine how hyposaline water in these condition affects the survival time of juvenile *A. pectinata* with no sediments. In addition, the present study examined a new statistical model that assumes a stepwise sharp increase of hazard (i.e., instantaneous mortality rate), not a classical exponential increase of hazard (Kleinbaum and Klein 2012), according to the decrease of salinity. This is because the survival time of some bivalves sharply changes according to a small change of salinity (Leyva-Valencia el al. 2001; Yamamoto and Handa 2011) and thus the stepwise increase of hazard appears to better fit such data.

Materials

Experimental specimens were artificially hatched in July 2014 at the Seikai National Fisheries Research Institute from the eggs of adult pen shells collected in the Ariake Sea, Japan. The orthodox species name at present for these pen shells is unclear owing to the taxonomic confusion of *Atrina* (Lemer et al. 2014). Yet, these pen shells all have been confirmed to be identical to "*Atrina pectinata* Lineage 2" (sensu Liu et al. 2011) by the analyses of the mtDNA COI region (Hashimoto, personal data). Therefore, we provisionally call our material "*Atrina pectinata* Lineage 2" or "*Atrina pectinata*".

In recent years pen shells in the Ariake Sea mainly consist of *A. pectinata* Lineage 2 (> 90% of specimens; Aramaki 2013; Hashimoto, personal data). Considering this, we judge that most pen shells collected from the Ariake Sea in recent studies (e.g., Yamamoto and Handa 2011; Koga and Aramaki 2013) consist of *A. pectinata* Lineage 2. Note that these pen shells were called "*Atrina lischkeana*" in some of recent studies (e.g., Yamamoto and Handa 2011).

Juvenile *A. pectinata* had been reared since its production in the laboratory of Seikai National Fisheries Research Institute. The specimens were kept in running seawater that had been sand-filtered. During the present study, the temperature ranged from 13.3 °C (February) to 18.6 °C (April), and the salinity from 33 and 34 ppt. The specimens were acclimated to the experimental water temperatures (either 12 or 24 °C) for > 10 days before the start of the experiment. The specimens were fed with *Isochrysis* sp., *Pavlova* sp. and *Chaetoceros gracilis* F. Schütt, 1895 before the start of the experiment but not during the experiment (120–140 h). The stopping of feeding during the experiment was necessary to keep the salinity of water as a predetermined level, because feeding the juveniles involves addition of phytoplankton culture medium to the water. Since juvenile *A. pectinata* survives experiments lasting for > 300 h without food (Nagasoe, personal communication), the stopping of feeding during our experiment ≤ 140 h is unlikely to significantly affect the survival time of the juvenile.

The hyposaline water to which experimental specimens were to be exposed was prepared by mixing distilled water at a predetermined dilution rate with the 100% seawater. The 100% seawater was what the specimens in the laboratory were being acclimated to. The salinity (ppt) of the 100% seawater varied between experimental days owing to the condition of the sea near the water intake. As a result, the salinity (ppt) somewhat varied among the replications of the same dilution rate. Considering the irregular variations of salinity, the salinity (ppt) attained for each replication was measured during the experiments with a salinometer EC300 (YSI Co.). The measurements for every experiment were regarded as an explanatory variable of the ratio scale (Upton and Cook 2005).

Experimental methods

Juvenile *A. pectinata* were exposed to various levels of low salinity at 12 and 24 °C. Salinity was either gradually or rapidly decreased (hereafter called "gradual-salinity-decrease experiment" and "rapid-salinity-decrease experiment", respectively). To prevent the influences of other specimens, each specimen was individually put into a cup with a different combination of salinity and temperature. The mean ± SD shell length (no. of individuals) of *A. pectinata* was 31.7 ± 2.8 mm (24) for the gradual-salinity-decrease experiment and 36.1 ± 3.5 mm (42) for the rapid-salinity-decrease experiment.

The gradual-salinity-decrease experiment was conducted in February 2015. The twenty-four specimens were allocated to the combinations of 12 dilutions × 2 temperatures. The dilutions were of 12 levels: 100, 80, 60, 55, 50, 45, 40, 35, 30, 20, 10, and 0% of the initial salinity that the specimens had been acclimated to. The salinities realized (ppt) were measured and analyzed as an explanatory variable in the present study.

Before the start of the gradual-salinity-decrease experiment, 24 experimental cups (radius 60 mm, height 147 mm) were prepared, which contained 800 mL of non-diluted, 0.2-μm filtered seawater. Half the cups were put in a water bath of 12 °C and the other half in another bath of 24 °C. One experimental specimen was put into each cup at the start of this experiment. Subsequently, the water in each cup (except for the two cups predetermined to contain 100% seawater) was in part exchanged with distilled water of the same temperature to dilute the salinity in a stepwise way: first, from 100% seawater to 80% seawater; second, from 80 to 60%; third, from 60 to 55%; fourth, from 55 to 50%; etc. The dilution in a cup was stopped when the salinity in the cup attained the predetermined level. One hour was consumed from a dilution to the next dilution. Therefore, cups predetermined for lower salinities needed a longer time before reaching its determined level. The gradual-salinity-decrease experiment lasted for 120 h, and the water in each cup was changed with new water of the same temperature and salinity every day.

During the gradual-salinity-decrease experiment each specimen was checked for survival intermittently 19 times in total. Specimens were regarded as dead that had opened valves and, when stimulated with a glass rod, did not move the valves or soft tissue. The remaining specimens, which had not been regarded as dead throughout the experiment, were confirmed to be alive in a post-experimental check in which these specimens were found to excrete and/or move the valves in 100% seawater containing the phytoplankton *Isochrysis* sp.

The rapid-salinity-decrease experiment was conducted in April 2015 in a similar manner to the gradual-salinity-decrease experiment, but all specimens were exposed to rapid, sharp decreases of salinity from 100% seawater to $\leq$ 100% seawater. In the rapid-salinity-decrease experiment, 42 specimens were allocated to the combinations of 21 dilutions (100, 95, 90, … , 5, and 0% seawater) × two temperatures (12 and 24 °C). These specimens were transferred to the water of these conditions directly from tanks containing 100% seawater of the same temperatures. The rapid-salinity-decrease experiment lasted for 140 h, and the survival of specimens was checked 28 times in total in a similar manner to the gradual-salinity-decrease experiment.

Statistical modelling

The survival time of juvenile *A. pectinata* was recorded as not a time point but a time range ("interval-censored data" sensu Kleinbaum and Klein 2012). That is, if a specimen was found to be alive at the *i*th observation and to be dead at the $(i + 1)$th observation, the survival time was recorded as the time range covering the two observations. In the case that the *i*th observation was the last observation of the experiment, the survival time was recorded as the time range later than the last observation. The minimum and maximum values in the range of a survival time are hereafter called the "min. survival time" and "max. survival time".

The effect of low salinity on the survival time of juvenile *A. pectinata* was analyzed with specific statistical models for each experiment and temperature. The models were modified from the parametric survival analysis (Dobson and Barnett 2008; Kleinbaum and Klein 2012). The models assume that a specimen exposed to a low salinity would follow one of the six patterns (Fig. 1) of the relationship between low salinity and instantaneous mortality rate (namely, hazard; "h" hereafter).

i. Null-constant model. The equation is:

$$h = c$$

where c denotes a constant > 0. The survival time under this model follows an exponential distribution (Upton and Cook 2005), the rate parameter of which is a constant. The biological meaning of this model is that there is neither hyposaline stress nor temporally increasing stress.

ii. Null-accumulation model. The equation is:

$$h = c \cdot t$$

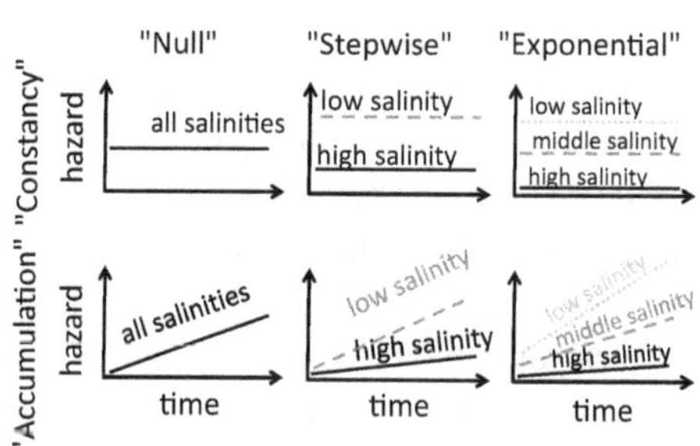

Fig. 1 Six statistical models for possible relationships of hazard (i.e., instantaneous mortality rate) against time at different salinities. Hazard is assumed to be: independent of salinity (the null models); dependent on whether or not salinity exceeds a threshold (the stepwise models); or exponentially variable according to salinity (the exponential models). Hazard is assumed to be either independent of time (the constancy models) or dependent (the accumulation models)

where c denotes a constant (> 0) and t the exposure time. The survival time under this model follows a Rayleigh distribution (Upton and Cook 2005), the scale parameter of which is a constant. The biological meaning of this model is that there is no hyposaline stress and some other stresses increase over time.

iii. Exponential-constant model. The equation is:

$$h = c \cdot \exp(-k \cdot s)$$

where c denotes a constant (> 0), k the slope (> 0), and s the salinity. The survival time under this model follows an exponential distribution, the rate parameter of which is a function of salinity. The biological meaning of this model is that stress exponentially increases with decreased salinity and is temporally constant.

iv. Exponential-accumulation model. The equation is:

$$h = c \cdot \exp(-k \cdot s) \cdot t$$

where c denotes a constant (> 0), k the slope (> 0), s the salinity, and t the exposure time. The survival time under this model follows a Rayleigh distribution, the scale parameter of which is a function of salinity. The biological meaning of this model is that stress exponentially increases with decreased salinity and increases over time.

v. Stepwise-constant model. The equation is:

$$h = c_1 + c_2 \cdot f(s, \tau)$$

where c_1 and c_2 denote a constant (> 0), s salinity, τ threshold, and f a step function (1 when $s < \tau$, and 0 when $s \geq \tau$). The biological meaning of this model is that stress increases in a stepwise manner when salinity was decreased below a threshold and is temporally constant.

vi. Stepwise-accumulation model. The equation is:

$$h = \{c_1 + c_2 \cdot f(s, \tau)\} \cdot t$$

where c_1 and c_2 denote a constant (> 0), s salinity, τ threshold, f the foregoing step function, and t the exposure time. The biological meaning of this model is that stress increases in a stepwise manner when salinity was decreased below a threshold and increases over time.

For each experiment and temperature, the parameters of the above six models were determined by the maximum likelihood estimation (Dobson and Barnett 2008) as follows. (1) The initial values of parameters were arbitrarily chosen. (2) From the parameters, the hazard for each specimen was calculated and the specimen's survival curve was determined. (3) From the survival curve, the likelihood of the specimen dying between the min. and max. survival time recorded was calculated. (4) The logarithm of the likelihood was summed across all the specimens. (5) The procedures 2–4 were repeated with the values of parameters changed by the Nelder–Mead simplex method (Upton and Cook 2005) to maximize the sum of the logarithm of the likelihood. This maximization algorithm was performed with the function "optim" in R (R Core Team 2015).

From the maximized sum of the logarithm of the likelihood and the number of parameters, the Bayesian information criterion (BIC) was calculated for each model (Upton and Cook 2005). The model with the lowest BIC was regarded as the best model.

Results

Gradual-salinity-decrease experiment

At both temperatures the survival time and expected LT50 (median lifetime) of juvenile *A. pectinata* sharply decreased owing to a small decrease (< 5 ppt) in salinity, as follows.

For the high temperature (24 °C) all juveniles survived the 83-h observation at salinities ≥ 21.1 ppt, whereas all juveniles died at salinities ≤ 18.6 ppt (Fig. 2) with the max. survival time < 20 h except for one specimen. In a sense, such a sharp difference in the max. survival time between high and low salinities did not correspond with the result of model selection (Table 1): the exponential-constant model showed a better, i.e. lower, BIC (60.2) than the stepwise models (64.7 and 73.7). The sharp difference, however, corresponded with the expected LT50 of the selected, exponential-constant model, which sharply decreased with the decrease of salinity from 20 to 15 ppt (Fig. 3).

For the low temperature (12 °C) all specimens survived the 83-h observation at salinities ≥ 17.2 ppt, whereas all specimens died at salinities ≤ 15.1 ppt (Fig. 2) with the max. survival time < 33.6 h. Such a sharp difference in the max. survival time between salinities corresponded with a result that the stepwise-accumulation model showed the best BIC, 26.3, among the six models (Table 1). This selected model estimated the lethal salinity threshold to be 15.1–17.1 ppt. At salinities < 15 ppt, the max. survival time and expected LT50 (18.3 h) were longer for lower temperature (Fig. 3).

Rapid-salinity-decrease experiment

At both temperatures, again, the survival time and expected LT50 sharply decreased owing to a small decrease in salinity < 5 ppt.

For the high temperature, all specimens except one survived the 140-h observation at salinities ≥ 17.7 ppt, whereas all specimens died at salinities ≤ 15.9 ppt (Fig. 4). At salinities ≤ 15.9 ppt the max. survival time varied among specimens from 13 to 20 h, neither increasing nor decreasing according to salinity. Such a sharp difference in the max. survival time between high and low salinities corresponded with a result that the stepwise-accumulation model showed the best BIC, 61.4 (Table 2). This model estimated the lethal salinity threshold to be 15.9–17.7 ppt and the LT50 below the threshold to be 12.4 h.

At the low temperature all specimens survived the 140-h observation at salinities ≥ 22.3 ppt, whereas all specimens died at salinities ≤ 21.3 ppt (Fig. 4). Such a difference in survival time corresponded with a result that the stepwise-accumulation model showed the best BIC, 90.7 (Table 2). This model estimated the lethal salinity threshold to be 18.6–21.3 ppt and the LT50 below the threshold to be 18.0 h. At salinities < 15.0 ppt the max. survival time (Fig. 4) and expected LT50 (Fig. 3) were somewhat longer for lower temperatures, as found in the gradual-salinity-decrease experiment. Expected LT50 for the rapid-salinity-decrease experiment

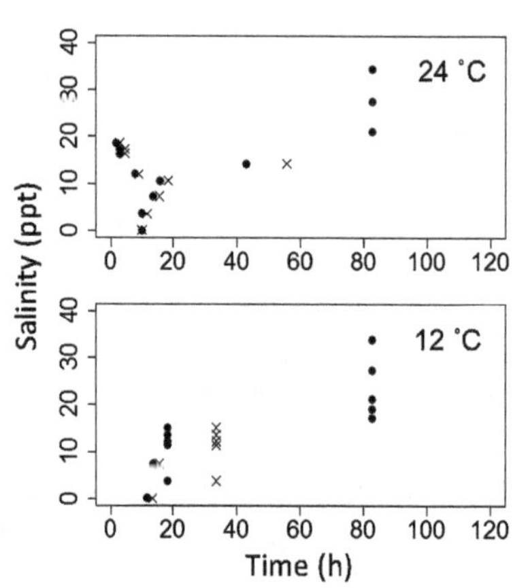

Fig. 2 Results of the gradual-salinity-decrease experiment at 24 and 12 °C. Filled circles: the last time a specimen was found to be alive. Crosses: the first time a specimen was found to be dead

Table 1 Results of statistical model selection for the gradual-salinity-decrease experiment

Model	Aspect	24 °C	12 °C
Null-Constant	No. of parameters	1	1
$h = c$	Data size	12	12
	BIC	81.2	48.0
	Parameter c	0.0244	0.0122
Null-Accumulation	No. of parameters	1	1
$h = c \cdot t$	Data size	12	12
	BIC	103.5	54.2
	Parameter c	0.0008	0.0004
Exponential-Constant	No. of parameters	2	2
$h = c_1 \cdot \exp(- c_2 \cdot s)$	Data size	12	12
	BIC	**60.2**	31.7
	Parameter c_1	1.578E+00	4.091E−01
	Parameter c_2	2.463E−01	2.349E−01
Exponential-Accumulation	No. of parameters	2	2
$h = c_1 \cdot \exp(- c_2 \cdot s) \cdot t$	Data size	12	12
	BIC	76.3	27.2
	Parameter c_1	1.726E+00	2.020E−01
	Parameter c_2	4.177E−01	3.802E−01
Stepwise-Constant	No. of parameters	3	3
$h = c_1 + c_2 \cdot f(s, \tau)$	Data size	12	12
	BIC	64.7	31.4
	Parameter c_1	4.830E−10	1.567E−09
	Parameter c_2	9.123E−02	5.871E−02
	Parameter τ	18.6–21.1	15.0–17.1
Stepwise-Accumulation	No. of parameters	3	3
$h = \{c_1 + c_2 \cdot f(s, \tau)\} \cdot t$	Data size	12	12
	BIC	73.7	**26.3**
	Parameter c_1	7.554E−11	0.000E+00
	Parameter c_2	6.100E−03	4.160E−03
	Parameter τ	18.6–21.1	15.0–17.1

h hazard or instantaneous mortality rate, c, c_1, c_2 parameters for the effect of salinity, τ threshold of salinity (determined as a range), s salinity, $f(s, \tau)$ step function (1 when $s < \tau$, and 0 when $s \geq \tau$). The lowest BIC for each temperature is shown in bold numericals

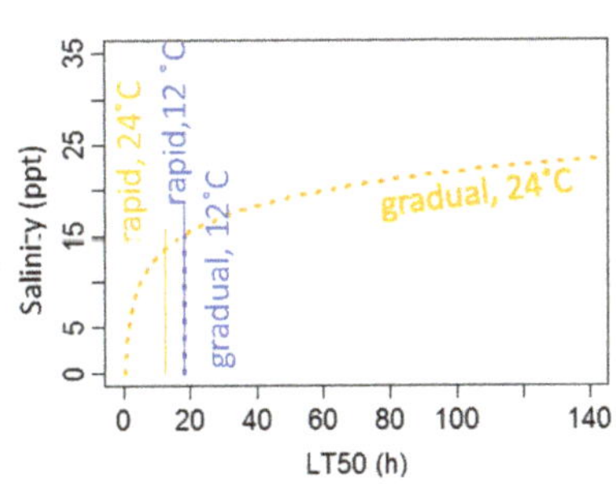

Fig. 3 Estimates of LT50 (median life time) for each experiment (gradual-salinity-decrease experiment: dotted lines, rapid-salinity-decrease experiment: continuous lines) at each temperature (24 °C: orange lines, 12 °C: blue lines). Except for the gradual-salinity-decrease experiment at 24 °C, only the LT50 at salinities below a threshold is presented because the LT50 at salinities exceeding a threshold was longer than the experimental durations

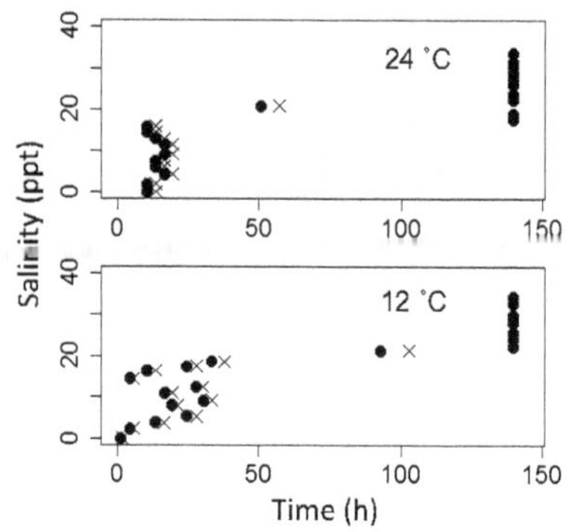

Fig. 4 Results of the rapid-salinity-decrease experiment at 24 and 12 °C. Filled circles: the last time a specimen was found to be alive. Crosses: the first time the specimen was found to be dead

Table 2 Results of statistical model selection for the rapid-salinity-decrease experiment

Model	Aspect	24 °C	12 °C
Null-Constant	No. of parameters	1	1
$h = c$	Data size	21	21
	BIC	109.0	125.7
	Parameter c	0.0069	0.0090
Null-Accumulation	No. of parameters	1	1
$h = c \cdot t$	Data size	21	21
	BIC	138.6	159.4
	Parameter c	0.0001	0.0002
Exponential-Constant	No. of parameters	2	2
$h = c_1 \cdot \exp(- c_2 \cdot s)$	Data size	21	21
	BIC	79.9	93.4
	Parameter c_1	1.710E−01	2.483E−01
	Parameter c_2	1.840E−01	1.818E−01
Exponential-Accumulation	No. of parameters	2	2
$h = c_1 \cdot \exp(- c_2 \cdot s) \cdot t$	Data size	21	21
	BIC	78.1	91.5
	Parameter c_1	3.373E−02	4.755E−02
	Parameter c_2	2.975E−01	2.802E−01
Stepwise-Constant	No. of parameters	3	3
$h = c_1 + c_2 \cdot f(s, \tau)$	Data size	21	21
	BIC	73.8	93.2
	Parameter c_1	6.899E−04	8.162E−05
	Parameter c_2	6.771E−02	0.0401171
	Parameter τ	15.9–17.7	21.3–22.3
Stepwise-Accumulation	No. of parameters	3	3
$h = \{c_1 + c_2 \cdot f(s, \tau)\} \cdot t$	Data size	21	21
	BIC	**61.4**	**90.7**
	Parameter c_1	1.012E−05	1.210E−05
	Parameter c_2	9.046E−03	4.245E−03
	Parameter τ	15.9–17.7	18.6–21.3

h hazard or instantaneous mortality rate, c, c_1, c_2 parameters for the effect of salinity, τ threshold of salinity (determined as a range), s salinity, $f(s, \tau)$ step function (1 when $s < \tau$, and 0 when $s \geq \tau$). The lowest BIC for each temperature is shown in bold numericals

were either similar (the low temperature) or longer (the high temperature) compared to those for the gradual-salinity-decrease experiment (Fig. 3).

Discussion

Based on the previous and present experiments in which *A. pectinata* were exposed to low salinities with no sediment, it is possible to compare hyposaline endurance between juveniles vs. adults, 12 vs. 24 °C, and gradual vs. sharp decreases in salinity, as follows.

Juvenile *A. pectinata* are likely to be more vulnerable to low salinity than adult *A. pectinata*. We found that most juvenile *A. pectinata* died within 60 h at salinities below a threshold existing between 15 and 21 ppt (water temperature 12 and 24 °C; rapid and gradual salinity decrease). On the other hand, Yamamoto and Handa (2011) found that adult *A. pectinata* died at salinities < 15 ppt but not at salinities ≥ 15 ppt in an experiment of 96 h at 28 °C with a rapid decrease in salinity. That is, juvenile *A. pectinata* died earlier than adult *A. pectinata* at higher salinities. Such a difference between juveniles and adults might have been greater, if juvenile *A. pectinata* had been exposed to low salinities at the high temperature, 28 °C, as with adult *A. pectinata*, because the hyposaline endurance of juvenile *A. pectinata* is likely to be weakened at higher temperatures (Fig. 3). Such a vulnerability of juvenile bivalves against low salinities was also reported for *Perna viridis* and *Mytella charruana* (Yuan et al. 2016).

What is the mechanism explaining the difference in hyposaline endurance between juvenile and adult *A. pectinata*? Bivalves protect themselves against environmental hyposaline water by means of: (1) preventing the hyposaline water from flowing on their soft tissue through valve closure and through ceasing pumping behavior; and/or (2) decreasing the osmotic pressure in their soft tissue cells to the osmotic pressure in the environmental water (Gosling 2003). *A. pectinata* is reported to adopt the former method at least (Yamamoto and Handa 2011). This method leads to a decreased supply of oxygen to the cells (Yamamoto and Handa 2011). The decreased supply of oxygen is likely to be more stressful for juvenile *A. pectinata* than for adult *A. pectinata*, since the juveniles die more quickly under hypoxia than the adults (Nagasoe et al. 2017; Nagasoe et al., in preparation). Hypoxia as an outcome of hyposaline water is suggested from the result of the statistical model selection. The selected models in general showed that the hazard (i.e., instantaneous mortality rate) of juvenile *A. pectinata* soars when salinity decreases below a threshold. This result may mean that the juveniles close their valves and stop water intake in salinities below the threshold, which results in suffering from gradual hypoxia. It is unclear which, juveniles or adults, are more adversely affected when they adopt the second method, namely, regulation of osmotic pressure. This is because there has been no study on osmoregulation for *Atrina* to our knowledge.

The difference found for LT50 between water temperatures indicates that the hyposaline endurance of juvenile *A. pectinata* with no sediments is weaker at 24 °C than 12 °C. Such a dependence of hyposaline endurance on water temperature is also reported for *Aequipecten opercularis* (Gosling 2003), *Perna viridis* and *Mytella charruana* (Yuan et al. 2016). A possible explanation for the decreased hyposaline endurance of juvenile *A. pectinata* at the higher temperature is related with metabolism. *A. pectinata* would increase their metabolism at the higher temperature, considering that animals (Schmidt-Nielsen 1997), including bivalves (Gosling 2003), increase their metabolism as the ambient temperature increases within a temperature range of their suitable habitat. To address the increased metabolism, *A. pectinata* would more strongly require oxygen. Such a stronger demand would not be met if *A. pectinata* stop their respiration at low salinities (Yamamoto and Handa 2011), which is likely to be more stressful. An additional, possible explanation for the decreased hyposaline endurance at the higher temperature is related with the van't Hoff's law. According to this law, osmotic pressure increases with temperature, so that the difference in osmotic pressure between the inside and outside of *A. pectinata* cells should increase, which may be more stressful for *A. pectinata*.

The LT50 estimated for the rapid-salinity-decrease experiment were either similar (the low temperature) or longer (the high temperature) compared to those for the gradual-salinity-decrease experiment. This is an unexpected result, considering that a more rapid decrease in salinity is likely to lead to a more acute effect of osmotic pressure on cells (Shumway et al. 1977) and to be more stressful. The result is unlikely due to the artifact that experimental period was longer for the rapid-salinity-decrease experiment (140 h) than the gradual-salinity-decrease experiment (120 h). This is because no specimen died between 120 and 140 h in the rapid-salinity-decrease experiment (Fig. 4), and thus the estimates of LT50 for this experiment would not change much even if the experimental period was reduced from 140 to 120 h. In the gradual-salinity-decrease experiment, perhaps, water in rearing cups was frequently agitated owing to the little-by-little dilution of salinity, which might have been stressful for juvenile *A. pectinata*.

Our data give suggestions for the stock enhancement of *A. pectinata* in the Ariake Sea. It should take into account that juvenile *A. pectinata* are vulnerable to low salinities < 21 ppt especially for 24 °C, at least when the specimens are not embedded in sediment. Therefore, juvenile *A. pectinata* should not be transplanted onto the tidal flat of the Ariake Sea in the rainy season and neighborhood of rivers leading to salinities < 21 ppt (Fisheries Research Agency and Saga University 2012). In addition, juvenile *A. pectinata* should not be kept in rearing systems where incidental intake of hyposaline water may occur. Some rearing systems near the Ariake Sea use underground seawater, which occasionally has salinity < 20 ppt (Kurihara, personal data). The above-mentioned transplantation and rearing should be avoided especially for seasons with the water temperature ≈ 24 °C. In the present study, juvenile *A. pectinata* were not embedded in sediments. Future study should examine the possibility that juvenile *A. pectinata* surrounded by sediments are affected by the salinities of not only water column but also the interstice water in sediments.

Acknowledgements We thank Dr. Junya Hiroi and Dr. Masahiko Matsuda for their information on hyposaline endurance of aquatic organisms. We thank Dr. Sou Nagasoe for his information on tolerance to starvation and hypoxia in *A. pectinata*. We thank Fukuoka Fisheries and Marine Technology Research Center and Saga Prefectural Ariake Fisheries Research and Development Center for providing adult *Atrina* spp. for the artificial hatchery. This study was funded by the "Nimaigai Kinkyu Zoshoku Taisaku Jigyo" from Japan Fisheries Agency.

References

Aramaki H (2013) The relationship between an effect of DNA analysis and proportion of pen-shells, *Atrina* spp. in the innermost area of Ariake Bay. Ariake Fish Promot Center Res Rep 26:89–91

Aramaki H, Tsukuda M (2014) The transplantation of pen-shells, *Atrina* spp. to tidal flats of the inner part of Ariake Bay in Saga Prefecture. Ariake Fish Promot Center Res Rep 27:1–8

Dobson AJ, Barnett A (2008) An introduction to generalized linear models, 3rd edn. Chapman & Hall, Florida

Fisheries Research Agency, Saga University (2012) Investigation for a plan to restore the ecosystem of Ariake Bay (evaluation of ecological functions of bivalves). In: Report of the results for contract research of Ministry of the Environment

Fisheries Research Agency, Seikai National Fisheries Research Institute, National Research Institute of Fisheries and Environment of Inland Sea, Fukuoka Fisheries and Marine Technology Research Center, Saga Ariake Fisheries Promotion Center, Nagasaki Prefectural Institute of Fisheries, Kumamoto Fisheries Research Center (2017) Nimaigai Shigen Kinkyu Zosyoku Taisaku Itaku Jigyo Hokokusyo. Fisheries Agency

Funakoshi S, Suzuki T, Wada K (1985) Salinity tolerances of marine bivalves. In: Sindermann CJ (ed) Environmental quality and aquaculture systems. Proceedings of the 13th US–Japan meeting aquaculture, Mie, 24–25 October 1984. NOAA technical report NMFS 69. US Department of Commerce, NOAA, National Marine Fisheries Service, Springfield, pp 15–18

Gillanders BM, Kingsford MJ (2002) Impact of changes in flow of freshwater on estuarine and open coastal habitats and the associated organisms. Oceanogr Mar Biol 40:233–309

Gosling E (2003) Bivalve molluscs (biology, ecology and culture). Fishing News Books, a division of Blackwell Publishing, Oxford. http://malacolog.com/files/Gosling_2004.pdf. Accessed 22 Jun 2017

Holmes SP, Miller N (2006) Aspects of the ecology and population genetics of the bivalve *Corbula gibba*. Mar Ecol Prog Ser 315:129–140

Jansen JM, Koutstaal A, Bonga SW, Hummel H (2009) Salinity related growth rates in populations of the European clam *Macoma balthica* and in field transplant experiments along the Baltic Sea salinity gradient. Mar Freshw Behav Physiol 42:157–166

Yamamoto K, Handa T (2011) Effect of low salinity in ambient water on ventilation in the pen-shell *Atrina* (*Servatrina*) *lischkeana*. Suisan Zosyoku (Aquacult Sci) 59:535–540

Kinne O (1966) Physiological aspects of animal life in estuaries with special reference to salinity. Neth J Sea Res 3:222–244

Kleinbaum DG, Klein M (2012) Survival analysis: a self-learning text. Springer, New York

Koga H, Aramaki H (2013) History of the pen shell, *Atrina* spp. fishery and the factors relevant to the fishing ground formation in Saga Ariake Bay—especially, some considerations about the factors of big catch in 2009 fiscal fishing period. Ariake Fish Promot Center Res Rep 26:13–24

Kurihara T (2017) Tolerance of the bivalve *Trapezium liratum* (Reeve, 1843) to decrease in salinity. Plankton Benthos Res 12:44–52

Lemer S, Buge B, Bemis A, Giribet G (2014) First molecular phylogeny of the circumtropical bivalve family Pinnidae (mollusca, bivalvia): evidence for high levels of cryptic species diversity. Mol Phylogen Evol 75:11–23

Leyva-Valencia I, Maeda-Martinez AN, Sicard MT, Roldan L, Robles-Mungaray M (2001) Halotolerance, upper thermotolerance, and optimum temperature for growth of the penshell *Atrina maura* (Sowerby, 1835) (Bivalvia: Pinnidae). J Shellfish Res 20:49–54

Liu J, Li Q, Kong LF, Zheng XD (2011) Cryptic diversity in the pen shell *Atrina pectinata* (Bivalvia: Pinnidae): high divergence and hybridization revealed by molecular and morphological data. Mol Ecol 20:4332–4345

Marsden ID (2004) Effects of reduced salinity and seston availability on growth of the New Zealand little-neck clam *Austrovenus stutchburyi*. Mar Ecol Prog Ser 266:157–171

Matsuda M, Shinagawa A, Higano J, Fujii A, Hirano K, Ishimatsu A (2008) Effects of low salinity on survival, hemolymph osmolality and tissue water content of the manila clam *Ruditapes philippinarum*. Suisan Zosyoku (Aquacult Sci) 56:127–136

McLeod RJ, Wing TR (2008) Influence of an altered salinity regime on the population structure of two infaunal bivalve species. Estur Coast Shelf S 78:529–540

Nagasoe S, Tokunaga T, Matsuyama Y (2017) Effect of hypoxic conditions on the survival of juvenile pen shell *Atrina* cf. *japonica* (an East Asian species). Suisan Zosyoku (Aquacult Sci) 65

Newell RC (1976) Adaptation to environment: essays on the physiology of marine animals. Butter Worths, London

Paul JD (1980) Salinity-temperature relationships in the queen scallop *Chlamys opercularis*. Mar Biol 56:295–300

R Core Team (2015) R: a language and environment for statistical computing. R Foundation for Statistical Computing, Vienna. http://www.R-project.org/. Accessed 22 Jun 2017

Rupp GS, Parsons GJ (2004) Effects of salinity and temperature on the survival and byssal attachment of the lion's paw scallop *Nodipecten nodosus* at its southern distribution limit. J Exp Mar Biol Ecol 309:173–198

Schmidt-Nielsen K (1997) Animal physiology: adaptation and environment, 5th edn. Cambridge University Press, Cambridge

Shumway SE, Gabbott PA, Youngson A (1977) Effect of fluctuating salinity on concentrations of free amino-acids and ninhydrin-positive substances in adductor muscles of 8 species of bivalve mollusks. J Exp Mar Biol Ecol 29:131–150

Upton G, Cook I (2005) Dictionary of statistics, 2nd edn. Oxford University Press, Oxford

Yoshida M, Ito T, Chikushi Y (2007) The study of transplantation of the pen-shell (*Atrina pectinata japonica*). Bull Fukuoka Fish Mar Tech Res Center 17:1–5

Yuan W, Walters LJ, Schneider KR, Hoffman EA (2010) Exploring the survival threshold: a study of salinity tolerance of the nonnative mussel *Mytella charruana*. J Shellfish Res 29:415–422

Yuan WS, Walters LJ, Brodsky SA, Schneider KR, Hoffman EA (2016) Synergistic effects of salinity and temperature on the survival of two nonnative bivalve molluscs, *Perna viridis* (Linnaeus 1758) and *Mytella charruana* (d'Orbigny, 1846). J Mar Biol. https://doi.org/10.1155/2016/9261309

6

Characterization of carotenoid pigments from bacterial symbionts of soft-coral *Sarcophyton* sp. from North Java Sea

Lia Kusmita · Erlita Verdia Mutiara · Handung Nuryadi · Petrick Ariska Pratama · Awang Surya Wiguna · Ocky Karna Radjasa

Abstract Marine bacteria produce natural pigments; however, the ability of marine bacterial symbionts to produce natural pigments has been less studied. Marine bacteria associated with soft-coral *Sarcophyton* sp. collected from Karimunjawa Island were successfully isolated and screened to synthesize the carotenoid pigments. This approach has allowed the use of these symbionts as an environmental friendly source of new natural pigments. Out of 33 bacterial isolates, only 4 bacterial symbionts (CBSCP 2-2, CBSCP 2-3, CBSCP 1-1, and CBSCP 2-4), positively contain carotenoid pigments. Molecular identification based on 16S rDNA method showed that bacterial symbionts CBSCP 2-2, CBSCP 2-3, CBSCP 1-1, and CBSCP 2-4 were closely related to *Pseudoalteromonas shioyasakiensis*, *Pseudoalteromonas rubra*, *Virgibacillus salaries*, and *Pseudoalteromonas spongiae*. Pigment analysis showed that the pigments have been categorized within the groups of carotenoid pigments. Antioxidant activity of pigment extracts was done by measuring inhibitory concentration (IC_{50}) of pigment extract against 2,2-diphenyl-1-picrylhydrazyl (DPPH) solution. The antioxidant activity measurement of CBSCP 1-1, CBSCP 2-2, CBSCP 2-3, and CBSCP 2-4 extract, and β-carotene was 2015, 5017, 2520, 4213, and 1980 mg l^{-1}, respectively.

Keywords Bacterial symbionts · Soft coral · Carotenoid · Antioxidant activity

Introduction

The marine environment is the largest habitat on Earth, representing more than 70% of the surface of our planet. These diverse marine environments still remain largely unexplored, understudied, and underexploited in comparison with terrestrial ecosystems and organisms (Joint et al. 2010). Indonesia is the global epicenter of marine biodiversity and is one of the mega diverse countries that harbor the majority of the Earth's species. Indonesia is well known for housing a unique and diverse sponge fauna. Natural products from marine invertebrates greatly expand the chemical diversity available for biotechnological exploitation. The diversity of marine invertebrates has not been utilized optimally, including soft corals (Radjasa et al.

L. Kusmita · E. V. Mutiara · P. A. Pratama · A. S. Wiguna
STIFAR "Yayasan Pharmasi", Letjend Sarwo Edie Wibowo KM 1, Semarang 50193, Indonesia

H. Nuryadi
Tropical Marine Biotechnology Laboratory, Diponegoro University, Semarang 50275, Indonesia

O. K. Radjasa (✉)
Faculty of Fisheries and Marine Science, Diponegoro University, Semarang, Central Java 50275, Indonesia
e-mail: ocky_radjasa@yahoo.com

2007a, 2011, 2013). Variation of the color within soft corals indicates that the organisms have diverse pigment contents, commonly found in yellow, orange, and red colors, which identify the carotenoid pigments.

One of the most serious bottlenecks in developing natural products from marine sources during the past decades has been the availability of biomass and/or of optimised cultivation conditions to gain sufficient amounts of compound for preclinical and clinical studies (Raghukumar 2008; Pan et al. 2008; Schulz et al. 2008). The supply of marine metabolites tested pre-clinically and in the clinic can be provided by several methods, including open aquaculture of invertebrates, total synthesis, semi-synthesis, and fermentation of the producing microbes. Many recognize that fermentation is the most appropriate method for sustainable production of natural products (Salomon et al. 2004). Symbionts that can be cultivated in the laboratory and still produce the bioactive metabolite can then be subjected to enhanced fermentation to produce large amounts of the targeted compound (Hildebrand et al. 2004).

Since marine diversity also reflects chemical diversity, the isolation of the under-exploited bacterial symbionts from Indonesian soft corals offers a great opportunity for discovering novel bioactive pigment compounds based on screening against various disease targets with significant impact and strong potential for treatment of diseases.

The exploitation of symbiotic bacteria as a source for novel secondary metabolites is considered to be in its infancy; however, the discovery rate of novel active metabolites from marine bacterial symbionts could surpass that of their terrestrial counterparts. Among the unusual niches for novel microbes are marine invertebrates, including soft corals, which host hundreds of different bacterial groups and contain diverse symbionts (Radjasa et al. 2007a, 2011).

Carotenoids have important function in photosynthetic process. They play role to absorb light energy in the range of 300–600 nm, beyond absorption area of chlorophyll. The light energy will be transferred to chlorophyll for driving photosynthetic process. Moreover, carotenoids have ability to photoprotect photosynthetic organisms from harmful of excess light during light harvesting in photosynthetic process (Vilches et al. 2011). In biological aspect, carotenoids are well known as an antioxidant and pro-vitamin A to protect the retina from macular degeneration (Dutta et al. 2005). These compounds can inhibit the chain reaction of free radicals that if not inhibited by antioxidants can cause degenerative diseases, such as cancer, coronary heart disease, diabetes, and others (Burrati et al. 2005). Recently, there have been reported novel activities of two specific carotenoids, Siphonaxanthin and Fucoxanthin, which have the potential to inhibit viability of human leukemia cell (Sugawara et al. 2014) and to act as an anti-obesity substance (Gammone and D'Ozario 2015), respectively. In commercial point of view, carotenoids can increase commercial value and quality of many industrial products. Having such potential and great benefits of carotenoids, exploration of the source of carotenoids mainly derived from marine organisms is quite abundant to be exploited.

It has been very well established for more than half a century (Kelecom 2002) that terrestrial bacteria and fungi are sources of valuable bioactive metabolites. It has also been noted that the rate at which new compounds are being discovered from traditional microbial resources, however, has diminished significantly in recent decades as exhaustive studies of soil microorganisms repeatedly yield the same species which in turn produce an unacceptably large number of previously described compounds. Therefore, it is reasonable to expect that exploration of untapped marine microbial diversity and resources will improve the rates at which new classes of secondary metabolites are discovered. In particular, the studies regarding screening of secondary metabolite-producing bacterial symbionts of soft corals are important for understanding their biotechnological potential, including as a sustainable source of marine carotenoids.

Materials and methods

Sampling and isolation of bacterial symbionts

Samples of soft coral were taken from a depth of approximately 2 m by scuba diving from three sampling sites at Cemara Besar, Karimunjawa Islands, North Java Sea (Fig. 1). Upon collection, the soft-coral colonies were then put into plastic containers containing sterile seawaters and then were stored temporarily in a cool box. The samples were rinsed 3× with sterile seawaters to clean the bacteria that temporarily attached to the surface waters (Burgess et al. 2003). The samples were homogenated and were serially diluted. A total of

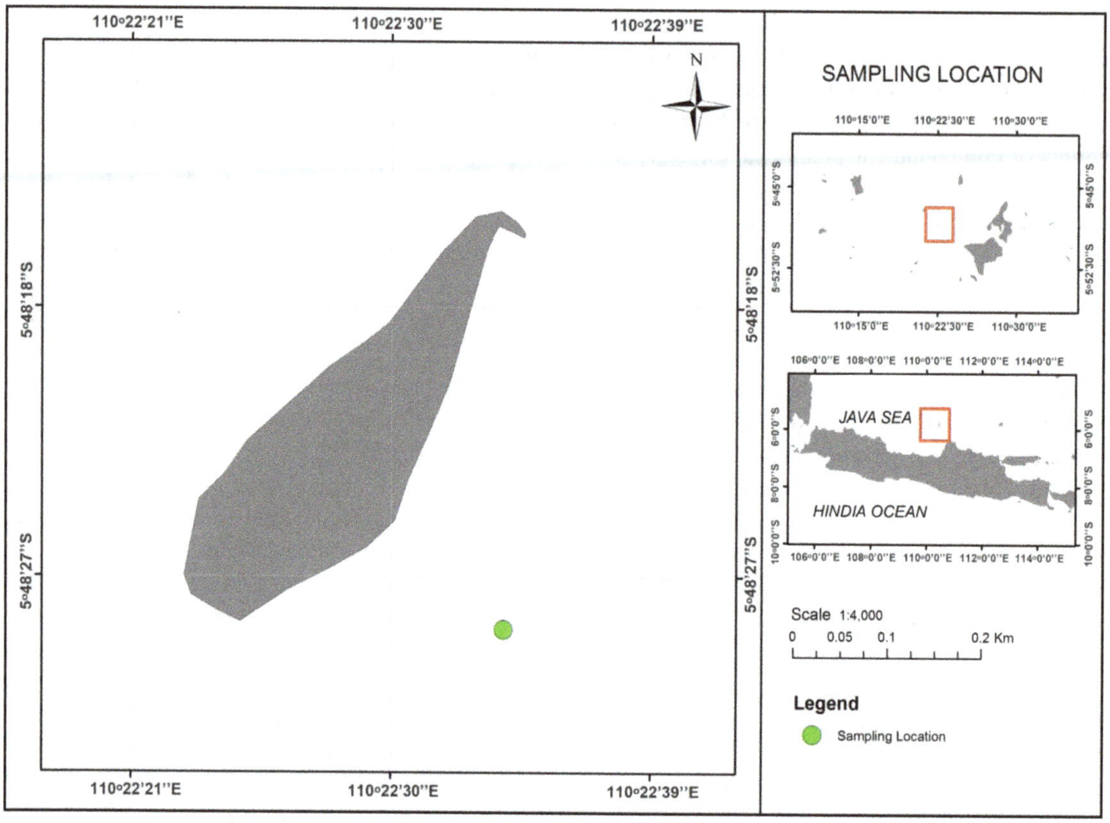

Fig. 1 Location of sampling of soft-coral in Cemara Besar island, Karimunjawa

100 μl were spread on the surface of Zobell 2216E marine medium using glass spreader. Petri dishes were then incubated at 30 °C for 2 days. The appeared colonies with pigment colors were then selected and purified. Unpurified bacteria were picked and streaked into Zobell 2216E medium using four quadrant streaking method. It was incubated at 30 °C for 2 days. A single colony was picked and streaked into new Zobell 2216E medium. The pure isolate was used for further steps.

DNA extraction and 16S rDNA PCR

Genomic DNA of the bacterial symbionts was isolated by the freeze and thaw method (Radjasa et al. 2007b). 1 ml suspension of bacterial cell was centrifuged at 14,000 rpm for 10 min, and supernatant was then discarded. 100 μl nuclease free water was added for resuspending of cell pellet. Bacterial suspensions were frozen at −20 °C and rapidly thawed at 65 °C for 10 min. That freeze and thaw procedure was repeated 4 times and followed by centrifugation at 14,000 rpm for 10 min. Supernatant containing genome DNA was transferred into a new sterile tube. Genomic DNA was used as DNA template in amplification process.

In PCR amplification, primers used for 16S rDNA PCR were universal primers 27F (5′-AGAGTTT-GATCMTGGCTCAG-3′) and Eubacteria-specific primer 1492R (5′-TACGGYTACCTTGTTACGACTT-3′) (Long and Azam 2001). The temperature cycle of amplification was as follows: the initial denaturation at a temperature of 94 °C for 2 min, and then successive denaturation (94 °C for 1 min), annealing (55 °C for 1 min), and extension (72 °C for 2 min). A series of denaturation, annealing, and extension were repeated 45 times (Radjasa et al. 2007c). Electrophoresis was done on 2% agarose. Sequencing was done according to Radjasa et al. (2007c). DNA sequencings, homology search, and phylogenetic analyses were carried out according to Atschul et al. (1997) and Radjasa et al. (2013).

Extraction of bacterial pigments

Bacterial symbionts producing pigments were cultured on Zobell 2216E broth medium. They were then centrifuged to obtain bacterial pellet. Pellets were taken and then were extracted using methanol. The extraction process was repeated several times until all pigments were removed and the residues became pale yellow. Pigment extracts were then concentrated using a vacuum evaporator at 30 °C.

UV–Vis spectrophotometer analysis

The extracted pigments were re-dissolved in 10 ml methanol 100%. The spectra were recorded in the λ 300–600 nm range with a Varian Cary spectrophotometer.

Thin-layer chromatography (TLC) analysis

TLC analysis for identification of β-carotene content was done using silica gel GF254 as Stationary phase. Mobile phase used combination of acetone:hexane (6:4 v/v). Retention factor (R_f) value was calculated to determine β-carotene compound.

High-performance liquid chromatography (HPLC) analysis

Carotenoid analysis was carried out by HPLC method. The method involved a Shim-Pack VP-ODS C18 column with methanol:acetonitrile (7:3) as the mobile phase for isocratic elution. Analysis was carried out for 10 min, and flow rates were 1 ml/min. Detection was with a photodiode array detector (PDA). Results of HPLC and spectrum PDA analysis were compared to the literature (Maeda et al. 2005).

Antioxidant activity

Carotenoid pigments from bacterial symbionts and β-carotene marker were dissolved in methanol at several concentrations. Antioxidant activity was measured using spectrophotometer at 517 nm wavelength. Assays were done according to the method reported by Panovska et al. (2005).

DNA sequences

DNA sequences of all bacterial isolates have been deposited in the DNA Database Bank of Japan (DDBJ) in the following accession numbers: NR026223, NR041270, NR043172, and NR125458.

Results and discussion

Results

Screening of bacterial symbionts of soft-coral Sarcophyton sp. resulted in four isolates that are capable of producing pigments. A representative pigment-producing isolates is shown in Fig. 2.

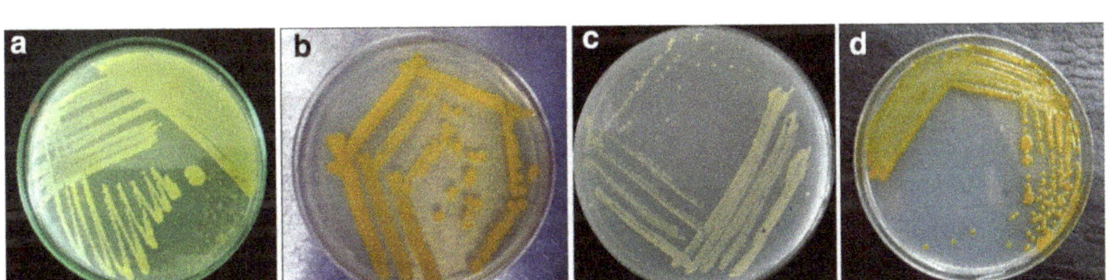

Fig. 2 Bacterial colonies of symbiont of soft-coral *Sarcophyton* sp. **a** CBSCP 2-2, **b** CBSCP 2-3, **c** CBSCP 1-1, and **d** CBSCP 2-4

Table 1 Molecular identification of isolate bacterial symbiont soft coral

No.	Strain	Length (bp)	Closest Relative	Accession number	Homology (%)
1	CBSCP 2-2	1392	*Pseudoalteromonas shioyasakiensis*	LC131142	96
2	CBSCP 2-3	1419	*Pseudoalteromonas rubra*	LC122513	98
3	CBSCP 1-1	1250	*Virgibacillus salarius*	LC131143	92
4	CBSCP 2-4	1393	*Pseudoalteromonas spongiae*	LC131144	97

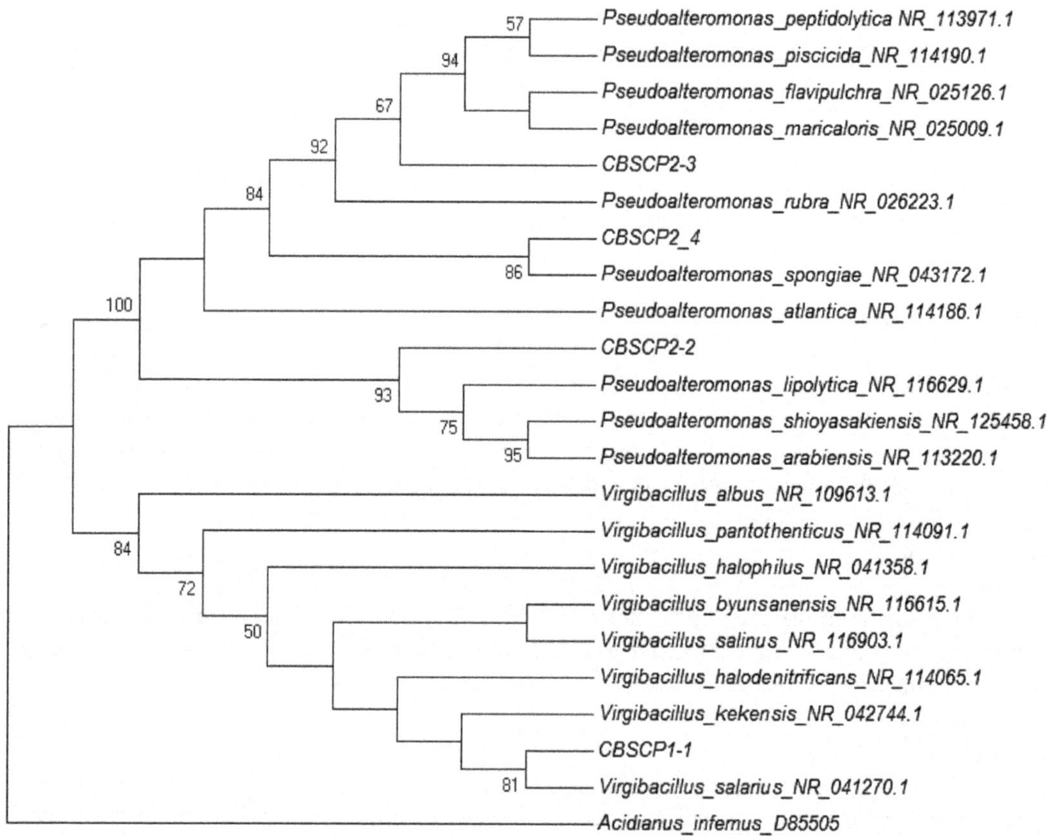

Fig. 3 Phylogenetic tree of pigment-producing bacterial symbionts of *Sarcophyton* sp.

Molecular identification (Table 1) and phylogenetic tree (Fig. 3) reveal that two genera were recognized from four bacterial symbionts, namely, Pseudoalteromonas and Virgibacillus. Three isolates were observed in the genus Pseudoalteromonas, and one isolate was detected in the genus Virgibacillus.

Figure 4 indicates the maximum absorption, which appears on each chromatogram. The results of the spectra obtained at each peak indicate that the pigments produced by all four bacterial symbionts are carotenoids. Bacterial isolate CBSCP 2-2 has three types of carotenoid pigments, isolate CBSCP 2-3 has two to three types of carotenoids, isolate CBSCP 1-1 has two types of carotenoids, and isolate CBSCP 2-4 has one type of carotenoid. Identification for each type of carotenoid produced can be seen from the absorption spectra and the maxima listed in Table 2.

The chromatogram of TLC analyses from β-carotene is shown in Fig. 5. The R_f value of each spot was calculated and then compared with β-carotene marker. The results of TLC have clearly confirmed the results of HPLC, in which the bacterial symbionts contain β-carotene.

The antioxidant activity from pigments produced by each bacterial symbiont is presented in Fig. 6. IC_{50} values from activity antioxidant assay are shown in Fig. 6.

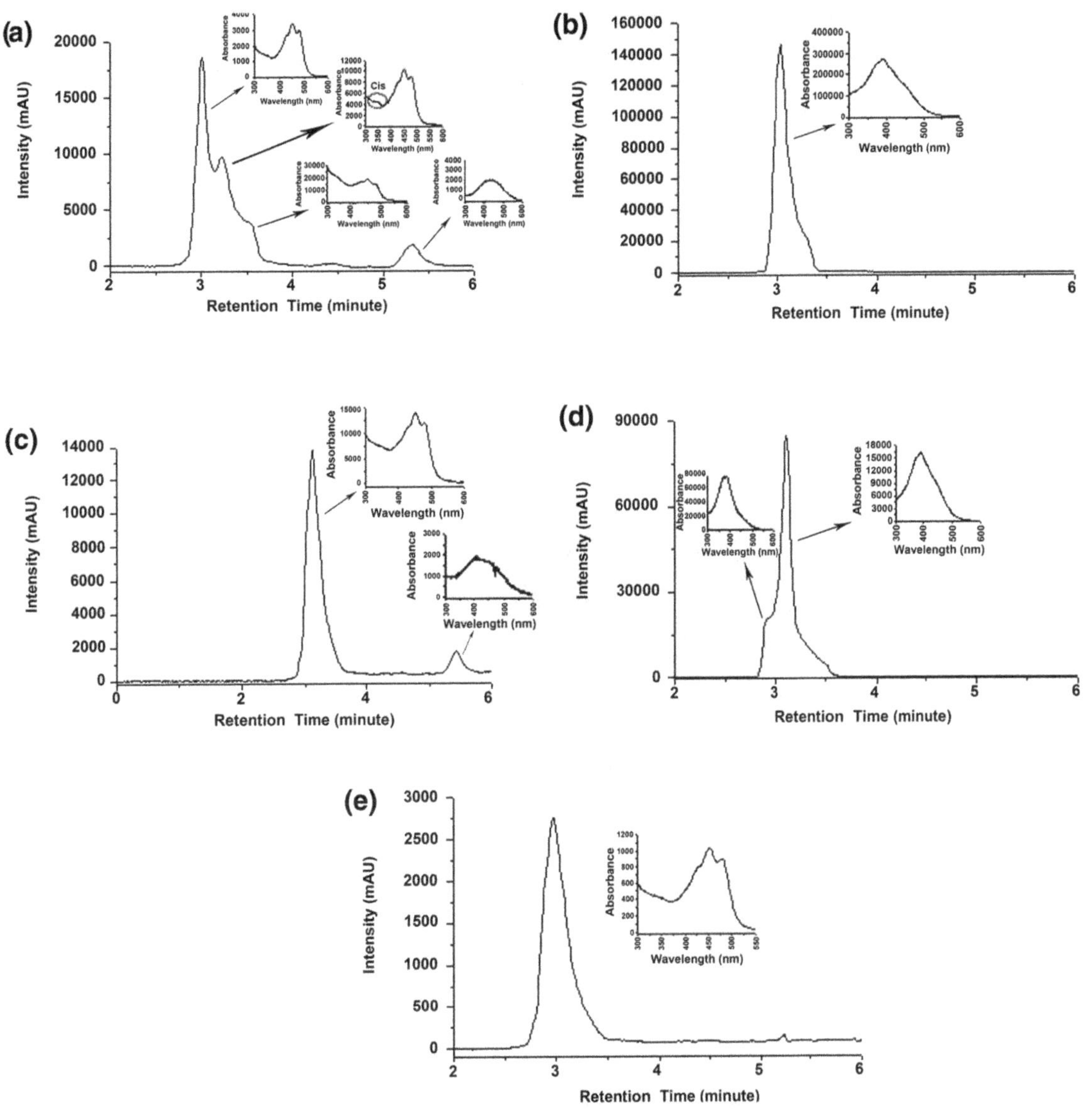

Fig. 4 HPLC profile of the extracts of **a** CBSCP 2-2, **b** CBSCP 2-3, **c** CBSCP 1-1, **d** CBSCP 2-4, and **e** β-carotene marker. The detection was carried out at 450 nm with a flow rate of 1 ml per min^{-1}

Discussion

As an effort to explore the potential of bacterial symbionts of reef invertebrates, especially soft corals, we successfully screened bacterial symbionts of soft-coral *Sarcophyton* sp. collected from Karimunjawa islands in the North Java Sea for producing carotenoid pigments and exhibiting antioxidant activity. The results have prompted the search for novel carotenoid pigment compounds from bacteria associated with soft corals in a sustainable manner, without harming the precious coral reefs. These findings were also in accordance with the highlights of microbial associates of invertebrates with medical potential (Radjasa et al. 2011).

Molecular phylogenetic analysis shows that four bacterial symbionts capable of producing carotenoid pigments are belonging to genera of *Pseudoalteromonas* and *Virgibacillus*. *Pseudoalteromonas* species are frequently found in association with eukaryotic hosts in the marine environment and are well known to produce active metabolites (Holmström and Kjelleberg 1999). *Pseudoalteromonas* is genus that can be divided relatively cleanly into pigmented and non-pigmented species clades and that pigmentation correlates with their proclivity for natural products (Bowman 2007). Furthermore, pigmented *Pseudoalteromonas* species possess a

Table 2 Carotenoid identification of bacterial symbionts of soft-coral *Sarcophyton* sp.

Sample	tr	Component	λ_{max} (nm)			
CBSCP 2-2	2.99	All trans β-carotene		430	451	480
	3.24	Cis β-carotene	(349)	430	451	480
	5.31	Not identified			433	
CBSCP 2-3	3.04	Not identified		371	386	
CBSCP 1-1	3.05	All trans β-carotene		430	451	480
	5.40	Not identified			416	
CBSCP 2-4	2.91	Not identified		370	387	
	3.10	Not identified			389	
β-Carotene marker	2.97	All trans β-carotene		430	451	480

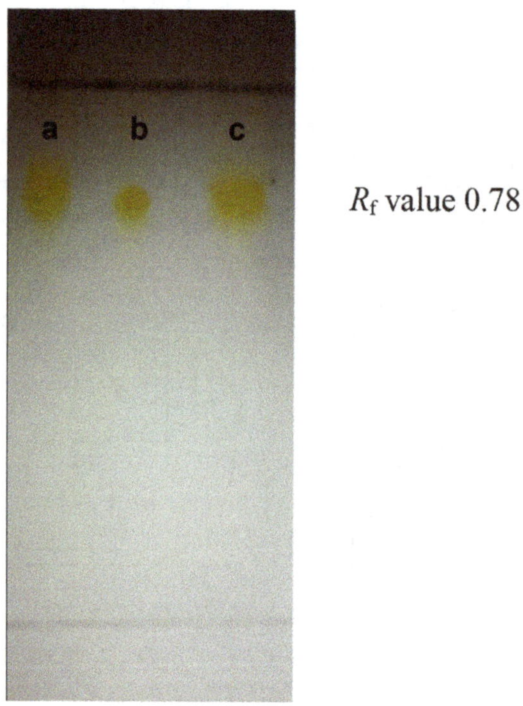

Fig. 5 Result from β-carotene of **a** CBSCP 2-2 and **b** CBSCP 1-1 compared with **c** β-carotene marker by TLC, using silica gel GF_{254} and mobile phase acetone:hexane (6:4)

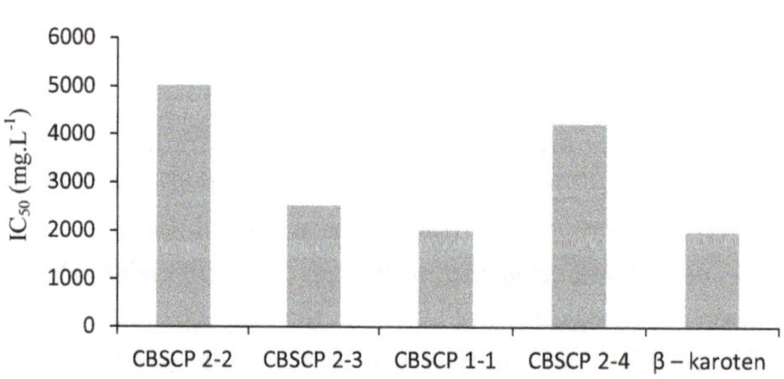

Fig. 6 IC_{50} of pigments of bacterial symbionts CBSCP 2-3, CBSCP 1-1, CBSCP 2-2, CBSCP 2-4, and β-carotene marker using DPPH method

broad range of bioactivity associated with the secretion of extracellular compounds, several of which include pigment compounds (Holmström and Kjelleberg 1999).

On the other hand, one bacterial isolate belonged to the genus *Virgibacillus*. Furthermore, Marhaeni et al. (2011) found that bacterial symbiont *Virgibacillus* sp. from seagrass *Enhalus acoroides* inhibited the growth of biofilm-forming bacteria. A study by Kanagasabhapathyt et al. (2005) demonstrated the inhibition of growth of fouling bacteria by sponge *Pseudoceratina purpure*-associated bacterium *Virgibacillus* sp.

There has been increase in the search for novel pigments especially from bacterial pigments. As an alternative to synthetic pigments, bacterial pigments due to their better biodegradability and higher compatibility with the environment offer promising avenues for various applications. The industry is now able to produce some bacterial pigments for applications in food, pharmaceuticals, cosmetics, and textiles (Venil et al. 2013).

Bacterial symbionts CBSCP 2-2 and CBSCP 1-1 contain carotenoids which have absorptions at 430, 451, and 480 nm. According to Jeffery et al. (1997), carotenoids which have absorptions at a wavelength of 433, 455, and 482 nm are β-carotene.

The chromatogram CBSCP 2-2 shows that there are two β-carotenes. A peak with retention time at 2.99 min^{-1} was identified as all trans-β-carotenes and retention times at 3.24 min^{-1} were identified as cis-isomer of β-carotene by its absorption spectrum: the subsidiary peak in the near ultraviolet (349 nm) is a characteristic for a cis-isomer. The cis-isomer can be identified from the appearance of new peak at 300–375 nm (Gross 1991; Rodriguez-Amaya and Kimura 2004). Identification of all trans-β-carotenes was shown by the similarity of the retention time and Rf value with comparison of β-carotene marker. Judging from the spectrum λ_{max} produced, it is shown that the types of carotenoids produced by bacteria CBSCP 2-3 and 2-4 are almost the same.

IC_{50} concentration of CBSCP 2-2 was 5017 mg l^{-1}, while CBSCP 2-3, CBSCP 1-1, CBSCP 2-4, and β-carotene were 2520, 2015, 4213, and 1980 mg l^{-1}, respectively. β-carotene was used as standard, because it is a natural antioxidant from the carotenoids. IC_{50} value CBSCP 1-1 is in a similar value with β-carotene marker. The smaller the IC_{50} value indicates the higher antioxidant activity.

Arunachalam and Appadorai (2013) reported that an antioxidant activity of bacterial symbiont *Vigibacillus* sp. associated with sponge *Callyspongia diffusa*. Park et al. (2009) also showed the antioxidant activity of pigment-producing marine bacterium *Pseudoalteromonas psicida* TA20.

In conclusion, bacterial symbionts of soft-coral *Sarcophyton* sp. from Karimunjawa islands represent the untapped richness of an under-utilized group of marine microorganisms and the possibility of environmentally friendly secondary metabolite producers with medical potential in particular as antioxidant compounds.

Acknowledgements The authors thank the Directorate of Research and Community Services, Ministry of Research, Technology and Higher Education, and Republic of Indonesia for financial support to this research through "PEKERTI" research grant scheme (Contract No. 034/O06.2/PP/SP/2015).

References

Arunachalam K, Appadorai RAJ (2013) Antioxidant potential and biochemical evaluation of metabolites from the marine bacteria *Virgibacillus* sp. associated with the sponge *Callyspongia diffusa*. Free Radicals Antioxid 3(1):47–51

Atschul SF, Madden TL, Schaffer AA, Zhang J, Zhang Z, Miller W, Lipman DJ (1997) Gapped BLAST and PSI-BLAST: a new generation of protein database search programs. Nucl Acid Res 25:3389–3402

Bowman JP (2007) Bioactive compound synthetic capacity and ecological significance of marine bacterial genus *Pseudoalteromonas*. Mar Drugs 5:220–241

Burgess JG, Boyd KG, Armstrong E, Jiang Z, Yan L, Berggren M, May U, Pisacane T, Granmo A, Adams DR (2003) The development of a marine natural product-based antifouling paint. Biofouling 19:197–205

Burrati S, Pellegrini N, Brenna OV, Mannino S (2005) Rapid electrochemical method for the evaluation of the antioxidant power some lipophilic food extracts. J Agric Food Chem 49:4151–5136

Dutta D, Chaudhuri UR, Chakraborty R (2005) Structure, health benefit, antioxidant property and processing and storage of carotenoid. Afr J Biotechnol 4:1510–1520

Gammone MA, D'Ozario N (2015) Anti-obesity activity of the marine carotenoid Fucoxanthin. Mar. Drugs. 13:2196–2214

Gross J (1991) Pigment in vegetables: chlorophyll and carotenoid. Van Nostrand Reinhold, New York

Hildebrand M, Waggoner LE, Lim GE, Sharp KH, Ridley CP, Haygood MG (2004) Approaches to identify, clone, and express symbiont bioactive metabolite genes. Nat Prod Rep 21:122–142

Holmström C, Kjelleberg S (1999) Marine *Pseudoalteromonas* species are associated with higher organisms and produce active extracellular agents. FEMS Microbiol Ecol 30:285–293

Jeffrey SW, Mantoura RFC, Wright SW (1997) Phytoplankton pigments in oceanography: guidelines to modern method. UNESCO Publishing, Paris, p 661

Joint I, Mühling M, Querellou J (2010) Culturing marine bacteria—an essential prerequisite for biodiscovery. Microb Biotechnol 3:564–575

Kanagasabhapathyt M, Sasaki H, Nakajima K, Nagata K, Nagata S (2005) Inhibitory activities of surface associated marine sponge *Pseudoceratina purpurea*. Microb Environ 20(3):178–185

Kelecom A (2002) Secondary metabolites from marine microorganisms. An Acad Bras Ciênc 74:151–170

Long RA, Azam F (2001) Antagonistic interactions among marine pelagic bacteria. Appl Environ Microbiol 67(11):4975–4983

Maeda H, Masashi H, Tokute S, Katsura F, Kazuo M (2005) Fucoxanthin from edible seaweed, *Undaniria pinnatifida*, shows antiobesity effect through UCP1 expression in white adipose tissues. Biochem Biophys Res Commun 332:392–397

Marhaeni B, Radjasa OK, Khoeri MM, Bengen DG, Sabdono A, Sudoyo H (2011) Antifouling activity of bacterial symbionts of seagrasses against marine biofilm-forming bacteria. J Environ Protect 2:1245–1249

Pan JH, Jones EBG, She ZG, Pang JY, Lin YC (2008) Review of bioactive compounds from fungi in the South China Sea. Bot Mar 51:179–190

Panovska TK, Svetlana K, Marina S (2005) In vitro antioxidant activity of some *Teucrium* species (Lamiace). Acta Pharm 55:207–214

Park J-S, Dong-Woo J, Myung-Hee K (2009) The physicochemical stabilities and biological activities of pigment extract from marine bacterium *Pseudoalteromonas psicida* TA20. J Life Sci 19(8):1132–1138

Radjasa OK, Salasia SIO, Sabdono A, Wiese J, Imhoff JF, Lammler C, Risk MJ (2007a) Antibacterial activity of marine bacterium *Pseudomonas* sp. Associated with soft coral *Sinularia polydactyla* against *Streptococcus equi* subsp. *zoopidemicus*. Int J Pharmacol 3(2):170–174

Radjasa OK, Martens T, Grossart H-P, Brinkoff T, Sabdono A, Simon M (2007b) Antagonistic activity of a marine bacterium *Pseudoalteromonas luteoviolacea* TAB 4.2 associated with coral *Acropora* sp. J Biol Sci 7(2):239–246

Radjasa OK, Sabdono A, Junaidi Zocchi E (2007c) Richness of secondary metaboliteproducing marine bacteria associated with sponge *Haliclona* sp. Int J Pharmacol 3(3):275–279

Radjasa OK, Vaske YM, Navarro G, Vervoort HC, Tenney K, Linington RG, Crews P (2011) Highlights of marine invertebrate-derived biosynthetic products: their biomedical potential and possible production by microbial associants. Bioorg Med Chem 19:6658–6674

Radjasa OK, Khoeri MM, Darusallam CC, Trimasanto H, Sudoyo H (2013) Bacterial symbionts of reef invertebrates: screening for anti-pathogenic bacteria activity. Biodiversity 14:80–86

Raghukumar C (2008) Marine fungal biotechnology: an ecological perspective. Fungal Divers 31:19–35

Rodriguez-Amaya, DB, Mieko K (2004) Harvest plus handbook for carotenoid analysis. International Food Policy Research Institute (IFPRI) and International Center for Tropical Agriculture (CIAT), Cali

Salomon CE, Magarvey NA, Sherman DH (2004) Merging the potential of microbial genetics with biological and chemical diversity: an even brighter future for marine natural product drug discovery. Nat Prod Rep 21:105–121

Schulz B, Draeger S, Delacruz TE, Rheinheimer J, Siems K, Loesgen S, Bitzer J, Schloerke O, Zeeck A, Kock I, Hussain H, Dai JQ, Krohn K (2008) Screening strategies for obtaining novel, biologically active, fungal secondary metabolites from marine habitats. Botanica Marina 51:219–234

Sugawara T, Ganesan P, Li Z, Manabe Y, Hirata T (2014) Siphonaxanthin, a green algal carotenoid, as a novel functional compound. Mar Drugs 12:3660–3668

Venil CK, Zakaria ZA, Ahmad WA (2013) Bacterial pigments and their applications. Process Biochem 48:1065–1079

Vilches C, Forjan E, Cuaresma M, Bedmar F, Garbayo F, Vega JM (2011) Marine carotenoids: biological function and commercial applications. Mar Drugs 9:319–333

7

Cypermethrin induced stress and changes in growth of freshwater fish *Oreochromis niloticus*

Rajib Majumder · Anilava Kaviraj

Abstract Bioassays were conducted with technical grade and commercial formulation of cypermethrin using freshwater fish *Oreochromis niloticus* as the test fish. The technical grade cypermethrin contained 92% active ingredient (a.i.) and the commercial formulation was an emulsified concentrate (EC) containing 10% a.i. (10% EC). Based on the actual concentration in water (2 h), the commercial formulation was found to be more acutely toxic to *O. niloticus* (96-h $LC_{50} = 4.85$ μg/L) than the technical grade cypermethrin (96-h $LC_{50} = 9.74$ μg/L). Exposure to sub-lethal concentrations (1.25, 2.5 μg/L) of commercial cypermethrin for 96 h produced stress on the fish, which was evident from the reduction of hepatic glycogen, reduction in the activities of alkaline phosphatase, acetylcholinesterase and catalase in liver and elevation of plasma glucose level and activities of hepatic acid phosphatase, aspartate aminotransferase and alanine aminotransferase. Exposure to these concentrations of cypermethrin for 14–28 days produced anaemia in fish. Long-term exposure (90 days) of the fish to these concentrations reduced the growth and deposition of protein and lipid in the body of fish as compared to control.It is concluded from this study that even minute concentration (1.25 μg/L) of cypermethrin (10% EC) in water can produce stress on fish. Long term exposure to such concentration of cypermethrin may also affect growth of the fish.

Keywords Pesticide · Toxicity · Tilapia · LC_{50} · Enzymes · Growth

Introduction

Cypermethrin is a 4th generation halogenated type II pyrethroid (Kaviraj and Gupta 2014) and is extensively used in tropical countries to control insect pests of cotton, cabbage, okra, brinjal, sugarcane, wheat, sunflower and many other crops. As a result, cypermethrin has been the centre of attention of many researches dealing with susceptibility of fish to agricultural run-offs. Type II pyrethroids are extensively used for pest management because of their relatively low toxicity to birds and mammals. But information on the toxicity of these pesticides to non-target aquatic organisms is scarce. Cypermethrin has already been reported as highly toxic to fish (Kumar et al. 2007; Saha and Kaviraj 2008). Sporadically, it has also been reported that the toxicity of cypermethrin to fish varies with stereochemical structure (Polat et al. 2002), mixtures of their isomers (Yilmaz et al. 2004) and formulation.

R. Majumder · A. Kaviraj (✉)
Department of Zoology, University of Kalyani, Kalyani, W.B. 741235, India
e-mail: akaviraj@gmail.com

R. Majumder
Department of Zoology, Vivekananda Mahavidyalaya, Hooghly, W.B. 712405, India

Commercial formulations of cypermethrin have always been found to be more toxic than its active ingredient (Demetrio et al. 2014; Majumder and Kaviraj 2015) because of inert ingredients added to the commercial formulations of pesticides to improve their efficiency in field conditions (Puglis and Boone 2011).

Emulsified concentrate (EC), with 10–20% active ingredients, is the most common formulation of cypermethrin being marketed in India. Composition of inert synergists added to these formulations and their contribution to the toxicity of cypermethrin to non-target aquatic organisms are not properly known. Lethal concentrations of different formulations of cypermethrin to fish, reported so far, are based mostly on the nominal concentration of cypermethrin. In this study we determined lethal concentrations of both chemically pure (technical grade) and commercial formulation of cypermethrin to freshwater fish *Oreochromis niloticus* based on nominal as well as actual concentration of cypermethrin measured after 2 h of exposure. Comparison of the data generated was helpful to evaluate impact of inert ingredients, if any, onthe toxicity of the commercial formulation of cypemethrin to fish. Since liver is the main site of detoxification of toxicants entering into the body, researchers have made efforts to detect stress of cypermethrin on fish from the activities of hepatic enzymes related to various metabolic pathways (Tiwari et al. 2012; Adeyemi et al. 2014). Reviews by Kaviraj and Gupta (2014) have documented the effects of cypermethrin on liver glycogen, activities of alkaline and acid phosphatase, activities of enzymes related to nitrogen metabolism and lipid peroxidation of liver. Changes in haematological profile have also been used as biomarkers of cypermethrin toxicity to fish (Vani et al. 2012; Ojutiku et al. 2013). However, the effects of cypermethrin on the growth of fish are rather poorly documented (Carriquiriborde et al. 2007). Although the toxic potential of commercial cypermethrin (10% EC) gradually decreases once it is exposed to light (Saha and Kaviraj 2009a), long-term exposure (60 days) to sub-lethal concentration of this pesticide has been found to reduce the growth of the catfish *Heteropneustes fossilis* (Saha and Kaviraj 2013). A comprehensive study on the acute toxicity and the effects of sub-lethal concentrations of cypermethrin on the biochemical parameters as well as the growth of fish is lacking.

Oreochromis niloticus is a freshwater fish. Although the species is exotic to India, it is widely cultured throughout the country because of its capacity to adapt to local conditions and tolerance to a wide variety of environmental conditions. The main objectives of the present study were to determine the stress produced by sub-lethal concentrations of commercial cypermethrin (10% EC) on *O. niloticus* and the subsequent impact of the pesticide on growth of the fish. Accordingly, we first determined the acute toxicity of technical grade cypermethrin containing 92% active ingredient (92% a.i.) and an emulsified concentrate containing 10% active ingredient (10% EC) on *O. niloticus* based on both nominal and actual concentrations of cypermethrin in water and then evaluated the degree of difference in toxicity between the two forms of cypermethrin. Detailed studies on the stress and growth of the fish, however, were carried out with sub-lethal concentrations of an emulsified concentrate (10% EC) of cypermethrin because this form is used in agricultural fields.

Materials and methods

Experimental design

Four experiments were performed in this study. These included acute toxicity bioassay to determine lethal concentrations of cypermethrin to *O. niloticus* and experiments with sub-lethal concentrations of cypermethrin (10% EC) on biochemical parameters (96 h), haematological parameters (28 days) and growth of *O. niloticus* (90 days).

Specimens of *O. niloticus* (mean length 4.61 ± 0.24 cm and mean weight 2.64 ± 0.16 g) were obtained from Kulia fish farm, Kalyani, West Bengal, and were acclimatized to the test conditions for 96 h before use. During acclimatization, the fish were fed a balanced diet containing 30% crude protein formulated in the laboratory (Table 1). Proximate compositions of the ingredients and the diet were determined by AOAC method (Helrich1990). Technical grade cypermethrin (92% active ingredient of (*RS*)-α-cyano-3-phenoxy-benzyl (1*RS*)-*cis,trans*-3-(2,2-dichlorovinyl)-2,2-dimethyl cyclopropane carboxylate) was obtained from Krishi Rasayan Group of Companies, Kolkata-700020 (India), and the emulsified concentrate (10% EC) of cypermethrin was procured under the brand name Ustaad® from the United Phosphorus Ltd., Vapi-396195, Gujarat. Deep tube-well water stored in an overhead tank (temperature 31 ± 3 °C, pH 7.2 ± 0.1; free CO_2 3.37 ± 0.31 mg/L; dissolved oxygen 6.5 ± 0.2 mg/L; total alkalinity 123.28 ± 3.37 mg/L as $CaCO_3$; total hardness 141.44 ± 11.01 mg/L as $CaCO_3$) was used as the test medium in all experiments.

Table 1 Proximate composition of the ingredients and the experimental diet

Ingredients used for formulation	Proximate composition (% dry matter)					Proportion used in formulation (%)
	Crude protein	Crude lipid	Crude fibre	Moisture	Ash	
Rice bran	11.96	10.8	16.3	7.46	8.73	38.74
Mustard oil cake	31.63	12.6	8.3	7.58	9.55	38.74
Fishmeal	65.98	13.6	3.1	13.62	9.5	20.52
Vitamin premix[a]	–	–	–	–	–	1.0
Mineral premix[b]	–	–	–	–	–	1.0
Formulated diet (dry matter: 94%)	30.29	4.98	11.29	6.0	12.21	

[a] Vitamin mixture (%): (Ambiplex; Brihans Lab, Pune): Vit B_1: 7.14, Vit B_2: 2.55, Vit B_6: 1.02, VitB_{12}: 0.012, biotin: 0.025, calcium pantothenate: 2.55, niacin: 76.50, choline chloride (B4): 10.20

[b] Mineral mixture (%): (Agrimin; Glaxo India Ltd, Mumbai): copper 3.12, cobalt: 0.45, magnesium: 21.14, iron: 9.79, iodine: 1.56, zinc: 21.30, calcium: 30.00, phosphorous: 8.25

Acute toxicity bioassays

Acute toxicity bioassays were made in 15-L glass aquaria, each holding 10 L of water and five acclimatized fish. Separate bioassays were made with technical grade (92% a.i.) and formulation (10% EC) of cypermethrin. The range of concentrations of cypermethrin used in these bioassays is presented in Table 2. Three replicates were maintained for each concentration including a water control and a solvent control. The solvent control contained 0.1 ml/L acetone, because acetone was used to dissolve technical grade cypermethrin, and the maximum amount of acetone present in the highest concentration of technical grade cypermethrin tested was less than 0.1 ml/L. The experiment was carried out for 96 h. Mortality of the fish was recorded every 24 h and the dead fish were removed. No food was provided during the bioassay to avoid the interference of excretory products of the fish with the test chemical. Lethal concentrations of cypermethrin at which 50% mortality of the fish occurred (LC_{50}) and its 95% confidence limits were estimated for 96 h from the mortality data using EPA-Probit analysis version 1.5 statistical software based on the probit analysis method of Finney (1971). LC_{50} values between the active ingredient and formulation were compared following the criteria of Mayer and Ellersieck (1986), Schmuck et al. (1994) and Demetrio et al. (2014).

Residue analysis of cypermethrin

Residue analyses of cypermethrin were made in the samples of water collected from the aquaria used for acute toxicity bioassay. The samples of water were taken in a separating funnel for separation into water and organic phase. The organic phase was extracted with ethyl acetate and the concentration of cypermethrin in the extract,

Table 2 Range of concentrations of cypermethrin (μg/L) used in the experiments

Technical (92% a.i.)		Formulation (10% EC)	
Nominal	Actual[a]	Nominal	Actual[a]
0	0	0	0
5	4.31	1	0.92
10	8.61	3	2.77
15	12.92	5	4.61
20	17.22	7	6.45
25	21.53	9	8.30
30	25.83	14	12.91

[a] After 2 h of exposure

after centrifugation (11,200*g*) and filtration through a 0.2-μm nylon membrane filter, was determined in a gas chromatograph (Agilent 6890N) equipped with a wide bore HP column (HP-5, 30 m, 0.32 mm ID, 0.25 μm film thickness), an electron capture detector (ECD) and a 7683 B Series auto-injector. N_2 gas was used as a carrier. Concentration of cypermethrin was quantified from the calibration curve prepared from standard cypermethrin concentrations, using ChemStation software. Limit of detection (LOD) and limit of quantification (LOQ) of the instrument were 0.01 and 0.03 ppm, respectively. Nominal concentrations (*N*) used in the experiments and the respective actual concentrations (*A*) determined by the above method after 2 h of exposure are presented in Table 2.

Experiments on biochemical parameters

Experiments on biochemical parameters were also made in 15-L glass aquaria, each holding 10 L of water and five fish. Three sub-lethal concentrations of the cypermethrin formulation used for these experiments were as follows: control (0.0 μg/L), a low dose (1.25 μg/L) and a high dose (2.25 μg/L) of cypermethrin (10% EC). The low and high doses of cypermethrin were approximately 25 and 50% of the 96-h LC_{50} value of the nominal concentration of cypermethrin (10% EC) to *O. niloticus*, respectively. There were three replicates for each concentration. The fish specimens were sampled after 96 h of exposure, rinsed in deionized water and dried on blotting paper. Blood was collected directly from the heart of the sampled fish with a micro-syringe and was used for glucose estimation. Simultaneously, liver was dissected out from the sampled fish, weighed and homogenized, and the homogenate was used for the analysis of glycogen content, alkaline and acid phosphatase, aspartate and alanine aminotransferase, acetylcholinesterase and catalase activities of the liver. Plasma glucose was determined by the method of Hyvarinen and Nikkila (1962) and liver glycogen was determined by the method of Carroll et al. (1956). Protein in liver tissue was determined by the method of Lowry et al. (1951), while the activities of the enzymes acid and alkaline phosphatase, aspartate and alanine aminotransferase, acetylcholinesterase and catalase were determined, respectively, by the methods of Walter and Schutt (1974), Reitman and Frankel (1957), Ellman et al. (1961) and Luck (1974).

Experiments on haematological parameters

These experiments were also made in 15-L glass aquaria using the same treatments and experimental conditions as described above for the experiment on biochemical parameters. However, the experiment was carried out for 28 days. One half of the fish were sampled from each aquarium at the end of 14 days and the rest were sampled at the end (28 days). Blood was collected from the sampled fish by the technique described above and was subjected to the determination of different haematological parameters. Haemoglobin (Hb%) was determined by cyanomethemoglobin method following Dacie and Lewis (1968), packed cell volume (PCV) or haematocrit (Hct) was determined by Wintrobe method or Macromethod (Dacie and Lewis 1968), and total erythrocyte count (TEC) and total leucocyte count (TLC) were determined with Neubauer's improved double haemocytometer using RBC and WBC diluting fluids, respectively.

Experiments on growth

Bioassays were carried out in outdoor cement vat (diameter 90 cm and average depth 50 cm). Before stocking of fish, each vat was prepared with a 3-cm-thick layer of uncontaminated soil at the bottom and approximately 100 g cow dung was applied on it. The vats were then filled with water (400 L) and kept in this condition for 1 month prior to the start of the experiment to ensure the growth of planktons, which serve as natural food for the test fish. Fingerlings of *O. niloticus* were stocked in these experimental vats and acclimatized in this condition for 1 week before the start of the experiments. Thirty fish, irrespective of their sex, were stocked in each vat. Altogether, nine vats were arranged according to randomized block design so that the fingerlings could be reared in three replicates for each of the three test concentrations (0, 1.25 and 2.50 μg/L) of cypermethrin (10% EC). Treatments with cypermethrin were made on day 1 of the experiment (initial treatment). Then 20% of the test medium was renewed at 10-day interval with a pulse treatment of pesticides at 20% of the initial nominal concentration. Static renewal bioassay with renewal of the bioassay medium at regular interval is a standard method to compensate loss of water through evaporation, to reduce the possibility of depletion of dissolved oxygen in the test medium and to reduce the possibility of loss of toxicant through volatilization and/or adsorption to test vessels (USEPA 2002). The experiment was continued for 90 days. During the experiments, the fish were hand fed the formulated diet (Table 1) twice daily at an interval of 10 h up to apparent satiation of the fish. Daily survey was made on the behaviour and mortality of the experimental fish. Water samples were collected from each vat every 15 days and dissolved oxygen, free carbon dioxide, temperature, total hardness and total alkalinity of the sampled water were determined by standard method (APHA 1995). All fish from each outdoor vat were sampled at the end of 90 days and the lengths (cm) and weights (g) of the sampled fish were recorded. Three sampled fish from each vat were subjected to biochemical analyses to determine crude protein, crude lipid and ash contents of the fish following the AOAC method (Helrich 1990). Growth was determined by percent increase in weight, specific growth rate (SGR), feed conversion ratio (FCR) and apparent net protein utilization (ANPU) using standard formulae (Bagenal 1978; Castell and Tiews 1980; Adams and Mclean 1985; Steffens 1989).

Statistical methods

The data obtained from the experiments on haematological, biochemical and growth parameters were subjected to single-factor ANOVA followed by least significant difference (LSD) test to verify the significance of difference between treatments at 5% level of probability (Gomez and Gomez 1984).

Results

Acute toxicity

96-h LC_{50} values of technical grade (92% a.i.) and formulation (10% EC) of cypermethrin to *O. niloticus* are presented in Table 3. The formulation was found to be more toxic to *O. niloticus* than the technical grade cypermethrin. In both groups, fish exposed to acute doses exhibited abnormal behaviour. Initially, fish showed frequent surfacing, increased opercular movement and faster swimming activity though within 5–9 h of treatment gradual lethargy was noted in their movement. Deposition of thick mucus layer was found in the buccal cavity and gills of exposed fish probably to minimize the irritating effects of toxicant coming in close contact with body. Sign of heavy internal haemorrhage around pharynx was noticed in the dead fish.

Effects of sub-lethal concentrations of cypermethrin

Liver glycogen and plasma glucose

Changes in liver glycogen and plasma glucose levels of control and cypermethrin (10% EC)-treated *O. niloticus* are presented in Fig. 1. There was a gradual decrease in hepatic glycogen level from 24 to 96 h of exposure, while plasma glucose level increased during this period. The effects were dose dependent, with the treatment of 2.5 μg/L cypermethrin showing more acute effects than that of 1.25 μg/L of cypermethrin.

Hepatic enzyme activities

Effects of cypermethrin on hepatic enzyme activities of *O. niloticus* are summarized in Table 4. Activities of the enzymes alkaline phosphatase, acetylcholinesterase and catalase decreased significantly ($P < 0.05$) in

Table 3 96-h LC_{50} values (μg/L) of cypermethrin and 95% confidence limit (CL) to freshwater teleost *Oreochromis niloticus*

Cypermethrin	Nominal		Actual (2 h)	
	LC_{50}	95% CL	LC_{50}	95% CL
Technical (92% a.i.)	10.71	8.79–12.75	9.74	8.10–11.35
Formulation (10% EC)	5.25	4.35–6.16	4.85	4.02–5.68

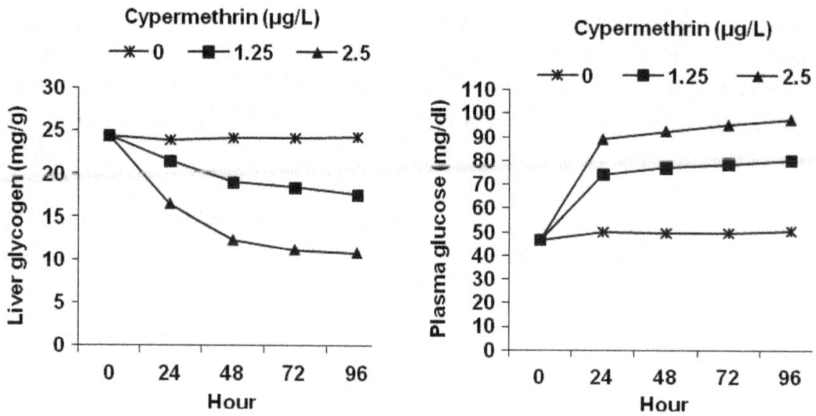

Fig. 1 Effects of cypermethrin (10% EC) on liver glycogen and plasma glucose of *Oreochromis niloticus*

Table 4 Changes in activities of hepatic enzymes of *O. niloticus* after 96-h exposure to cypermethrin (10% EC)

Parameter	0.0 μg/L	1.25 μg/L	2.5 μg/L
Alkaline phosphatase[1]	13.28 ± 0.07[a]	8.95 ± 0.08[b]	7.29 ± 0.07[c]
Acid phosphatase[1]	6.39 ± 0.06[a]	7.67 ± 0.12[b]	8.86 ± 0.07[c]
Aspartate aminotransferase[2]	2.28 ± 0.02[a]	3.04 ± 0.04[b]	3.58 ± 0.03[c]
Alanine aminotransferase[2]	5.71 ± 0.02[a]	6.98 ± 0.04[b]	9.55 ± 0.04[c]
Acetylcholinesterase[3]	0.73 ± 0.02[a]	0.48 ± 0.01[b]	0.38 ± 0.02[c]
Catalase[4]	13.37 ± 0.04[a]	11.06 ± 0.09[b]	9.36 ± 0.08[c]

Data are mean ± SD ($n = 3$); means with dissimilar superscripts in the same row indicate the least significant difference between the means at $P < 0.05$

[1] μg PNP/mg tissue protein/30 min

[2] μM pyruvate/mg tissue protein/h

[3] μM acetylthiocholine iodide/mg tissue protein/min

[4] μM H_2O_2 decomposed/mg tissue protein/min

cypermethrin treatments as compared to control. The higher the concentration of cypermethrin, the lower the activities of these enzymes. On the contrary, the activities of the enzymes acid phosphatase, aspartate aminotransferase and alanine aminotransferase increased significantly ($P < 0.05$) in both treatments of cypermethrin as compared to control. There were also significant differences in activities between the two sub-lethal concentrations (1.25 and 2.5 μg/L) of cypermethrin.

Haematological parameters

The changes in haematological parameters of *O. niloticus* exposed to control and cypermethrin are summarized in Table 5. The results indicated that TEC, haemoglobin concentration (Hb), PCV, mean cell volume (MCV), mean cell haemoglobin (MCH) and mean corpuscular haemoglobin concentration (MCHC) decreased significantly ($P < 0.05$) in both sub-lethal concentrations of cypermethrin (1.25 and 2.5 μg/L) as compared to control. On the other hand, total leucocyte count showed significant elevation in cypermethrin-treated *O. niloticus* than the control. There was no significant difference in these parameters between the exposure periods of 14 and 28 days.

Effects on growth

Long-term exposure (90 days) to sub-lethal concentrations of cypermethrin increased mortality and reduced growth of *O. niloticus* as compared to control. The growth parameters are summarized in Table 6. Fish exposed to 2.5 μg/L cypermethrin showed more reduction in weight gain (%), SGR, protein efficiency ratio

Table 5 Changes in haematological parameters of *O. niloticus* exposed to cypermethrin

Parameters	Days of exposure	0.0 μg/L	1.25 μg/L	2.5 μg/L
Total erythrocyte count (10^6/mm^3)	14	2.81 ± 0.03[a]	2.74 ± 0.04[b]	2.62 ± 0.03[c]
	28	2.86 ± 0.05[a]	2.68 ± 0.06[b]	2.51 ± 0.02[c]
Haemoglobin concentration (g/dl)	14	8.72 ± 0.05[a]	7.77 ± 0.04[b]	7.08 ± 0.09[c]
	28	8.83 ± 0.05[a]	7.34 ± 0.05[b]	6.74 ± 0.06[c]
Packed cell volume (%)	14	25.58 ± 0.16[a]	23.68 ± 0.07[b]	21.99 ± 0.69[c]
	28	25.67 ± 0.14[a]	22.53 ± 0.11[b]	20.96 ± 0.21[c]
Mean cell volume (fl/cell)	14	91.05 ± 1.12[a]	86.55 ± 1.13[b]	83.92 ± 0.64[c]
	28	89.66 ± 1.26[a]	84.18 ± 1.76[b]	83.51 ± 0.99[c]
Mean cell haemoglobin (pg/cell)	14	31.04 ± 0.17[a]	28.41 ± 0.45[b]	27.04 ± 0.25[b]
	28	30.86 ± 0.58[a]	27.43 ± 0.51[b]	26.84 ± 0.01[b]
Mean corpuscular haemoglobin concentration (g/dl)	14	34.09 ± 0.17[a]	32.82 ± 0.24[b]	32.22 ± 0.31[b]
	28	34.41 ± 0.36[a]	32.58 ± 0.33[b]	32.14 ± 0.40[b]
Total leucocyte count (10^3/mm^3)	14	23.55 ± 0.13[a]	24.74 ± 0.07[b]	25.96 ± 0.08[c]
	28	23.94 ± 0.09[a]	25.59 ± 0.15[b]	26.86 ± 0.07[c]

Data are mean ± SD ($n = 3$); means with dissimilar superscripts in the same row indicate the least significant difference between the means at $P < 0.05$

Table 6 Changes in growth parameters and mortality of *O. niloticus* exposed to cypermethrin (10% EC) for 90 days

Parameters	0.0 μg/L	1.25 μg/L	2.5 μg/L
Initial weight (g)	2.64 ± 0.16	2.64 ± 0.16	2.64 ± 0.16
Weight gain (%)[1]	128.39 ± 8.36[a]	74.72 ± 10.89[b]	51.79 ± 9.45[c]
FCR[2]	3.08 ± 0.21[a]	4.19 ± 0.53[b]	4.95 ± 0.98[c]
SGR (%/day)[3]	0.92 ± 0.04[a]	0.62 ± 0.07[b]	0.46 ± 0.07[c]
PER[4]	1.09 ± 0.07[a]	0.81 ± 0.12[b]	0.70 ± 0.12[c]
HSI[5]	1.10 ± 0.04[a]	1.57 ± 0.09[b]	1.83 ± 0.12[c]
ANPU (%)[6]	20.85 ± 1.12[a]	13.61 ± 1.60[b]	10.74 ± 1.54[c]
Mortality (%)	4.04 ± 1.22[a]	17.39 ± 1.26[b]	21.75 ± 1.30[c]

Values are the mean of three replicates ± SD; means with dissimilar superscripts in the same row indicate the least significant difference between the means at $P < 0.05$

[1] Weight gain (%) = [{final wt (g) − initial wt (g)}/initial wt (g)] × 100

[2] Feed conversion ratio = food given/weight gain

[3] Specific growth rate (%/day) = $\{(\log_e W_2 - \log_e W_1)/T\} \times 100$, where W_1 = initial body weight (g), W_2 = final body weight (g) and T = days of exposure

[4] Protein efficiency ratio = increase in weight of fish (wet weight)/weight of protein in feed (dry weight)

[5] Hepatosomatic index = [{wet weight of liver (g) without gall bladder}/wet body weight] × 100

[6] Apparent net protein utilization (%) = (net increase in carcass protein/amount of protein consumed) × 100

(PER) and ANPU than those treated by 1.25 μg/L cypermethrin. FCR and hepatosomatic index (HSI) increased significantly ($P < 0.05$) in fish treated with 2.5 μg/L cypermethrin than in those treated with 1.25 μg/L cypermethrin treatment and control. Proximate compositions of carcass of the sampled fish are presented in Table 7. The results clearly indicated that crude protein, crude lipid and ash contents increased from their initial values. However, the rate of increase was significantly lower in cypermethrin-treated fish in comparison to control. Among the cypermethrin-treated fish, the values reduced with the increase of cypermethrin concentration.

Table 7 Proximate composition (% wet weight basis) of carcass of *O. niloticus* exposed to cypermethrin (10% EC) for 90 days

	Initial	Final		
		0.0 μg/L	1.25 μg/L	2.5 μg/L
Crude protein	11.13 ± 0.27	15.40 ± 0.24^{a}	13.56 ± 0.24^{b}	12.43 ± 0.24^{c}
Crude lipid	2.73 ± 0.14	4.78 ± 0.12^{a}	3.63 ± 0.14^{b}	3.22 + 1.12^{c}
Ash	1.89 ± 0.08	3.01 ± 0.07^{a}	2.49 ± 0.06^{b}	2.29 ± 0.07^{c}

Values are the mean of three replicates ± SD; means with dissimilar superscripts in the same row indicate the least significant difference between the means at $P < 0.05$

Discussion

Acute toxicity

96-h LC_{50} values of technical grade (92% a.i.) and commercial (10% EC) cypermethrin to *O. niloticus* as observed in the present investigation are, respectively, close to 96-h LC_{50} values of technical grade (98%) alpha-cypermethrin to *Poecilia reticulata* (9.43 μg/L; Yilmaz et al. 2004) and commercial cypermethrin (10% EC) to freshwater fish *Colisa fasciatus* (6.0 μg/L; Singh et al. 2010). The present results indicate that the emulsified concentrate (EC) of cypermethrin is more toxic than the technical grade (active ingredient) cypermethrin. This has also been confirmed by the toxicity studies of cypermethrin on various species of invertebrates (Demetrio et al. 2014; Majumder and Kaviraj 2015). Several studies have demonstrated that the inert ingredients added to the formulated products increase its toxicity (Puglis and Boone 2011). Quotient (LC_{50} of X/LC_{50} of Y) is used to interpret and compare the acute toxicity data between technical grade (X) and formulation (Y) of pesticides. Mayer and Ellersieck (1986) assumed the formulation as more toxic when the quotient was more than 1, while Schmuck et al. (1994) observed that there was a natural variability of quotient between 0.5 and 2.0 and considered the formulation as more toxic when the quotient was more than 2. Evaluating the results of the present study, it was revealed that the formulation was more toxic than the technical grade cypermethrin according to the criteria of Mayer and Ellersieck (1986) and Schmuck et al. (1994). Since only one single point (LC_{50}) for the concentration–effect function was considered under both criteria, Demetrio et al. (2014) proposed to accept the criteria as valid only when the concentration effect lines were parallel. Log-probit model regression lines for log concentration versus probit values of mortality for *O. niloticus* following 96-h exposure to technical grade (T) and formulation (F) of cypermethrin were parallel to each other (Fig. 2).

Biochemical parameters

Reduction of hepatic glycogen content in *O. niloticus* due to cypermethrin treatment as observed in the present study is a common symptom of stress exhibited by fish when exposed to type II pyrethroids (Kaviraj and Gupta 2014). Similar effects of cypermethrin have been reported for *Clarias batrachus* (Begum 2005), *H. fossilis* (Saha and Kaviraj 2009b) and *Ophiocephalus punctatus* (Shruti et al. 2011) and *Labeo rohita* (Tiwari et al. 2012). Carbohydrate serves as the instant source of energy during stress. Accordingly, glycogen of liver is broken down (glycogenolysis) to meet the energy demand raised during cypermethrin stress, resulting in the reduction of hepatic glycogen content. On the other hand, an increase in plasma glucose level due to cypermethrin treatment, as observed in the present study, has also been observed on *H. fossilis* (Saha and Kaviraj 2009b) and *O. niloticus* (Firat et al. 2011) probably due to an increase in hepatic glucose-6-phosphatase activity which enhances glycogenolysis and glucose synthesis from extrahepatic tissue proteins and amino acids (Firat et al. 2011). Gluconeogenesis is another pathway to increase plasma glucose level during the stress of pesticide to fish (Saravanan et al. 2011). This is initiated by aminotransferase enzymes, which provide strategic links between protein and carbohydrate metabolism to meet the demand of energy under stressed condition (Neelima et al. 2013). Increased activities of hepatic aspartate aminotransferase (AST) and alanine aminotransferase (ALT) as observed in the present study indicate active catabolism of amino acid to meet the immediate energy demand under cypermethrin stress (Kumar et al. 2011). In the first

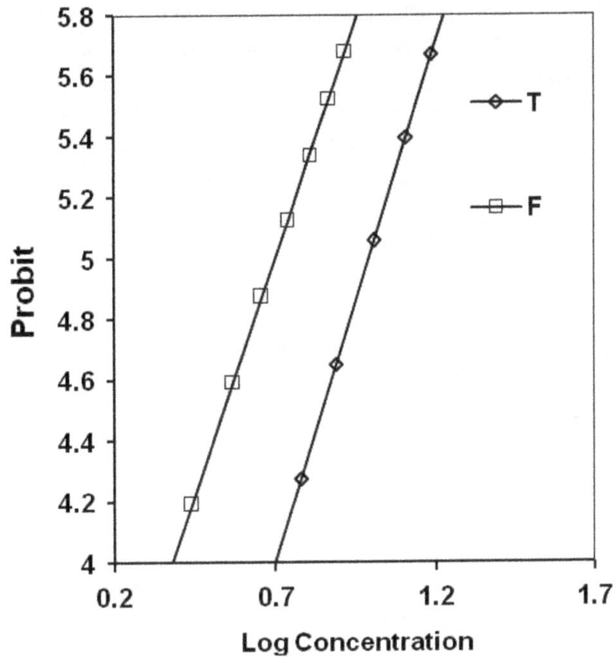

Fig. 2 Log-probit model regression lines for log concentration (actual) versus probit values for *Oreochromis niloticus* following 96-h exposure to technical grade (*T*) and formulation (*F*) of cypermethrin

phase of catabolism, α keto acids are formed through transamination. Gluconeogenesis is promoted by the incorporation of keto acids into tricarboxylic acid (TCA) cycle via active transdeamination (Neelima et al. 2013). Studies on cypermethrin-induced increase in the activities of hepatic AST and ALT in *C. batrachus* (Begum 2005), *Cirrhinus mrigala* (Prasanth and Neelagund 2008), *Channa punctatus* and *C. batrachus* (Kumar et al. 2011), *L. rohita* (Tiwari et al. 2012), *Catla catla* (Vani et al. 2012) and *Prochilodus lineatus* (Loteste et al. 2013) indicate that fish utilize amino acids to overcome the stress of cypermethrin.

Changes in glucose metabolism were also linked with the activities of hepatic alkaline phosphatase, which were decreased in cypermethrin-treated *O. niloticus* as compared to control. Similar effects of cypermethrin on alkaline phosphatase activities were found in the liver of *Heterobranchus bidorsalis* (Gabriel et al. 2011), *H. fossilis* (Saha and Kaviraj 2009b) and *L. rohita* (Tiwari et al. 2012). Liver alkaline phosphatase promotes glycogen synthesis by inactivating the enzyme phosphorylase. Reduced activities of hepatic alkaline phosphatase may lead to the breakdown of glycogen in order to get necessary energy to overcome the stress. Some researchers reported increased activities of alkaline phosphatase in the serum of fish exposed to cypermethrin, which might be due to the efflux of enzymes from the liver into bloodstream (Firat et al. 2011; Meenambal et al. 2012). On the other hand, the activity of acid phosphatase was elevated in cypermethrin-exposed *O. niloticus* as compared to control. These results are in agreement with the findings of Kumar et al. (2014) on *C. batrachus*. Acid phosphatase is a lysosomal enzyme that hydrolyses the ester linkage of phosphate esters and cause autolysis of cell after its death. Increased acid phosphatase activity may be due to hepatocellular damages in *O. niloticus* exposed to cypermethrin.

Acetylcholinesterase (AChE) is distributed in synaptic region and induces nerve impulse propagation by converting acetylcholine into acetic acid and choline. Inhibition of this enzyme can cause hyperexcitability. This reaction was evident in the acute toxicity experiment with cypermethrin in the present study. Cypermethrin probably interacts with the hydrophobic aromatic surface region of acetylcholinesterase and inhibits the activities of this enzyme as observed in *L. rohita* (Tiwari et al. 2012), *C. batrachus* (Kumar et al. 2014), *Channa striatus* (Ahmed et al. 2015) and *Cyprinus carpio* (Neelima et al. 2015).

Cypermethrin also produces oxidative stress in fish. This was evident from the decreased activity of catalase in the liver of *O. niloticus* exposed to sub-lethal concentrations of cypermethrin, as observed in the present study. Catalase is an enzymatic antioxidant. Reactive oxygen species (ROS) like superoxide anion, hydrogen peroxide and hydroxyl radicals are formed in fish upon exposure to pollutants. Catalase scavenges reactive oxygen species, converts them into less reactive species and prevents lipid peroxidation. Accordingly, the activity of malondialdehyde (MDA) is increased and that of catalase is decreased as oxidative defense in fish (Marigoudar et al. 2013; Kaviraj and Gupta 2014). Catalase activity thus serves as a potential oxidative stress biomarker in fish toxicity studies. Tripathi and Bandooni (2011) observed decreased activity of catalase in the liver of *C. batrachus* treated with alphamethrin, a racemic mixture of two isomers of cypermethrin. Cypermethrin-induced reduction in the catalase activity of spermatozoa has also been observed in *Oncorhynchus mykiss* (Kutluyer et al. 2016).

Haematological parameters

Decline in total erythrocyte count, Hb content, PCV and MCHC values as observed in the present study is an indication of anaemia. Anaemia is probably caused by the inhibition of erythropoiesis and chemosynthesis as well as the destruction of erythrocytes (Adhikari et al. 2004; Jee et al. 2005; Akinrotimi et al. 2012). In addition, WBC is liberated from the spleen into the blood to counter the cypermethrin-induced stress (Akinrotimi et al. 2012), resulting in an increase in leucocyte count (Adhikari et al. 2004; Ojutiku et al. 2013) as also observed in the present study.

Effects on growth and biochemical composition

Results of the present study indicated that the growth of *O. niloticus* is reduced when exposed to sub-lethal concentrations of cypermethrin for 90 days. This was due to poor efficiency in feed conversion, net protein utilization and subsequent poor deposition of protein and lipid in the muscle. There are evidence that cypermethrin can reduce the protein level of serum (Vani et al. 2012; Kannan et al. 2014), gills and liver (Begum 2005), muscle and kidneys (Begum 2007) in several species of fish. Gijare et al. (2011) also found that the lipid content was decreased in *O. punctatus* due to cypermethrin treatment. The toxic impact of cypermethrin is quickly removed from water (Saha and Kaviraj 2009a). But the results of the present study indicate that the stress produced by 1.25–2.5 μg/L cypermethrin (10% EC) on biochemical and haematological parameters is carried and reflected on feed conversion and deposition of protein.

Conclusion

It is concluded from the present study that cypermethrin (10% EC) is highly toxic to *O. niloticus.* Concentration of cypermethrin as low as 1.25 μg/L is capable of inducing changes in biochemical and haematological parameters of the fish after short-term exposure.These changes are indication of stress on the fish. The stress also affects growth of the fish, if itis exposed to cypermethrin for long period.

Acknowledgements Partial research grants received from the University of Kalyani through the Department of Science and Technology-Promotion of University Research and Scientific Excellence (DST-PURSE) and Personal Research Grant are thankfully acknowledged.

Conflict of interest There is no conflict of interest.

References

Adams SM, McLean RB (1985) Estimation of the large mouth bass, *Micropterus salmoides* Lacepede, growth using the liver somatic index physiological variables. J Fish Biol 26:111–126

Adeyemi JA, Adewale OO, Oguma AY (2014) Mortality, oxidative stress and hepatotoxicities in juvenile African catfish, *Clarias gariepinus* Burchell, exposed to lead and cypermethrin. Bull Environ Contam Toxicol 92:529–533

Adhikari S, Sarkar B, Chatterjee A, Mahapatra CT, Ayyappan S (2004) Effects of cypermethrin and carbofuran on certain haematological parameters and prediction of their recovery in a freshwater teleost, *Labeo rohita* (Hamilton). Ecotoxicol Environ Saf 58:220–226

Ahmed O, Mastan SA, Rabiabanu S, Indira P (2015) Sub lethal effect of cypermethrin on acetylcholinesterase (AChE) activity and acetylcholine (Ach) content in selected tissues of *Channa striatus* (Bloch.). J Toxicol Environ Health Sci 7(4):31–37

Akinrotimi OA, Gabriel UU, Ariweriokuma SV (2012) Haematotoxicity of cypermethrin to African catfish (*Clarias gariepinus*) under laboratory conditions. J Environ Eng Technol 1(2):13–19

APHA (American Public Health Association) (1995) Standard methods for the examination of water and wastewater. American Public Health Association, American Water Works association and Water Pollution Control Federation, Washington, DC

Bagenal T (1978) Methods for assessment of fish production in fresh water, 3rd edn. Blackwell Scientific Publication, Oxford

Begum G (2005) In vivo biochemical changes in liver and gill of *Clarias batrachus* during cypermethrin exposure and following cessation of exposure. Pestic Biochem Phys 82(3):185–196

Begum G (2007) Cypermethrin induced biochemical perturbations in freshwater fish *Clarias batrachus* at sublethal exposure and after released into freshwater. Drug Chem Toxicol 30:55–65

Carriquiriborde P, Díaz J, Mugni H, Bonetto C, Ronco AE (2007) Impact of cypermethrin on stream fish populations under field use in biotech-soybean production. Chemosphere 68:613–621

Carroll NV, Longley RW, Roe JH (1956) The determination of glycogen in liver and muscle by use of anthrone reagent. J Biol Chem 220:583–593

Castell JD, Tiews K (1980) Report of the EIFAC, IUNS and ICES working Group on the standardization of methodology in fish nutrition research. EIFAC Technical Paper 36

Dacie JV, Lewis SM (1968) Practical haematology, 4th edn. Churchill, London

Demetrio PM, Bonetto C, Ronco AE (2014) The effect of cypermethrin, chlorpyrifos and glyphosate active ingredients and formulations on *Daphnia magna* (Straus). Bull Environ Contam Toxicol 93:268–273

Ellman GL, Courtney KD, Andres V Jr, Featherstone RM (1961) A new and rapid colorimetric determination of acetylcholinesterase activity. Biochem Pharmacol 7:88–95

Finney DJ (1971) Probit analysis. Cambridge University Press, London

Firat Ö, Cogum HY, Yüzereroğlu TA, Gök G, Firat Ö, Kargin F, Kötemen Y (2011) A comparative study on the effects of a pesticide (cypermethrin) and two metals (copper, lead) to serum biochemistry of Nile tilapia, *Oreochromis niloticus*. Fish Physiol Biochem 37:657–666

Gabriel UU, Jack IR, Egobueze E, Edori OS (2011) Impact of cypermethrin on selected enzymes in tissues of *Heterobranchus bidorsalis*. West Afr J Appl Ecol 18:121–127

Gijare SS, Raja IA, Tantarpale VT, Kulkarni KM (2011) Lipid changes in the freshwater fish *Ophiocephalus punctatus* exposed to synthetic pyrethroid cypermethrin. Biosci Biotechnol Res Commun 4(1):52–54

Gomez KA, Gomez AA (1984) Statistical procedures for agricultural research, 2nd edn. Wiley Interscience, New York

Helrich K (1990) Official methods of analyses, vol I, 15th edn. Association of Official Analytical Chemists (AOAC), Washington, DC

Hyvarinen A, Nikkila E (1962) Specific determination of blood glucose with *O*-toluidine. Clin Chim Acta 7:140

Jee JH, Masroor F, Kang JC (2005) Responses of cypermethrin induced stress in haematological parameters of Korean rock fish, *Sebastes schlegeli* (Hilgendorf). Aquac Res 36:898–905

Kannan M, Muthusamy P, Venkatachalam U (2014) Response of synthetic pyrethroid cypermethrin (10% EC) induced stress in biochemical and haematological parameters of Indian major carp *Catla catla* (Hamilton, 1822). World J Pharm Res 3(4):1976–1996

Kaviraj A, Gupta A (2014) Biomarkers of type II synthetic pyrethroid pesticides in freshwater fish. BioMed Res Int. Article ID 928063. doi:10.1155/2014/928063

Kumar A, Sharma B, Pandey RS (2007) Preliminary evaluation of the acute toxicity of cypermethrin and λ-cyhalothrin to *Channa punctatus*. Bull Environ Contam Toxicol 79:613–616

Kumar A, Sharma B, Pandey RS (2011) Cypermethrin induced alterations in nitrogen metabolism in freshwater fishes. Chemosphere 83:492–501

Kumar A, Sharma B, Pandey RS (2014) λ-Cyhalothrin and cypermethrin induce stress in the freshwater muddy fish, *Clarias batrachus*. Toxicol Environ Chem 96(1):136–149

Kutluyer F, Benzer F, Erişir M, Öğretmen F, Inanan BE (2016) The in vitro effect of cypermethrin on quality and oxidative stress indices of rainbow trout *Oncorhynchus mykiss* spermatozoa. Pestic Biochem Phys 128:63–68

Loteste A, Scagnetti J, Simoniello MF, Campana M, Parma MJ (2013) Hepatic enzymes activity in the fish *Prochilodus lineatus* (Valenciennes, 1836) after sublethal cypermethrin exposure. Bull Environ Contam Toxicol 90:601–604

Lowry OH, Rosebrough NJ, Farr AL, Randall RJ (1951) Protein measurement with the folin-phenol reagent. J Biol Chem 193:265–275

Luck H (1974) Catalase. In: Bergmeyer HU (ed) Methods of enzymatic analysis, 2nd edn. Academic Press, New York, pp 885–890

Majumder R, Kaviraj A (2015) Variation in acute toxicity between technical grade and commercial formulation of cypermethrin to some non-target freshwater organisms. Int J Curr Res 7(06):16755–16759

Marigoudar SR, Ahmed RN, David M (2013) Ultrastructural responses and oxidative stress induced by cypermethrin in the liver of *Labeo rohita*. Chem Ecol 29(4):296–308

Mayer FL Jr, Ellersieck MR (1986) Manual of acute toxicity: interpretation and data base for 410 chemicals and 66 species of freshwater animals. US Department of the Interior, Fish and Wildlife Service. Resource Publication 160, Washington, DC

Meenambal M, Pugazhendy K, Vasantharaja C, Venkatesan S (2012) Ameliorative property of *Delonix elata* supplementary feed against cypermethrin induced serum biochemical changes in freshwater fish *Cyprinus carpio* (Linn). J Pharm Res 5(5):2489–2492

Neelima P, Kumar LCA, Rao NG, Rao JCS (2013) Studies on activity levels of aspartate amino transferase (ASAT) and alanine aminotransferase (ALAT) in the tissues of *Cyprinus carpio* (Linn.) exposed to cypermethrin (25% EC). Int J Pharm Pharm Sci 5(3):566–570

Neelima P, Rao KG, Rao NG, Jammu CSR (2015) Enzymatic alterations as biomarkers of cypermethrin (25% EC) toxicity in a freshwater fish, *Cyprinus carpio* (Linn.). Int J Fish Aquat Stud 3(1):149–158

Ojutiku RO, Asuwaju FP, Ayanda IO, Obande RA, Agbelege OO (2013) Effects of acute toxicity of cypermethrin on some biochemical parameters of juveniles of *Clarias gariepinus* (Burchell, 1822). Int J Eng Sci Invent 2(3):01–07

Polat H, Erkoç FÜ, Viran R, Koçak O (2002) Investigation of acute toxicity of beta-cypermethrin on guppies *Poecilia reticulata*. Chemosphere 49:39–44

Prasanth MS, Neelagund SE (2008) Impact of cypermethrin on enzyme activities in the freshwater fish *Cirrhinus mrigala* (Hamilton). Casp J Environ Sci 6(2):91–95

Puglis HJ, Boone MD (2011) Effects of technical-grade active ingredient vs. commercial formulation of seven pesticides in the presence or absence of UV radiation on survival of green frog tadpoles. Arch Environ Contam Toxicol 60:145–155

Reitman S, Frankel S (1957) A colorimetric method for determination of serum glutamic oxaloacetic transaminase and serum glutamic pyruvic transaminase. Am J Clin Pathol 28:649–654

Saha S, Kaviraj A (2008) Acute toxicity of synthetic pyrethroid cypermethrin to some freshwater organisms. Bull Environ Contam Toxicol 80:49–52

Saha S, Kaviraj A (2009a) Effect of ambient temperature and daylight on the survival of freshwater catfish *Heteropneustes fossilis* (Bloch, 1974) exposed to cypermethrin. Environ Eng Sci 26(2):459–462

Saha S, Kaviraj A (2009b) Effects of cypermethrin on some biochemical parameters and its amelioration through dietary supplementation of ascorbic acid in freshwater catfish *Heteropneustes fossilis*. Chemosphere 74(19):1254–1258

Saha S, Kaviraj A (2013) Dietary ascorbic acid as a means to counter the stress of cypermethrin on the growth of freshwater catfish *Heteropneustes fossilis*. Toxicol Ind Health 29(5):468–473

Saravanan M, Karthika S, Malarvizhi A, Ramesh M (2011) Ecotoxicological impacts of clofibric acid and diclofenac in common carp (*Cyprinus carpio*) fingerlings: hematological, biochemical, ionoregulatory and enzymological responses. J Hazard Mater 195:188–194

Schmuck R, Pfleeger W, Grau R, Hollihn U, Fischer R (1994) Comparison of short-term aquatic toxicity: formulation vs active ingredients of pesticides. Arch Environ Contam Toxicol 26:240–250

Shruti SG, Raja IA, Tantarpale VT, Kulkarni KM (2011) Modulation in glycogen of the freshwater fish *Ophiocephalus punctatus* exposed to cypermethrin. Bioscan 6(1):129–130

Singh SK, Singh SK, Yadav RP (2010) Toxicological and biochemical alterations of cypermethrin (synthetic pyrethroids) against freshwater fish *Colisa fasciatus* at different season. World J Zool 5(1):25–32

Steffens RR (1989) Principles of fish nutrition. Ellis Horwood, Chichester

Tiwari S, Tiwari R, Singh A (2012) Impact of cypermethrin on fingerlings of common edible carps (*Labeo rohita*). Sci World. Article ID 291395. doi:10.1100/2012/291395

Tripathi G, Bandooni N (2011) Impact of alphamethrin on antioxidant defense (catalase) and protein profile of a catfish. Environmentalist 31:54–58

USEPA (2002) Methods for measuring the acute toxicity of effluents and receiving waters to freshwater and marine organisms. US Environment Protection Agency, Washington, DC. EPA-821-R-02-012

Vani T, Saharan N, Roy SD, Ranjan R, Pal AK, Siddaiah GM, Kumar R (2012) Alteration in haematological and biochemical parameters of *Catla catla* exposed to sub-lethal concentration of cypermethrin. Fish Physiol Biochem 38:1577–1584

Walter K, Schutt C (1974) Method of enzymatic analysis. Academic, New York

Yilmaz M, Gül A, Erbaşlı K (2004) Acute toxicity of alpha-cypermethrin to guppy (*Poecilia reticulata* Pallas, 1859). Chemosphere 56:381–385

8

Annual variations in hepatic antioxidant defenses and lipid peroxidation in a temperate fish, common bream *Abramis brama* (L.)

Alexey A. Morozov · Grigorii M. Chuiko · Victoria V. Yurchenko

Abstract Alternation of physiological states during the annual cycle (pre-spawning mobilization, spawning energy loss, post-spawning accumulation of energy reserve, declined metabolic activity during the wintering) occurring in the wide temperature range requires high plasticity of fish defense systems. The article presents a 2-year study of hepatic antioxidant defenses and lipid peroxidation in the temperate cyprinid fish, common bream *Abramis brama* (L.), in relation to basic annual rhythms, such as photoperiodic and temperature cycling, reproduction and feeding activity. The objective of the study was to evaluate the plasticity of antioxidant defenses in bream under normal conditions throughout the annual cycle, by measuring hepatic antioxidants (superoxide dismutase, catalase, glutathione-*S*-transferase activities and reduced glutathione concentrations) and products of lipid peroxidation (thiobarbituric acid reactive substances concentrations). The results suggested that hepatic levels of antioxidants in bream varied dramatically during the annual cycle allowing bream to develop necessary adaptation and implement reproduction program. The lowest values of all studied antioxidants were witnessed in the middle of winter. In early spring, antioxidant enzyme activities increased rapidly reaching their maxima in late spring and summer. Toward the end of autumn, enzyme activities were declining toward the levels observed before the ice breakup. Spring–summer up-regulation of antioxidant defenses along with increased lipid peroxidation indicated bream to be experiencing mild oxidative stress.

Keywords Antioxidants · Lipid peroxidation · Oxidative stress · Seasonal fluctuations

Introduction

Dynamic equilibrium between an organism and its environment is maintained by the effective regulation of metabolic processes in response to the external factors. In the temperate-boreal climatic zone, environmental conditions change dramatically through the year demanding a large adaptive potential from fish. Alternation of physiological states during the annual cycle (pre-spawning mobilization, spawning energy loss, post-spawning accumulation of energy reserve, declined metabolic activity during the wintering) occurring in the wide temperature range requires high plasticity of fish defense systems.

Energetic life support in aerobic organisms is maintained by the complex combination of processes, in which the key role is played by the oxidative reactions involving molecular oxygen as the main oxidizing

A. A. Morozov (✉) · G. M. Chuiko · V. V. Yurchenko
Laboratory of Physiology and Toxicology of Aquatic Animals, I.D. Papanin Institute for Biology of Inland Waters Russian Academy of Sciences, IBIW RAS, 152742 Borok, Russia
e-mail: aleksey.a.morozov@gmail.com

agent or electron acceptor. Due to electrons escaping the mitochondrial electron-transport chain to molecular oxygen, reactive oxygen species (superoxide anion radical, hydrogen peroxide, hydroxyl radical, and hydroxyl anion) are produced. The amount of electrons escaping the electron-transport chain depends on the physiological state of the organism and oxygen availability. Living organisms possess a multilevel and complicated antioxidant defense system operating to prevent the formation of reactive oxygen species (and reactive oxygen species modified molecules) and to eliminate them, or minimize their negative effects (Winston and Di Giulio 1991; Menshchikova et al. 2006; Lushchak 2014, 2016).

Primary antioxidant enzymes use reactive oxygen species as substrates, thus performing direct detoxification. Superoxide dismutase (SOD, EC 1.15.1.1) converts superoxide anion radical to hydrogen peroxide and molecular oxygen. Catalase (CAT, EC 1.11.1.6) destroys hydrogen peroxide to molecular oxygen and water. Secondary antioxidant enzymes assist the primary ones. For instance, glutathione-*S*-transferase (GST, EC 2.5.1.18) protects cells by reducing hydroperoxyl groups of oxidized phospholipids directly in the membranes. A no less important component of the antioxidant defense is a non-enzymatic antioxidant—reduced glutathione (GSH), which can be used as a cofactor by primary and secondary antioxidant enzymes, such as glutathione-dependent peroxidase (EC 1.11.1.9) and GST, respectively. GSH is involved in reduction and conjugation reactions and can directly neutralize many reactive oxygen species and stabilize cell membranes (Hayes et al. 2005; Menshchikova et al. 2006; Di Giulio and Meyer 2008; Forman et al. 2009; Lushchak 2014).

Normally, antioxidant defenses effectively maintain very low steady-state levels of reactive oxygen species, protecting cells from the oxidative damage. However, under certain circumstances, an imbalance between production and elimination of reactive oxygen species can occur, causing oxidative stress (Lushchak 2014; Sies 2015) which has specific physiological consequences including lipid peroxidation (Cajaraville et al. 2003; Di Giulio and Meyer 2008). Changes in water temperature, oxygen deficiency as well as reoxygenation may affect steady-state levels of reactive oxygen species and antioxidant defenses in fish (Martínez-Álvarez et al. 2005). It is also known that release of sexual products during spawning leads to tissue degradation coupled with oxidation processes (Shulman and Love 1999), therefore, spawning could induce natural oxidative stress (Soldatov et al. 2007).

Useful information on the state of the organism could be obtained by accessing antioxidant enzyme activities (Sies 2015). The objective of our study was to evaluate the plasticity of antioxidant defenses in the temperate cyprinid fish, the common bream (*Abramis brama* L.), under normal conditions throughout the annual cycle, by measuring hepatic antioxidants (SOD, CAT, GST activities, and GSH concentrations) and products of lipid peroxidation (concentrations of thiobarbituric acid reactive substances, TBARS).

Materials and methods

Fish and sampling

Common bream is a bottom dwelling fish. It inhabits slow-flowing waters and forms shoals. The bream is a typical benthivore, feeding on macrozoobenthos including oligochaetes, chironomid larvae, and bivalves. Most actively, it feeds during the summer time. The bream spawns when the water temperature reaches 12–16 °C (in the study area, it usually happens in May). Under normal weather conditions, spawning is of short duration (Nikolsky 1978; Yakovlev et al. 2001).

Bream was collected during 2 years, January 2008–December 2009. In February, November, December 2008, January–April and November 2009 sampling could not been conducted due to legal reasons. Mature fish was captured using a seine in the mouth area of the Sutka River (58°1.18′N 38°15.976′E, Upper Volga basin). Based on hydrobiological and hydrochemical parameters of the water, the river is characterized as mesotrophic water object. Concentrations of the main ions (ppm) in the sampling area varied in the following ranges 18.1–76.1 (Ca^{2+}), 5.3–23.3 (Mg^{2+}), 2.4–13.1 (Na^{+}), 1.1–3.0 (K^{+}), 1.8–12.4 (Cl^{-}), 2.9–23.4 (SO_4^{2-}), and 73.8–344.4 (HCO_3^{-}); pH values varied from 7.40 to 8.09 (Tsvetkov et al. 2015). Sampling site may be considered as a reference site as it is free of industrial discharges, surrounded by forests and paludified lands.

Table 1 Length, weight, sex, and age of the studied bream, *Abramis brama*

Sampling date	n (♂, ♀)	Standard length (cm)	Total weight (g)	Eviscerated weight (g)	Age
16 Jan 2008	9 (3, 6)	34.7 ± 0.6 cd	833 ± 40 c	742 ± 34 c	7–10
1 Apr 2008	16 (7, 9)	24.6 ± 0.5 a	218 ± 15 a	199 ± 14 a	4–7
24 Apr 2008	11 (3, 8)	35.0 ± 0.5 cd	861 ± 39 c	741 ± 23 c	7–10
5 May 2008	11 (9, 2)	38.4 ± 0.6 fg	825 ± 34 c	738 ± 30 c	9–12
19 June 2008	10 (3, 7)	40.5 ± 0.8 g	1053 ± 67 d	937 ± 62 d	11–15
3 July 2008	10 (4, 6)	36.8 ± 0.7 def	794 ± 47 c	694 ± 40 c	8–10
7 Aug 2008	10 (4, 6)	36.0 ± 1.1 cde	887 ± 76 c	775 ± 65 c	6–11
25 Sep 2008	10 (5, 5)	30.1 ± 1.8 b	279 ± 12 a	251 ± 11 a	4–10
30 Oct 2008	10 (4, 6)	23.2 ± 0.3 a	245 ± 5 a	226 ± 5 a	3–4
20 May 2009	10 (5, 5)	34.9 ± 0.5 cd	847 ± 37 c	740 ± 34 c	7–10
25 June 2009	10 (2, 8)	38.4 ± 0.8 fg	1077 ± 79 d	939 ± 75 d	8–12
16 July 2009	10 (4, 6)	36.9 ± 0.6 def	1072 ± 51 d	938 ± 43 d	9–12
13 Aug 2009	10 (5, 5)	30.2 ± 0.4 b	478 ± 37 b	399 ± 15 b	5–8
24 Sep 2009	10 (4, 6)	29.2 ± 1.2 b	504 ± 52 b	454 ± 43 b	4–8
16 Oct 2009	10 (7, 3)	37.2 ± 0.3 ef	1112 ± 28 d	969 ± 17 d	9–10
30 Dec 2009	10 (4, 6)	33.9 ± 0.7 c	890 ± 62 c	786 ± 51 c	7–9

Standard length, total weight, eviscerated weight values are given as mean ± SEM. Age values are given as min–max. The differences in length and weight among the sampling events were examined by Kruskal–Wallis test with Mann–Whitney *U* post hoc test; means with same letters do not have significant differences

After fish was sacrificed by cervical transection, the abdominal cavity was opened on ice, and the liver was excised. Each liver sample was immediately frozen at 20 °C until analyzed. A total of 167 bream were analyzed. Sampling dates, number of individuals analyzed, fish size, age and sex are presented in Table 1.

The duration of daylight at each sampling date and additional dates to build a photoperiod curve was calculated using the day length calculator (http://www.solartopo.com). Water temperature was measured at depth 0.5 m at each sampling event.

Biochemical assays

Each liver sample was homogenized in ice-cold potassium phosphate buffer pH 7.5 at a 1:5 ratio (weight/volume) for enzyme activity assays (1 g tissue in 5 ml buffer) and a 1:1 ratio for GSH assay (1 g tissue in 1 ml buffer) using an Ultra-Turrax T10 Basic homogenizer (IKA). Homogenates were centrifuged at 12,000*g* for 15 min at 4 °C. After centrifugation, lipid phase was removed, and supernatants were collected. Biochemical assays were carried out using a Lambda 25 spectrophotometer (PerkinElmer). Each sample was measured in duplicate. Protein contents were measured using the method of Bradford (1976) with bovine serum albumin as a standard.

SOD activity was measured by following the inhibition of nitroblue tetrazolium reduction in reaction with phenazine methosulfate and NADH under basic conditions (Spitz and Oberley 1989). Measurement was performed in 100 mM potassium phosphate buffer pH 7.8 at 20 °C. The reaction mixture contained 0.33 mM EDTA, 0.6 mM phenazine methosulfate, 1.35 mM nitroblue tetrazolium, 78 mM NADH, and 0.3 ml sample aliquot. SOD activity determination was performed at 540 nm and reported as the concentration of nitroformazan produced per mg protein in the sample per minute of reaction time (μM/mg protein/min).

CAT activity was measured by following the rate of H_2O_2 decomposition, which can be determined directly by the decrease in absorbance at 240 nm, using a molar extinction coefficient of $22.2 \times 10^3\ M^{-1}\ cm^{-1}$. Additional dilution in the potassium phosphate buffer at a 1:9 ratio was applied to the liver supernatant. An aliquot of the resultant solution was added to a quartz cuvette with a reactive mixture of 100 mM potassium phosphate buffer pH 7.0 and 1.2 mM H_2O_2. Change of absorbance per minute was measured and reported as μM/mg protein/min (Aebi 1984).

GST activity was determined by the method of Habig et al. (1974) by monitoring the conjugation of GSH with 1-chloro-2,4-dinitrobenzene used as a substrate. The reaction mixture was composed of 100 mM potassium phosphate buffer pH 6.5, 100 mM GSH, 50 mM 1-chloro-2,4-dinitrobenzene, and 0.1 ml of the sample aliquot. Absorbance was measured at 340 nm, 20 °C. The increase in absorbance was recorded for 3 min. Enzyme activity was measured using a molar extinction coefficient of $9.6 \times 10^3\ M^{-1}\ cm^{-1}$ and reported as the concentration of 1-GS-2,4-dinitrobenzene produced per mg protein in the sample per minute of reaction time (μM/mg protein/min).

GSH concentrations were detected in reaction with 5,5′-dithiobis(2-nitrobenzoic acid). Trichloroacetic acid (25%) was added to homogenate aliquot, mixed thoroughly, and placed on ice for 1 h. The resulting precipitate was separated by centrifugation at 5000 *g* for 10 min at 4 °C. Each sample cuvette contained 100 mM potassium phosphate buffer pH 8.0, 0.6 mM 5,5′-dithiobis(2-nitrobenzoic acid), and 0.5 ml supernatant fraction. The colored complex was registered at 412 nm. GSH content was calculated using a molar extinction coefficient of $13.6 \times 10^3\ M^{-1}\ cm^{-1}$ and reported as nM/mg protein (Moron et al. 1979).

TBARS concentrations were assayed by color reaction with 2-thiobarbituric acid (Buege and Aust 1978). Trichloroacetic acid solution at 30%, 0.5 M HCl solution, and 0.75% 2-thiobarbituric acid solution were successively added to tissue homogenate. The mixture was stirred vigorously and heated for 15 min in a boiling water bath. After cooling, samples were separated by centrifugation. Ethanol solution of ionol was added to supernatant (final concentration 5 mM) in order to prevent lipid peroxidation in the sample. Samples were measured at 532 nm. A molar extinction coefficient of $1.56 \times 10^5\ M^{-1}\ cm^{-1}$ was used for calculations. TBARS concentration was reported as nM/mg protein.

Statistical analysis

The data are presented as the means with the standard errors (mean ± SEM). The statistical analyses were performed using Statistica software. The data were assessed for normality using the Shapiro–Wilk test. As the test results suggested that the data did not come from a normal distribution, nonparametric methods were used. The differences in the levels of antioxidants and TBARS, length and weight of the bream among the sampling points were examined by Kruskal–Wallis test with Mann–Whitney *U* post hoc test. The strength of the association between the studied variables was estimated using the Spearman rank correlation coefficient (*R*). The graphs were constructed using Microsoft Office Excel software.

Results

No significant influence of the fish size on the antioxidant activities and TBARS levels was detected. Weak relationships were registered between the standard length and SOD ($R = 0.38$, $p < 0.05$), CAT ($R = 0.31$, $p < 0.05$), and TBARS ($R = 0.24$, $p < 0.05$).

Annual variations of SOD activity are shown in Fig. 1. The lowest SOD activities were found in winter and fall with the minimum registered in January 2008 (12.76 ± 1.35 μM/mg protein/min). In spring, by the time

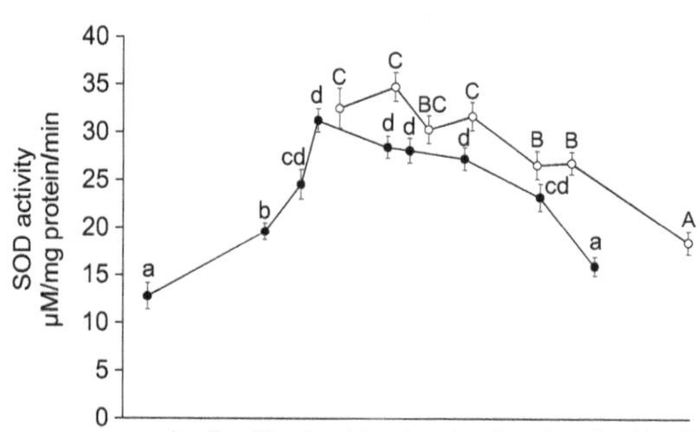

Fig. 1 Superoxide dismutase (SOD) activity (μM/mg protein/min) in the liver of bream in the years 2008 (*filled circle*) and 2009 (*open circle*). Data are presented as mean ± SEM. Means with *different letters* are significantly different (Kruskal–Wallis test with Mann–Whitney *U* post hoc test, *lower case* the year 2008, *upper case* the year 2009)

of spawning, a fast significant increase in activity was observed. The greatest value was registered in May 2008 (31.23 ± 1.23 μM/mg protein/min), and then a slow decrease in activity was observed. In 2009, higher SOD levels were witnessed. Two small peaks were registered in summer with the maximum in June (34.77 ± 1.49 μM/mg protein/min).

CAT activity increased rapidly in spring (Fig. 2) reaching its maximum of 50.19 ± 1.97 μM/mg protein/min by May 2008 (spawning period). In 2009, the greatest value was registered in May as well (52.43 ± 1.69 μM/mg protein/min). Both years, since June, CAT activities were decreasing to a lower winter level (24.41 ± 1.41 μM/mg protein/min, December).

Similar to CAT and SOD activities, GST level increased significantly in spring by the time of spawning (Fig. 3). Slight decreases in GST activity followed by new increases were observed in summer months. The greatest GST values were registered in August of each year (3.26 ± 0.12 and 3.13 ± 0.1 μM/mg protein/min). In 2008, the period of high enzyme activity was more prolonged, GST level stayed high from June to the late September, while the next September GST activity was significantly lower as compared with the values registered in August (Mann–Whitney test, $p < 0.05$).

GSH pattern differed from those witnessed for SOD, CAT, and GST activities (Fig. 4). Spring values were significantly greater as compared with winter level, being relatively small on a whole year scale though. Pronounced peaks were observed in August of each year (11.21 ± 0.84 and 8.12 ± 0.33 nM/mg protein). Late autumn and winter levels were equal to those registered in spring and early summer.

TBARS content did not peak through the annual cycle (Fig. 5), though May–August levels were significantly greater than those seen in April, October, and December ($p < 0.05$). TBARS level started to rise during spawning and continued to elevate slowly reaching the annual maximum in August of each year (1.83 ± 0.1 and 1.79 ± 0.09 nM/mg protein).

Moderate relationships were observed between SOD and CAT ($R = 0.61$, $p < 0.05$), SOD and GST ($R = 0.55$, $p < 0.05$). Weaker associations were evident between SOD and TBARS ($R = 0.43$, $p < 0.05$), CAT and GST ($R = 0.43$, $p < 0.05$). Other relationships were much weaker or non-significant.

Water temperature values registered at the sampling dates are given in Fig. 6, showing an increase from April (after the breakup) to July. The photoperiod curve for the sampling site is also presented in Fig. 6.

Statistically significant correlations (Spearman coefficient, $p < 0.05$) were observed between all studied parameters and photoperiod (R_p) and water temperature (R_t), except for GSH. The strength of association varied from moderate to strong. Correlation coefficient values were as follows SOD $R_p = 0.83$, $R_t = 0.78$; CAT $R_p = 0.82$, $R_t = 0.6$; GST $R_p = 0.59$, $R_t = 0.7$; TBARS $R_p = 0.77$, $R_t = 0.88$.

Discussion

It is well established by numerous studies that seasonal rhythms govern various vital functions of fish (Bulow et al. 1981; Eliassen and Vahl 1982; Zapata et al. 1992; Gozdowska et al. 2006; Prihartato et al. 2015; etc.). Along with the alternation of seasons, temperate fish experience changes in mean temperature, amounts of

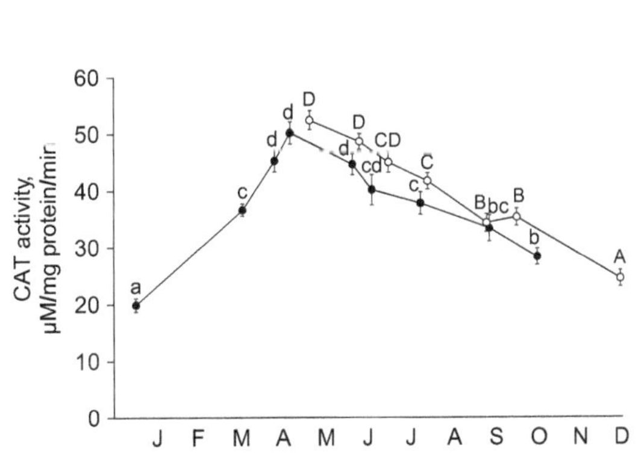

Fig. 2 Catalase (CAT) activity (μM/mg protein/min) in the liver of bream in the years 2008 (*filled circle*) and 2009 (*open circle*). Data are presented as mean ± SEM. Means with *different letters* are significantly different (Kruskal–Wallis test with Mann–Whitney *U* post hoc test, *lower case* the year 2008, *upper case* the year 2009)

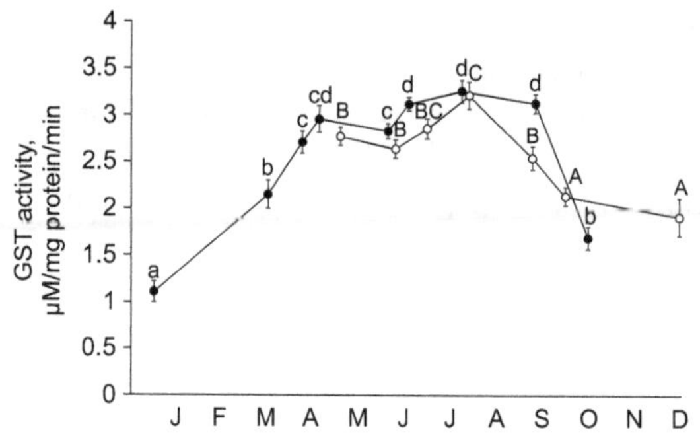

Fig. 3 Glutathione-*S*-transferase (GST) activity (μM/mg protein/min) in the liver of bream in the years 2008 (*filled circle*) and 2009 (*open circle*). Data are presented as mean ± SEM. Means with *different letters* are significantly different (Kruskal–Wallis test with Mann–Whitney *U* post hoc test, *lower case* the year 2008, *upper case* the year 2009)

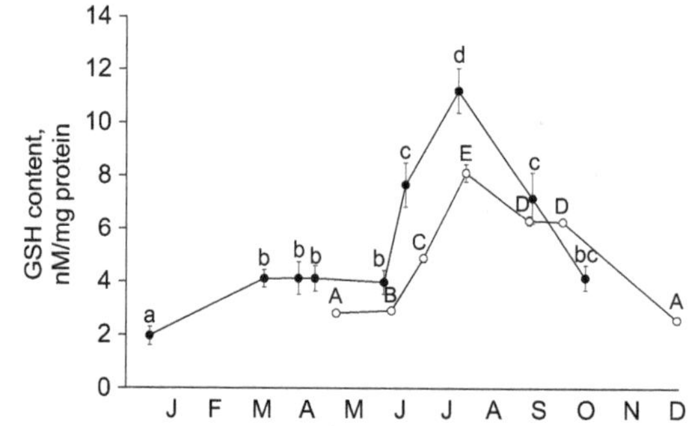

Fig. 4 Reduced glutathione (GSH) concentrations (nM/mg protein) in the liver of bream in the years 2008 (*filled circle*) and 2009 (*open circle*). Data are presented as mean ± SEM. Means with *different letters* are significantly different (Kruskal–Wallis test with Mann–Whitney *U* post hoc test, *lower case* the year 2008, *upper case* the year 2009)

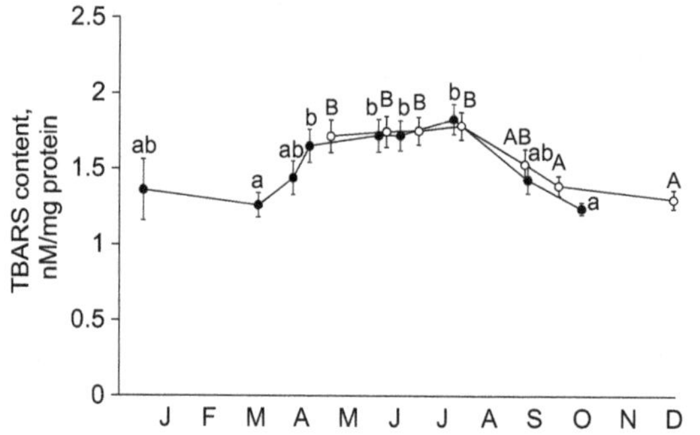

Fig. 5 Thiobarbituric acid reactive substances (TBARS) concentrations (nM/mg protein) in the liver of bream in the years 2008 (*filled circle*) and 2009 (*open circle*). Data are presented as mean ± SEM. Means with *different letters* are significantly different (Kruskal–Wallis test with Mann–Whitney *U* post hoc test, *lower case* the year 2008, *upper case* the year 2009)

oxygen, food availability large enough to present organisms with a physiological challenge (Clarke 1993). Yet, these fluctuations have a constant regime for which certain functional adaptations exist.

The lowest values of all studied antioxidants were seen in the middle of winter. This fact reflects a general decline in fish metabolic activity during the wintering period (Shulman and Love 1999).

Enzyme activities and GSH level increased in early spring before the ice breakup and water temperature elevation. At that, the TBARS level decreased as compared with January level. In the obvious absence of temperature compensation on the enzymatic activities, one might consider a cueing role of photoperiod. Temperate fish's fundamental metabolic processes are driven to maintain seasonality in reproduction; photoperiod is an important synchronizing factor inducing changes in trends and rates of metabolism (Shulman

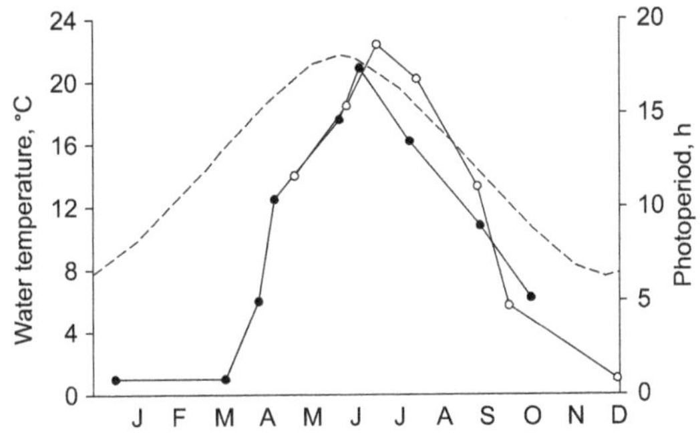

Fig. 6 Photoperiod, h (*dashed line*) and water temperature courses, °C, in the years 2008 (*filled circle*) and 2009 (*open circle*)

and Love 1999; Wang et al. 2010). The second possible explanation of the observed elevation of the antioxidant levels might be the fish's response to hypoxia termed "preparation to oxidative stress" (Hermes-Lima et al. 1998). This response is considered as an evolutionary adaptation playing a protective role during oxygen deficient conditions, which usually occur by the end of ice-covered period. Hypoxia activates antioxidant defences, allowing fish to cope with oxidative stress that might arise from tissue reoxygenation (Martínez-Álvarez et al. 2005). The possibility of such post-breakup oxidative stress is also illustrated by the fact that the bream inhabiting the studied area experiences spring (post-breakup) increase of plasma concentrations of triiodothyronine, known to enhance tissue oxygen uptake (Bolotovskiy and Levin 2015).

By the end of April, when the water temperature started to grow, SOD, CAT, and GST activities were at their rapid rise, while the GSH level was still relatively low and equal to that registered before the ice breakup. TBARS concentrations heightened before the beginning of spawning. At the period from May to August, increased TBARS concentrations along with elevated levels of the studied antioxidants indicated a state of oxidative stress. According to the intensity-based classification of oxidative stress proposed by Lushchak (2014), a low-intensity/mild oxidative stress is that is witnessed here. The spawning may be considered as a physiological state causing naturally induced oxidative stress in aquatic organisms (Soldatov et al. 2007). Thus, Soldatov et al. (2007) observed increased glutathione peroxidase and CAT activities along with elevated TBARS level in bivalve mollusk *Mytilus galloprovincialis* Lam, during the spawning period.

In the present study, the signs of oxidative stress were registered prior to spawning and throughout the summer. Endogenous compounds, both involved in gametogenesis and resorption of unspawned gametes, may be considered as effectors inducing oxidative stress. For instance, metabolism of steroid hormones involves hepatic cytochrome P450-dependent enzymes (Whyte et al. 2000). In the cytochrome P450 metabolic pathways, reactive oxygen species such as superoxide anion, hydrogen peroxide, and hydroxyl radical can be generated, evoking certain responses (Menshchikova et al. 2006).

It was evident that SOD and CAT activities had a tendency to decline after spawning. Apart from these enzymes, maximum of GST activity was observed in August. Although there were no direct associations between the enzyme activities and water temperature, SOD, CAT, and GST were likely to be under relatively strong temperature impact. Enhanced temperature can be considered as a natural pro-oxidant factor during summer months (Gorbi et al. 2005). High ambient temperature can disturb metabolic processes causing elevation of steady-state concentrations of reactive oxygen species, which required adjustments in the antioxidant defense. Each year, while SOD and CAT levels were decreasing after a spring rise, GST activity increased to its new high. At that, the TBARS levels, although relatively high, were almost even throughout the summer indicating elevated but steady level of oxidative modification.

An important antioxidant action of GSH is that of being a substrate to antioxidant enzymes such as glutathione-dependent peroxidase and GST. Thus, the highest levels of GSH were timed to the GST activity maxima. However, in a whole year scale, their fluctuations were not in complete agreement. No changes of the GSH concentrations were found during the spawning period when GST activity was at significant rise. Summer GSH peaks appear to be not only GST-related. As the spawning is accomplished, the bream leaves the nearshore zones and moves to the deep in search for areas rich with benthos (Poddubny 1971). Active feeding governed by the need for energy reserve accumulation that will cover vital functions later when food consumption is ceased, stimulates intensive digestion. A sharp summer increase of the GSH levels might be related to the intensification of feeding and digestion. The uptake of amino acids formed in the result of proteolysis is controlled by passive or active mechanisms functioning via diffusion or requiring cellular energy (Forrest et al. 2011; Hediger et al. 2013). Presumably, GSH may act as an amino acid transporter in the γ-glutamyl cycle (Orlowski and Meister 1970; Meister 1974). According to the initial γ-glutamyl cycle hypothesis, a series of catalytic events involves the synthesis and degradation of GSH, and the coupled uptake and release of free amino acids. GSH is known to play multiple roles in animal metabolism (Wu 2013). Therefore, the up-regulation of GSH synthesis witnessed in summer may be associated with not only antioxidant functions of GSH but also its non-antioxidant contributions to metabolic and regulation processes as well.

As observed in 2008 and 2009, spring–summer up-regulation of the studied antioxidants along with relatively intensified but steady lipid peroxidation indicate bream's adaptation to periodically occurring natural oxidative stress. Such high plasticity of antioxidant defenses provides restraining of lipid peroxidation and thus surviving under oxidative stress and realization of reproduction program.

Toward the end of autumn, after abrupt decline of water temperature before the ice-covered period, the bream ceases feeding (Poddubny 1971). At this time, fish metabolism is adjusted to a stable state preparing for the wintering period (Shulman and Love 1999). In the present study, in autumn, enzyme activities along with TBARS concentrations were approaching the levels observed prior to the ice breakup.

As shown in this study, hepatic levels of antioxidants in bream vary dramatically during the annual cycle. Spring–summer levels significantly exceed those registered in autumn and winter. Driven by endogenous and exogenous factors, the greatest levels of antioxidants are timed to the periods of spawning and active feeding. In spring prior to spawning and further to the end of summer, bream seems to be experiencing mild oxidative stress.

Acknowledgements The authors are sincerely grateful to A. V. Elagin and A. V. Orlov for their help in the field. We are immensely thankful to three anonymous reviewers for their perceptive criticism. This work was supported by the Russian Foundation for Basic Research (RFBR), Research Project No. 16-34-00854.

References

Aebi H (1984) Catalase in vitro. Method Enzymol 105:121–126. doi:10.1016/S0076-6879(84)05016-3

Bolotovskiy AA, Levin BA (2015) Thyroid hormone divergence between two closely related but ecologically diverse cyprinid fish species (Cyprinidae). Biochem Syst Ecol 59:305–310. doi:10.1016/j.bse.2015.02.012

Bradford MM (1976) A rapid and sensitive method for the quantitation of microgram quantities of protein utilizing the principal of protein-dye binding. Anal Biochem 72:248–254. doi:10.1016/0003-2697(76)90527-3

Buege JA, Aust SD (1978) Microsomal lipid peroxidation. Method Enzymol 52:302–310. doi:10.1016/S0076-6879(78)52032-6

Bulow FJ, Zeman ME, Winningham JR, Hudson WF (1981) Seasonal variations in RNA–DNA ratios and in indicators of feeding, reproduction, energy storage, and condition in a population of bluegill, *Lepomis macrochirus* Rafinesque. J Fish Biol 18:237–244. doi:10.1111/j.1095-8649.1981.tb03766.x

Cajaraville MP, Hauser L, Carvalho G, Hylland K, Olabarrieta I, Lawrence AJ, Lowe D, Goksøyr A (2003) Genetic damage and the molecular/cellular response to pollution. In: Lawrence A, Hemingway K (eds) Effects of pollution on fish: molecular effects and population responses. Blackwell Science Ltd., Oxford, pp 14–82

Clarke A (1993) Seasonal acclimatization and latitudinal compensation in metabolism: do they exist? Funct Ecol 7:139–149. doi:10.2307/2389880

Di Giulio RT, Meyer JN (2008) Reactive oxygen species and oxidative stress. In: Di Giulio RT, Hinton DE (eds) The toxicology of fishes. CRC Press, Taylor & Francis Group, Boca Raton, pp 273–324

Eliassen JE, Vahl O (1982) Seasonal variations in biochemical composition and energy content of liver, gonad and muscle of mature and immature cod, *Gadus morhua* (L.) from Balsfjorden, northern Norway. J Fish Biol 20:707–716. doi:10.1111/j.1095-8649.1982.tb03981.x

Forman HJ, Zhang H, Rinna A (2009) Glutathione: overview of its protective roles, measurement, and biosynthesis. Mol Aspects Med 30:1–12. doi:10.1016/j.mam.2008.08.006

Forrest LR, Krämer R, Ziegler C (2011) The structural basis of secondary active transport mechanism. BBA Bioenerg 1807:167–188. doi:10.1016/j.bbabio.2010.10.014

Gorbi S, Baldini C, Regoli F (2005) Seasonal variability of metallothioneins, cytochrome P450, bile metabolites and oxyradical metabolism in the European eel *Anguilla anguilla* L. (Anguillidae) and Striped mullet *Mugil cephalus* L. (Mugilidae). Arch Environ Con Tox 49:62–70. doi:10.1007/s00244-004-0150-9

Gozdowska M, Kleszczyńska A, Sokołowska E, Kulczykowska E (2006) Arginine vasotocin (AVT) and isotocin (IT) in fish brain: diurnal and seasonal variations. Comp Biochem Phys B 143:330–334. doi:10.1016/j.cbpb.2005.12.004

Habig WH, Pabst MJ, Jacoby WB (1974) Glutathione *S*-transferases: the first enzymatic step in mercapturic acid formation. J Biol Chem 249:7130–7139

Hayes JD, Flanagan JU, Jowsey IR (2005) Glutathione transferases. Annu Rev Pharmacol 45:51–88. doi:10.1146/annurev.pharmtox.45.120403.095857

Hediger MA, Clémençon B, Burrier RE, Bruford EA (2013) The ABCs of membrane transporters in health and disease (SLC series): introduction. Mol Aspects Med 34:95–107. doi:10.1016/j.mam.2012.12.009

Hermes-Lima M, Storey JM, Storey KB (1998) Antioxidant defenses and metabolic depression. The hypothesis of preparation for oxidative stress in land snails. Comp Biochem Phys B 120:437–448. doi:10.1016/S0305-0491(98)10053-6

Lushchak VI (2014) Free radicals, reactive oxygen species, oxidative stress and its classification. Chem Biol Interact 224:164–175. doi:10.1016/j.cbi.2014.10.016

Lushchak VI (2016) Contaminant-induced oxidative stress in fish: a mechanistic approach. Fish Physiol Biochem 42:711–747. doi:10.1007/s10695-015-0171-5

Martínez-Álvarez RM, Morales AE, Sanz A (2005) Antioxidant defenses in fish: biotic and abiotic factors. Rev Fish Biol Fish 15:75–88. doi:10.1007/s11160-005-7846-4

Meister A (1974) Glutathione; metabolism and function via the γ-glutamyl cycle. Life Sci 15:177–190. doi:10.1016/0024-3205(74)90206-9

Menshchikova EB, Lankin VZ, Zenkov NK, Bondar IA, Krugovykh NF, Trufakin VA (2006) Oxidative stress. Prooxidants and antioxidants. Firma Slovo, Moscow

Moron MS, Depierre JW, Mannervik B (1979) Levels of glutathione, glutathione reductase and glutathione *S*-transferase activities in rat lung and liver. BBA Gen Subj 582:67–78. doi:10.1016/0304-4165(79)90289-7

Nikolsky GV (1978) The ecology of fishes. TFH Publications, Neptune City

Orlowski M, Meister A (1970) The γ-glutamyl cycle: a possible transport system for amino acids. Proc Natl Acad Sci USA 67:1248–1255

Poddubny AG (1971) Environmental topography of fish populations in reservoirs. Nauka, Leningrad

Prihartato PK, Aksnes DL, Kaartvedt S (2015) Seasonal patterns in the nocturnal distribution and behavior of the mesopelagic fish *Maurolicus muelleri* at high latitudes. Mar Ecol Prog Ser 521:189–200. doi:10.3354/meps11139

Shulman GE, Love RM (1999) The biochemical ecology of marine fishes. Academic Press Ltd., London

Sies H (2015) Oxidative stress: a concept in redox biology and medicine. Redox Biol 4:180–183. doi:10.1016/j.redox.2015.01.002

Soldatov AA, Gostyukhina OL, Golovina IV (2007) Antioxidant enzyme complex of tissues of the bivalve *Mytilus galloprovincialis* Lam. under normal and oxidative-stress conditions: a review. Appl Biochem Microbiol 43:556–562. doi:10.1134/S0003683807050092

Spitz DR, Oberley LW (1989) An assay of superoxide dismutase activity in mammalian tissue homogenates. Anal Biochem 179:8–18. doi:10.1016/0003-2697(89)90192-9

Tsvetkov AI, Krylov AV, Bolotov SE, Otyukova NG (2015) Physico-chemical characteristics of water in distinguished zones of tributary's mouth. In: Krylov A (ed) Hydroecology of lowland water reservoir tributaries' mouth. Publisher's bureau "Filigran", Yaroslavl, pp 56–75

Wang N, Teletchea F, Kestemont P, Milla S, Fontaine P (2010) Photothermal control of the reproductive cycle in temperate fishes. Rev Aquac 2:209–222. doi:10.1111/j.1753-5131.2010.01037.x

Whyte JJ, Jung RE, Schmitt CJ, Tillitt DE (2000) Ethoxyresorufin-*O*-deethylase (EROD) activity in fish as a biomarker of chemical exposure. Crit Rev Toxicol 30:347–570. doi:10.1080/10408440091159239

Winston GW, Di Giulio RT (1991) Prooxidant and antioxidant mechanisms in aquatic organisms. Aquat Toxicol 19:137–161. doi:10.1016/0166-445X(91)90033-6

Wu G (2013) Amino acids: biochemistry and nutrition. CRC Press, Taylor & Francis Group, Boca Raton

Yakovlev VN, Slyn'ko YV, Kiyashko VI (2001) Annotated catalog of cyclostomata and fish from water bodies of the Upper Volga basin. In: Kopylov AI (ed) Ecological problems of the Upper Volga. YaGTU, Yaroslavl, pp 52–68

Zapata AG, Varas A, Torroba M (1992) Seasonal variations in the immune system of lower vertebrates. Immunol Today 13:142–147. doi:10.1016/0167-5699(92)90112-K

9

Effects of dietary protein and essential amino acid deficiencies on growth, body composition, and digestive enzyme activities of silvery-black porgy (*Sparidentex hasta*)

Mansour Torfi Mozanzadeh · Morteza Yaghoubi · Jasem G. Marammazi · Omid Safari · Enric Gisbert

Abstract A 6-week feeding trial was conducted to evaluate the effects of protein-free (PF) and essential amino-acid-deficient (EAAD) diets on the growth performance and digestive enzyme activities of silvery-black porgy (*Sparidentex hasta*) juveniles. Three experimental diets were formulated: a control diet in which 60% of dietary nitrogen was provided by intact protein (fish meal) and 40% by crystalline AA [(blends of essential amino acids (EAA) and none essential amino acids (NEAA)]; an essential amino-acid-deficient diet in which 60% of dietary *N* was provided by intact protein, whereas the rest was provided by NEAA and a protein-free (PF) diet, based on carbohydrate sources. Survival rates in fish fed the EAAD and PF diets were lower than in fish fed the control diet. Weight gain in fish fed with EAAD was 5.6 g lower than fish fed with the control diet. Furthermore, fish fed the PF diet lost weight and their final body weight was 12.2 g lower than the control group. Fish fed the PF diet had the highest whole-body moisture, but the lowest whole-body protein, lipid, threonine, aspartic acid, glutamic acid, and serine levels. Digestive pancreatic enzyme activities including trypsin, lipase, α-amylase, and carboxypeptidase A significantly decreased in fish fed EAAD and PF diets. The information obtained from this study testing two extreme diets (EAAD and PF) may serve for better understanding the impact of protein and EAA nutritional imbalances in fish performance.

Keywords Amino acid profile · Feed efficiency · Nitrogen retention · Sparidae · Digestive enzymes

Introduction

The optimization of the dietary protein and essential amino acid (EAA) profile according to the requirements of a given fish species maximizes dietary EAA and non-EAA utilization and increases protein synthesis resulting in a proper fish growth and economic return. An absolute requirement for the 10 EAA (arginine,

M. T. Mozanzadeh (✉) · M. Yaghoubi · J. G. Marammazi
Agriculture Research, Education and Extension, Iran Fisheries Research Organization (IFRO), South Iran Aquaculture Research Center, Ahwaz, Iran
e-mail: mansour.torfi@gmail.com

O. Safari
Department of Fisheries, Faculty of Natural Resources and Environment, Ferdowsi University of Mashhad, Khorasan Razavi, Iran

E. Gisbert
Unitat de Cultius Experimentals, Centre de Sant Carles de la Ràpita (IRTA-SCR), IRTA, Crta. del Poble Nou Km 5.5, Sant Carles de la Ràpita, Spain

histidine, isoleucine, leucine, lysine, methionine, phenylalanine, threonine, tryptophan, and valine) has been demonstrated in most fish species studied to date (NRC 2011). Essential amino acids are key molecules for building proteins, as well as important regulators of key metabolic pathways including cell signaling, appetite stimulation, growth and development, energy utilization, immunity, osmoregulation, ammonia detoxification, antioxidative defense, metamorphosis, pigmentation, gut and neuronal development, stress responses, reproduction, and suppression of aggressive behavior in aquatic animals (Wu et al. 2013, 2014). In addition, the previous studies have reported that different EAA have a significant role on the digestive enzyme activities in fish (Tang et al. 2009, 2013, Chen et al. 2012, Li et al. 2015). In fact, the positive effects of dietary EAA supplementation on the activities of digestive enzymes may partially be attributed to the growth and structural integrity of pancreas and intestinal brush borders as it has been reported in the above-mentioned studies.

Silvery-black porgy (*Sparidentex hasta*) is recognized as one of the most promising fish species for aquaculture diversification in the Persian Gulf and the Oman Sea regions. In this regard, during recent years, nutritional studies focused on establishing the nutritional requirements for improving diet formulation for this species (Mozanzadeh et al. 2017). Previous studies have shown that the dietary protein requirement of *S. hasta* juvenile is estimated to be 48%, being fish meal the major source of protein in the diet (Mozanzadeh et al. 2017). Moreover, the optimal dietary EAA profile for *S. hasta* juveniles was estimated to be (g 16 g/N): arginine 5.3, lysine 6.0, threonine 5.2, histidine 2.5, isoleucine 4.6, leucine 5.4, methionine + cysteine 4.0 (in a diet containing 0.6 cysteine), phenylalanine + tyrosine 5.6 (in a diet containing 1.9 tyrosine), tryptophan 1.0, and valine 4.6 (Marammazi et al. 2017). In addition, Yaghoubi et al. (2017) reported that arginine, threonine, and lysine were the most limited EAA for humoral immunity in this species. Several studies used protein-free diets (PF) for estimating nitrogen utilization and loss as well as protein and EAA requirements for maintenance in different fish species (Kaushik et al. 1995, Abboudi et al. 2006, 2009). To our knowledge, there is no available information about the effects of dietary protein and EAA deficiencies on physiological responses in this sparid species. Thus, the present study aimed to evaluate effects of protein and EAA deficiencies on the growth performance, feed utilization, body composition, and pancreatic digestive enzyme activities to clarify how protein and EAA deficiencies can affect physiological responses in this species.

Materials and methods

Experimental diets

Three diets were formulated (Tables 1, 2): a control diet (74.7 g N/kg DM), a diet deficient in essential amino acids (EAAD 76.6 g N/kg DM), and a protein-free diet (PF 7.5 g N/kg DM). In the control diet, 60% of dietary N was provided by intact protein (fishmeal and gelatine) and 40% by crystalline AAs (CAA). The CAA mixture was prepared by a blend of EAA and non-essential amino acids (NEAA), which were coated with 1% agar in order to delay their absorption and to optimize their use for protein accretion (Peres and Oliva-Teles 2006). All the EAA and NEAA incorporated in the CAA mixture of the control diet simulated the AA profile of FM from *Clupeonella* sp. and matched the EAA nutritional requirements for this species (Marammazi et al. 2017). Agar was dissolved in boiled water (10 g/kg of the diet) and cooled to 40 °C before the addition of CAAs (Peres and Oliva-Teles 2006). Regarding the EAAD diet, 60% of dietary *N* was provided by fish meal, whereas the rest was incorporated as a mixture of NEAA (crystalline form). The PF diet was based on wheat middling, corn starch, and fish oil. In all diets, ingredients were finely ground (particle size: < 500 μm), mixed together in a Hobart mixer, and then, the pre-coated CAA mixture was added, followed by the fish oil and sufficient distilled water to form a soft dough that was extruded to obtain pellets of 2 mm in size. Pellets were dried in a convection oven at 25 °C and stored in re-sealable plastic bags at − 20 °C until their use.

Feeding trial

Experimental procedures were conducted in compliance with the guidelines of the international council for laboratory animals science for the use of laboratory animals. The study was carried out at the Mariculture Research Station of the South Iranian Aquaculture Research Center (SIARC) (Khuzestan, Sarbandar, Iran).

Table 1 Ingredient and proximate composition of the control, essential amino-acid-deficient (EAAD), and protein-free (PF) diets

	Diets		
	Control	EAAD	PF
Dietary ingredients (g kg^{-1} dry diet)			
Fish meal[a]	360	360	–
Gelatin[b]	40	40	–
Wheat middling[c]	70	70	275
Corn starch[d]	205	205	540
Fish oil[e]	110	110	155
Agar[f]	10	10	10
Vitamin Premix[g]	10	10	10
Mineral Premix[h]	10	10	10
L-arginine	8.5	–	–
L-lysine-HCl	11.5	–	–
L-threonine	8	–	–
L-histidine	5	–	–
L-isoleucine	8.5	–	–
L-leucine	14	–	–
L-methionine	6	–	–
L-phenylalanine	7.5	–	–
L-tryptophan	2	–	–
L-valine	9.5	–	–
NEAA mixture[i]	104.5	185	–
Proximate composition (%)			
Dry matter	92.9 ± 0.4	93.2 ± 0.1	91.5 ± 0.3
Crude protein	46.7 ± 0.6	47.9 ± 0.1	4.7 ± 2.5
Crude lipid	20.1 ± 0.0	20.0 ± 0.0	19.9 ± 0.6
NFE[j]	19.1 ± 0.5	18.3 ± 0.2	64.3 ± 1.7
Ash	6.8 ± 0.1	6.1 ± 0.1	1.6 ± 0.0
Gross energy (kJ g^{-1})[k]	20.9 ± 0.1	20.1 ± 0.0	19.5 ± 0.0

[a]Fish meal (*Clupeonella* sp.); Parskilka Mazandaran, Iran (63.5% crude protein, 17.7% crude lipid)

[b]Gelatine; Beyza feed mill, Shiraz, Iran. (85% crude protein, crude lipid, 4.2)

[c]Wheat meal; Beyza feed mill, Shiraz, Iran. 12% crude protein, 3% crude lipid)]

[d]Corn starch, Beyza feed mill, Shiraz, Iran

[e]Parskilka Mazandaran, Iran (*Clupeonella* sp.)

[f] Merck, Germany

[g]Vitamin premix (mg kg^{-1}) of premix: vitamin A, 5,000,000 IU; vitamin D3, 500,000 IU; vitamin E, 3000 mg; vitamin K3, 1500; vitamin B1, 6000; vitamin B2, 24,000; vitamin B5, 52,000; vitamin B6, 18,000; vitamin B12, 60,000; folic acid, 3000; nicotinamide 180,000; antioxidant, 500, Damloran pharmaceutical company, Broujerd, Iran

[h]Mineral premix (mg kg^{-1}) of premix: copper, 3000; zinc, 15,000; manganese, 20,000; Iron, 10,000; potassium iodate, 300. Microvit®, Razak laboratories, Tehran, Iran

Crystalline amino acids: Merck, Germany, except isoleucine (Sigma-Aldrich, USA)

[i]Dispensable amino acid mixture (% mixture): L-alanine: 13; L-aspartic acid: 20; sodium glutamate: 32; L-glycine: 15; L-serine: 10; and L-proline: 10, Merck, Germany

[j]Nitrogen-free extract = 100 − (protein + lipid + ash)

[k]Gross energy content was estimated as: total carbohydrate × 17.2 J kg^{-1}; fat × 39.5 Jkg^{-1}; and protein × 23.5 J kg^{-1}

Fish were adapted to the experimental conditions for 2 weeks and fed with the control diet during this period. 135 juveniles [initial body weight (BW_i) of 4.7 ± 0.1 g; mean ± standard error] were randomly distributed into 9 cylindrical polyethylene tanks (250 l of volume; 15 fish/tank; 3 replicates per diet). Fish were hand fed each of the experimental diets to visual satiation, three times per day (0800, 1200, and 1600 h), for 42 days.

Table 2 Amino acid profile of the control, essential amino acid deficient (EAAD), and protein-free (PF) diets ($n = 1$), g 100 g^{-1} Diet

Amino acid composition	Control	EAAD	PF
Arginine	2.6	1.7	0.2
Lysine	2.2	1.3	0.2
Threonine	1.9	1.9	0.2
Histidine	1.2	0.7	0.1
Isoleucine	2.1	1.2	0.3
Leucine	2.4	1.5	0.3
Methionine	1.4	0.8	0.1
Cysteine*	0.3	0.3	0.1
Phenylalanine	1.8	1.1	0.3
Tyrosine	0.9	0.9	0.2
Tryptophan*	0.5	0.3	0.1
Valine	2.1	1.3	0.3
Alanine	3.6	4.4	0.2
Aspartate	5.5	6.7	0.3
Glutamate	7.8	9.8	0.3
Glycine	5.0	6.0	0.3
Proline	2.3	3.0	0.2
Serine	2.0	2.7	0.3

*These amounts were calculated based on the ingredient cysteine and tryptophan composition

Uneaten pellets were drained off an hour after feeding and counted every day to calculate the total feed intake (FI) per tank using a mean dry pellet weight. Tanks were supplied with filtered seawater (flow rate 1 l/min; salinity 48.0 ± 0.5 ‰; oxygen 7.6 ± 0.2 mg/l; pH 7.8 ± 0.4; temperature of 29.0 ± 1.5 °C) and subjected to natural photoperiod (30°32′N, 49°20′E; June and July).

Sample collection

At the end of the trial, fish were fasted for 24 h before being anaesthetized (2-phenoxyethanol at 0.5 ml/l; Merck, Schuchardt, Germany) and individually weighed (BW_f). For blood collection, three specimens from each replicate tank ($n = 9$ per diet, $n = 3$ per experimental replicate) were anesthetized and blood (ca. 250–500 μl) was collected from the caudal vein with heparinized syringes, then centrifuged (4000×g, 10 min, 4 °C) and plasma separated. The vials containing plasma samples were transferred into liquid nitrogen and stored at − 80 °C until further analysis. For the assessment of digestive enzyme activities, three fish per tank ($n = 9$ fish per diet treatment, $n = 3$ fish per replicate) were randomly sampled, euthanized with the same anaesthetic, and immediately eviscerated on ice surface. The alimentary tract was dissected, adherent adipose and connective tissues carefully removed and placed in individually marked plastic test tubes and stored at − 80 °C until their analysis. For whole-body biochemical analyses, at the beginning of the trial, five fish from the initial stock were killed with an overdose of the anaesthetic and frozen at − 80 °C for subsequent analysis. At the end of the feeding trial, three fish per experimental replicate were sacrificed with overdose of the same anaesthetic for subsequent biochemical analyses.

Fish growth, condition, and feed efficiency parameters

The BW_f, liver, viscera, and intraperitoneal fat weights were measured to the nearest 0.1 g and standard length (SL) to the nearest 1 mm. The following formulae were used to assess growth performance, feed utilization, and other parameters: body weight gain (WG, g) = ($BW_f - BW_i$); specific growth rate (SGR, %/day) = [(ln BW_f − ln BW_i)/t] × 100, where t is experimental period (42 days); survival rate (SR, %) = (number of fish in each group remaining on day 42/initial number of fish) × 100; feed intake (FI) = total feed intake (g)/number of fish; feed conversion ratio (FCR, %) = (feed intake (g)/weight gain (g)); protein efficiency ratio

(PER) = weight gain (g)/protein intake (g); condition factor (K) = (body weight (g)/(body length (cm))^3) × 100; hepatosomatic index (HSI, %) = (liver weight (g)/whole-body weight (g)) × 100; viscerosomatic index (VSI, %) = (visceral weight (g)/whole-body weight (g)) and intraperitoneal fat index (IPF, %) = (IPF weight (g)/whole-body weight (g)) × 100. Daily N intake, gain and retention were estimated by following formulas:

$$\text{Mean metabolic weight gain (MBW)} = 0.5 \times \left(\text{BW}_f^{0.75} + \text{BW}_i^{0.75}\right),$$
$$N \text{ intake (g N/kg MBW/day)} = (D_i \times N_d)/(\text{MBW} \times t),$$
$$N \text{ gain (g N/kg MBW/day)} = \left(\text{BW}_f \times N_f - \text{BW}_i \times N_i\right)/(\text{MBW} \times t),$$
$$N \text{ retention (\% } N \text{ intake)} = (N \text{ gain}/N \text{ intake}) \times 100$$

where MBW is the average metabolic body weight; D_i is the dry feed intake (g); N_d is the N content of the experimental diets; t is the duration of the feeding period (42 days), N_i and N_f are the initial and final body N content, and BW_i and BW_f are the mean initial and final body weight, respectively.

Biochemical analyses

Proximate analyses of ingredients, diets and whole body were determined using the standard methods (AOAC 2005). Amino acids (except for tryptophan and cysteine because of the susceptibility of these AA to acid hydrolysis) were determined after hydrolysis. Samples were hydrolysed in 6 N HCl for 24 h at 110 °C in glass vials filled with nitrogen. O-phthalaldehyde (OPA) was used as a pre-column derivatization reagent according to Lindroth and Mopper (1979). Total AA levels were determined by HPLC (Knauer, Germany), using C18 column (Knauer, Germany), at a flow rate of 1 ml/min with fluorescence detector (RF-530, Knauer, Germany). Peak identification and integration was performed by the software Waters Empower 2 (Milford, MA) using an AA standard H (Pierce, USA) as an external standard. Plasma biochemical parameters including glucose, total protein, albumin, and creatinine were analyzed by means of an auto-analyser (Mindray BS-200, China) using commercial clinical investigation kits (Pars Azmoon Kit, Tehran, Iran).

Determination of digestive enzymes

Samples were homogenized in 30 volumes (w/v) of ice-cold mannitol (50 mM), Tris–HCl buffer (2 mM) pH 7.0, at a maximum speed for 30 s (IKA, Ultra-turrax®, USA); then, 100 µl of 0.1 M $CaCl_2$ was added to the homogenate, which was centrifuged at 10,000×g for 10 min at 4 °C and the supernatants collected and stored in aliquots at − 80 °C for further enzyme quantification (Gisbert et al. 2009). Trypsin activity was measured with N-α-benzoyl-dlarginine-p-nitroanilide (BAPNA) as substrate. BAPNA (1 mM in 50 mM Tris–HCl, pH 8.2, 20 mM $CaCl_2$) was incubated with the enzyme extract at 37 °C and absorbance was recorded at 410 nm (Erlanger et al. 1961). One unit of trypsin activity corresponded to 1 µmol of 4-nitroaniline liberated in 1 min per ml of extracellular enzymatic extract, based on the extinction coefficient of the substrate (ε410 = 8800 M/cm). The activity of α-amylase was determined at 25 °C using 1% starch (Sigma-Aldrich) (diluted in a buffer at pH 6.9, 0.02 M Na_2HPO_4 and 0.006 M NaCl) as substrate by the 3, 5-dinitrosalicylic acid method (Worthington 1991) and absorbance was recorded at 540 nm. The α-amylase activity (U) was expressed as µg maltose liberated from starch per mg protein in sample per hour. Non-specific lipase activity was determined at 25 °C (absorbance: 405 nm) using 0.4 mM 4-nitrophenyl-myristate (Sigma-Aldrich) as substrate and absorbance were recorded at 405 nm (Gawlicka et al. 2000). One unit of lipase was defined as the amount of enzyme, which liberated one micromole of 4-nitrophenyl-myristate per minute. Carboxypeptidase A activity was measured using the protocol of Folk and Schirmer (1963) using HPA (hippuryl-L-phenylalanine) as substrate (dissolved in 25 mM Tris–HCl, 10 mM $CaCl_2$ buffer, pH 7.8). One unit of enzyme activity was defined as 1 M of hippuryl hydrolysed per minute using a coefficient of molar extinction of 0.36 at 254 nm. The protein concentration of the enzyme extracts was determined by the Bradford method (Bradford 1976) using bovine serum albumin as a protein standard. All enzymes were analyzed following the recommendations provided by Solovyev and Gisbert (2016).

Statistical analysis

Data were analyzed using SPSS version 15.0 (Chicago, IL, USA). All data are presented as mean ± standard error of the mean. Equality of variances was tested with Levene's test and normality with Shapiro–Wilk's test. Arcsine transformations were conducted on all percentage data to achieve homogeneity of variance before statistical analysis. One-way analysis of variance was performed at a significance level of 0.05 following the confirmation of normality and homogeneity of variance. Tukey's test was used for multiple comparisons when statistical differences among groups were detected ($P < 0.05$).

Results and discussion

Fish growth, condition, and feed efficiency

In the present study, one interesting finding was the difference in the physiological responses between fish fed EAAD and PF diets. The information obtained from this study testing two extreme diets (EAAD and PF) may serve to better understand the impact of protein nutritional imbalances in fish performance, as well as generate reference values for a large series of physiological parameters under unsuitable nutritional conditions. Fish survival was significantly affected by experimental diets; thus, survival rates in fish fed the EAAD and PF diets were 8.9 and 25.6% lower, respectively, than in fish fed the control diet (Table 3). However, no external pathological signs were observed in fish, even in those fed the PF diet. Protein synthesis and accretion is a key component of the processes involved in growth response (Anthony et al. 2001). The lower survival rate in the PF and EAAD groups may have been related to disorders in different metabolic pathways, including (a) AA synthesis and catabolism, (b) generation of physiologically important low-molecular-weight peptides and nitrogenous substances, (c) intracellular protein turnover, (d) ammonia detoxification, (e) lipid and glucose metabolism, and (f) antioxidative reactions (Wu et al. 2014). In this study, weight gain in fish fed with EAAD was 5.6 g lower than fish fed with the control diet ($P < 0.05$). Furthermore, fish fed the PF diet lost weight and their final body weight was 12.2 g lower than the control group. Feed conversion ratio and PER values were negligible in fish fed with the PF diet, since those animals lost weight during the trial. All somatic indices, including the K, HSI, VSI, and IPF significantly decreased in fish fed the EAAD and PF diets ($P < 0.05$). In general, deficiency of most EAA in fish feeds leads to failure or loss of appetite, which results in a reduced FI and weight gain, as well as lower disease resistance (Wilson 2002). In addition, it has been reported that dietary deficiency in one or more EAAs results in the deamination of other AAs in the liver, which leads to an increased excretion of nitrogenous compounds and inefficient protein synthesis and somatic growth performance (Von der Decken and Lied 1993). In fact, imbalances in the dietary AA profile tend to lead to an increase in the oxidation of other EAAs and NEAAs present at normal levels in the feed, which results in reduced protein utilization (Rønnestad et al. 2007). In the current study, nitrogen retention extremely decreased in fish fed the EAAD and PF diets ($P < 0.05$). The N loss in fish fed the PF diet was 60 mg/kg MBW/day, indicating high whole-body protein breakdown in this group due to the lack of dietary AAs for protein synthesis, which impaired the turnover of tissue proteins. The highest plasma creatinine, which indicates high muscular AA catabolism for gluconeogenesis, in fish fed the PF diet also supported high whole-body AA breakdown in this group. However, fish fed the control and EAAD diets showed 240 and 60 mg/kg MBW/day *N* accretion, respectively. In this context, Marammazi et al. (2017) reported that deletion of a single EAA did not significantly affect *N* intake (g/kg MBW/day), whereas *N* gain (g/kg MBW/day) significantly decreased in all fish fed the EAA-deficient diets and the degree of *N* retention reduction was dependent upon the EAA deficiencies.

Body proximate composition and AA profile

Whole-body composition was significantly affected by different experimental diets (Table 4). As expected, fish fed the PF diet had the highest moisture, but the lowest protein, lipid, and gross energy contents ($P < 0.05$). The reduction in whole-body protein in the PF group was largely due to proteolysis, as an energy source to maintain lipid and carbohydrate utilization, and to synthesize the necessary enzymes and hormones

Table 3 Growth, feed utilization, and biometric parameters of *S. hasta* juvenile fed control, essential amino acid deficient (EAAD), and protein-free (PF) diets (mean ± SE, $n = 3$)

	Diets		
	Control	EAAD	PF
Growth performance			
BW_i (g)[a]	4.6 ± 0.1	4.6 ± 0.0	4.6 ± 0.0
BW_f (g)[b]	12.8 ± 0.5[a]	7.2 ± 0.3[b]	4.0 ± 0.1[c]
WG (g)[c]	8.2 ± 0.6[a]	2.6 ± 0.1[b]	−0.6 ± 0.0[c]
SGR (% body weight day^{-1})	2.4 ± 0.1[a]	1.0 ± 0.1[b]	−0.4 ± 0.1[c]
Survival (%)[d]	100.0 ± 0.0[a]	91.1 ± 3.1[b]	74.4 ± 2.2[c]
FI (g $fish^{-1}$)[e]	11.4 ± 0.2[a]	8.4 ± 0.2[b]	6.9 ± 0.1[c]
FCR[f]	1.4 ± 0.1[a]	3.2 ± 0.1[b]	–
PER[g]	1.6 ± 0.1[a]	0.5 ± 0.1[b]	–
K[h]	2.4 ± 0.2[a]	2.2 ± 0.1[b]	1.9 ± 0.2[c]
HSI (%)[i]	2.0 ± 0.0[a]	1.1 ± 0.1[b]	1.1 ± 0.4[b]
VSI (%)[j]	9.5 ± 0.4[a]	5.8 ± 0.3[b]	6.4 ± 0.5[b]
IPF (%)[k]	1.8 ± 0.2[a]	0.2 ± 0.1[b]	0.1 ± 0.0[b]
N intake (g kg^{-1} MBW day^{-1})	0.74 ± 0.0[a]	0.72 ± 0.03[a]	0.03 ± 0.0[b]
N gain (g kg^{-1} MBW day^{-1})	0.24 ± 0.01[a]	0.06 ± 0.0[b]	−0.06 ± 0.0[c]
N retention (% *N* intake)	32.2 ± 1.9[a]	7.9 ± 0.46[b]	–

A different superscript in the same row denotes statistically significant differences ($P < 0.05$)

[a] BW_i: initial body weight

[b] BW_f final body weight

[c] SGR: specific growth rate = ((ln final weight − ln initial weight)/*t*) × 100, where *t* is experimental period = 42 days

[d] Survival = (number of fish in each group remaining on day 60/initial number of fish) × 100

[e] FI: feed intake = feed intake/fish number

[f] FCR: feed conversion ratio = weight gain (g)/feed intake (g)

[g] PER: protein efficiency ratio = weight gain (g)/protein intake (g)

[h] K: condition factor = (body weight (g)/(body length (cm))^3) × 100

[i] HSI: hepatosomatic index = (liver weight (g)/whole-body weight (g)) × 100

[j] VSI: viscerosomatic index = (viscera weight (g)/whole-body weight (g)) × 100

[k] IPF: intraperitoneal fat = (IPF weight (g)/whole-body weight (g)) × 100

for intermediary metabolism, as also reported in Atlantic salmon (*Salmo salar*) fry fed a PF diet (Abboudi et al. 2006, 2009). In addition, during protein turnover, protein breakdown and resynthesize occurs, with oxidation of part of the amino acids from the protein breakdown. Thus, proteolysis must have occurred due to a lack of AAs from the diet. As intestinal metabolism may have a profound influence on whole-body AA composition, the differences in patterns of AAs between diets and body proteins have been reported to be

Table 4 Composition of whole-body proximate (% of wet weight) of *S. hasta* juvenile fed control, essential amino acid deficient (EAAD), and protein-free (PF) diets (mean ± SE, $n = 3$)

	Diets			
	Initial	Control	EAAD	PF
Moisture	72.4 ± 1.2	69.7 ± 0.9[c]	74.3 ± 1.2[b]	79.6 ± 0.7[a]
Crude protein	17.6 ± 0.6	16.5 ± 0.3[a]	15.7 ± 0.4[a]	14.1 ± 0.5[b]
Crude lipid	5.1 ± 0.3	7.6 ± 0.2[a]	3.3 ± 0.4[b]	2.5 ± 0.5[c]
Ash	4.3 ± 0.2	4.7 ± 0.2	5.1 ± 0.3	5.4 ± 0.1
Gross energy (MJ kg^{-1})	6.2 ± 0.2	7.0 ± 0.0[a]	5.2 ± 0.1[b]	3.9 ± 0.0[c]

A different superscript in the same row denotes statistically significant differences ($P < 0.05$)

particularly large for various AAs (Wu et al. 2014). The whole-body AA composition of fish is shown in Table 5. Fish fed the PF diet had the highest and lowest whole-body arginine and threonine levels, respectively ($P < 0.05$). The lowest whole-body threonine in fish fed PF diet might be linked to the lowest survival rate in this group. In this regard, Yaghoubi et al. (2017) reported that *S. hasta* fed threonine-deficient diet had the lowest survival rate than fish fed other EAA-deficient diets. It has been reported that threonine is involved in many physiological and biochemical processes, including somatic growth, feed efficiency, digestive and absorptive processes, gene expression regulation, antioxidant and immune functions in different fish species (Gao et al. 2014, Habte-Tsion et al. 2015). In addition, whole-body aspartate, glutamate, and serine in fish fed the PF diet were lower than in other groups. Non-essential AAs can be synthesized de novo in adequate amounts by the animal organism from EAAs to meet the requirements for maintenance, growth, development and health (Wu et al. 2013). Thus, in the present study, dietary deficiencies in EAA may have led to the reduction of NEAA in fish fed the PF diet. Other whole-body AAs were not affected by dietary treatments ($P > 0.05$).

Plasma biochemical parameters

In the present study, plasma total protein and albumin were significantly reduced in the EAAD and PF groups in comparison with the control, suggesting a malfunctioning of the liver in these groups (Table 6), since all the plasma proteins with the exception of immunoglobulins are synthesized in this organ (Rosalki and Mcintyre 1999). The results of this study showed that fish fed the EAAD and PF diets had the higher plasma glucose levels than the control group, results that might be attributed to a chronic stress condition in these groups (Barcellos et al. 1999, 2009). The highest plasma glucose in the PF group may have also been attributed to the high levels of carbohydrates in the PF diet as a consequence of limited ability of metabolizing glucose in most

Table 5 Amino acid profile of whole body (% of amino acids) of *S. hasta* juvenile fed control, essential amino acid deficient (EAAD), and protein-free (PF) diets (mean ± SE, $n = 3$)

	Diets			
	Initial	Control	EAAD	PF
Essential amino acid				
Arginine	6.4 ± 0.0	6.7 ± 0.0^{b}	6.8 ± 0.1^{b}	7.6 ± 0.1^{a}
Histidine	0.7 ± 0.0	0.6 ± 0.0	0.6 ± 0.0	0.7 ± 0.0
Isoleusine	6.3 ± 0.0	6.6 ± 0.2	6.5 ± 0.1	6.7 ± 0.1
Leucine	7.5 ± 0.0	7.3 ± 0.2	7.6 ± 0.1	7.8 ± 0.4
Lysine	1.4 ± 0.1	1.1 ± 0.0	1.2 ± 0.0	1.4 ± 0.1
Methionine	2.8 ± 0.0	2.9 ± 0.1	2.8 ± 0.0	3.1 ± 0.1
Phenylalanine	2.9 ± 0.0	3.0 ± 0.0	2.9 ± 0.0	3.0 ± 0.1
Threonine	6.7 ± 0.1	6.8 ± 0.1^{a}	7.2 ± 0.2^{a}	4.7 ± 0.5^{b}
Valine	7.7 ± 0.0	8.1 ± 0.2	8.1 ± 0.2	8.4 ± 0.2
Non-essential amino acid composition				
Aspartate	10.9 ± 0.4	10.2 ± 0.0^{a}	10.2 ± 0.3^{a}	9.3 ± 0.4^{b}
Glutamate	18.1 ± 0.43	17.5 ± 0.2^{a}	17.2 ± 0.1^{a}	16.0 ± 0.2^{b}
Serine	6.9 ± 0.1	7.4 ± 0.2^{a}	7.3 ± 0.1^{a}	6.5 ± 0.2^{b}
Glycine	8.3 ± 0.7	9.0 ± 0.7	9.6 ± 0.1	11.6 ± 1.7
Taurine	2.1 ± 0.1	1.7 ± 0.1	1.6 ± 0.0	1.7 ± 0.1
Alanine	7.0 ± 0.1	7.7 ± 0.1	7.4 ± 0.0	7.7 ± 0.1
Tyrosine	2.0 ± 0.0	2.0 ± 0.0	2.0 ± 0.0	2.1 ± 0.1
Total EAA	42.3 ± 0.0	43.1 ± 0.6	41.7 ± 0.5	43.4 ± 0.4
Total NEAA	55.2 ± 0.1	55.5 ± 0.7	57.4 ± 0.6	54.9 ± 0.2
Total AA	97.5 ± 0.0	98.6 ± 0.1	99.1 ± 0.1	98.3 ± 0.5

A different superscript in the same row denotes statistically significant differences ($P < 0.05$)

carnivorous fish species (Booth et al. 2013). On the other hand, the increase in plasma glucose in fish fed the EAAD and PF diets was associated with an increase in plasma creatinine. Creatinine is a waste product from muscle turnover and its plasmatic levels provide an accurate marker of the hepatic and renal function. In one side, the increase in plasma creatinine level indicated a high muscular AA catabolism for gluconeogenesis, which might have resulted in higher plasma glucose levels in fish fed the PF and EAAD diets.

Digestive enzyme activities

Fish growth is mainly associated with its digestive ability that correlates with digestive enzyme activity (Rønnestad et al. 2007). The specific activity of selected pancreatic digestive enzymes in *S. hasta* juveniles fed different diets is presented in Table 7. Trypsin, non-specific lipase, α-amylase, and carboxypeptidase A activities were highest in fish fed the control diet, whereas the lowest activity levels were found in fish fed the PF diet. The activities of trypsin and carboxypeptidase A in fish fed the control diet were 10 and 30 times higher than in fish fed the EAAD and PF diets, respectively. Moreover, the activity of non-specific lipase in fish fed the EAAD and PF diets was ca. four times lower than fish fed the control diet. Regarding α-amylase activity, fish fed the control diet had 14 and 35 times higher α-amylase activity than fish fed the EAAD and PF diets, respectively ($P < 0.05$). Amino acids have a main role in digestive function through chemical sensing via the G protein-coupled receptors in the gastrointestinal tract and possibly in other tissues (i.e., liver and brain), gastrointestinal emptying, as potent stimulators of hormone (i.e, cholecystokinin), and pancreatic enzymes secretion, and they (glycine and taurine) conjugate with bile salts, which emulsify diet fats. In addition, AAs modulate growth and metabolism, and regulate intestinal microbiota (Wu et al. 2013, 2014). For instance, glutamic and aspartic acids are major metabolic fuels for intestine to maintain its digestive function and protect the mucosal integrity (Wu et al. 2013). The lowest digestive enzyme activities in fish fed the PF diet supported the above-mentioned results. Furthermore, enzyme secretion closely relates to the development of the pancreas and several studies reported that various EAA can affect the development and structural integrity of pancreas in fish (Chen et al. 2012, Tang et al. 2013, Li et al. 2015). The results of this study showed that drastic malfunction of exocrine pancreas may as a consequence of disorders in the above-mentioned mechanisms.

Conclusion

The results of this study showed that dietary protein and EAA deficiencies had negative effects on metabolic and physiological responses in *S. hasta* juveniles. The lack of dietary AAs for the protein synthesis led to lower survival and growth rates in fish fed the NEAA and PF diets. In addition, drastic changes in plasma metabolite parameters including protein, albumin, glucose, and creatinine indicated disorders in the liver and kidney function, as well in the metabolic pathways of fish fed NEAA and PF diets. Digestive enzyme secretion was lower in the NEAA and PF groups, indicating a main role of AAs in the synthesis and secretion of these enzymes from the exocrine pancreas in fish. The results of the current study may serve as a basis for establishing the optimal nitrogen content with proper AAs profile in the diet for *S. hasta* and other sparid species, to increase fish productivity while preventing N waste and environmental damage. However, more

Table 6 Plasma biochemical parameters of *S. hasta* juvenile fed control, essential amino acid deficient (EAAD), and protein-free (PF) diets (mean ± SE, $n = 3$)

	Dietary treatments		
	Control	EAAD	PF
Total Protein (g l^{-1})	29.0 ± 3.0^{a}	4.3 ± 0.2^{b}	2.8 ± 0.1^{c}
Albumin (g dl^{-1})	19.6 ± 0.1^{a}	0.3 ± 0.0^{b}	0.2 ± 0.0^{b}
Glucose (mg dl^{-1})	134.0 ± 7.0^{c}	145.2 ± 8.0^{b}	231.6 ± 14.0^{a}
Creatinine (mg dl^{-1})	2.1 ± 0.1^{c}	10.8 ± 0.6^{b}	23.4 ± 1.1^{a}

A different superscript in the same row denotes statistically significant differences ($P < 0.05$)

Table 7 Pancreatic digestive enzyme activities (U mg protein^{-1}) of *S. hasta* juvenile fed control, essential amino acid deficient (EAAD), and protein-free (PF) diets (mean ± SE, $n = 3$)

	Diets		
	Control	EAAD	PF
Trypsin	3.1 ± 0.5^a	0.3 ± 0.2^b	0.1 ± 0.0^c
Lipase	2.2 ± 0.1^a	0.6 ± 0.1^b	0.5 ± 0.0^b
α-Amylase	1.4 ± 0.0^a	0.1 ± 0.0^b	0.04 ± 0.0^c
Carboxypeptidase A	3.0 ± 0.4^a	0.3 ± 0.0^b	0.1 ± 0.0^c

A different superscript in the same row denotes statistically significant differences ($P < 0.05$)

work is needed to evaluate the effects of dietary protein levels and AA profile on the utilization efficiency of limiting AAs, as well as to determine the influence of the duration of the experiments or fish size on their EAA requirements for maintenance.

Acknowledgements This article is extracted from the project recorded under code number of 92011610 and financially supported by Iran National Science Foundation. We are grateful to the director and staff of the Mariculture Research Station, Sarbandar, Iran for providing the necessary facilities for the experiment.

Author contributions MTM: Conducting the feeding trial, sampling and lab works. MY: Helping in lab works. JGM: Feed furmulation. OS: Supervising all experimental analyses and statistics. EG: Help in manuscript preparation.

Conflict of interest There is no conflict of interest between authors in the publication of this paper.

References

Abboudi T, Mambrini M, Ooghe W, Larondelle Y, Rollin X (2006) Protein and lysine requirements for maintenance and for tissue accretion in Atlantic salmon (*Salmo salar*) fry. Aquaculture 261:369–383

Abboudi T, Mambrini M, Larondelle Y, Rollin X (2009) The effect of dispensable amino acids on nitrogen and amino acid losses in Atlantic salmon (*Salmo salar*) fry fed a protein-free diet. Aquaculture 289:327–333

Anthony TG, Reiter AK, Anthony JC, Kimball SR, Jefferson LS (2001) Deficiency of dietary EAA preferentially inhibits mRNA translation of ribosomal proteins in liver of meal-fed rats. Am J Physiol Endocrinol Metabol 281:430–439

AOAC (2005) Official methods of analysis of association of official analytical chemists. Association of Official Analytical Chemists, Gaithersburg

Barcellos L, Nicolaiewsky S, De Souza S, Lulhier F (1999) Plasmatic levels of cortisol in the response to acute stress in Nile tilapia, *Oreochromis niloticus*(L.), previously exposed to chronic stress. Aquac Res 30:437–444

Barcellos LJG, Kreutz LC, Quevedo RM, da Rosa JGS, Koakoski G, Centenaro L, Pottker E (2009) Influence of color background and shelter availability on jundiá (*Rhamdia quelen*) stress response. Aquaculture 288:51–56

Booth MA, Moses MD, Allan GL (2013) Utilisation of carbohydrate by yellowtail kingfish *Seriola lalandi*. Aquac 376–379:151–161

Bradford MM (1976) A rapid and sensitive method for the quantitation of microgram quantities of protein utilizing the principle of protein-dye binding. Anal Biochem 72:248–254

Chen G, Feng L, Kuang S, Liu Y, Jiang J, Hu K, Jiang W, Li S, Tang L, Zhou X (2012) Effect of dietary arginine on growth, intestinal enzyme activities and gene expression in muscle, hepatopancreas and intestine of juvenile Jian carp (*Cyprinus carpio* var. Jian). Br J Nutr 108:195–207

Erlanger BF, Kokowsky N, Cohen W (1961) The preparation and properties of two new chromogenic substrates of trypsin. Arc Biochem Biophys 95:271–278

Folk J, Schirmer E (1963) The porcine pancreatic carboxypeptidase A system I. Three forms of the active enzyme. J Biol Chem 238:3884–3894

Gao YJ, Yang HJ, Liu YJ, Chen SJ, Guo DQ, Yu YY, Tian LX (2014) Effects of graded levels of threonine on growth performance, biochemical parameters and intestine morphology of juvenile grass carp *Ctenopharyngodon idella*. Aquaculture 424:113–119

Gawlicka A, Parent B, Horn MH, Ross N, Opstad I, Torrissen OJ (2000) Activity of digestive enzymes in yolk-sac larvae of Atlantic halibut (*Hippoglossus hippoglossus*): indication of readiness for first feeding. Aquaculture 184:303–314

Gisbert E, Giménez G, Fernández I, Kotzamanis Y, Estévez A (2009) Development of digestive enzymes in common dentex *Dentex dentex* during early ontogeny. Aquaculture 287:381–387

Habte-Tsion HM, Ge X, Liu B, Xie J, Ren M, Zhou Q, Miao L, Pan L, Chen R (2015) A deficiency or an excess of dietary threonine level affects weight gain, enzyme activity, immune response and immune-related gene expression in juvenile blunt snout bream (*Megalobrama amblycephala*). Fish Shellfish Immunol 42:439–446

Kaushik S, Doudet T, Médale F, Aguirre P, Blanc D (1995) Protein and energy needs for maintenance and growth of Nile tilapia (*Oreochromis niloticus*). J App Ichthyol 11:290–296

Li W, Feng L, Liu Y, Jiang WD, Kuang SY, Jiang J, Li SH, Tang L, Zhou XQ (2015) Effects of dietary phenylalanine on growth, digestive and brush border enzyme activities and antioxidant capacity in the hepatopancreas and intestine of young grass carp (*Ctenopharyngodon idella*). Aquacult Nutr 21:913–925

Lindroth P, Mopper K (1979) High performance liquid chromatographic determination of subpicomole amounts of amino acids by precolumn fluorescence derivatization with o-phthaldialdehyde. Anal Chem 51:1667–1674

Marammazi JG, Yaghoubi M, Safari O, Peres H, Mozanzadeh MT (2017) Establishing the optimum dietary essential amino acids pattern for silvery-black porgy (*Sparidentex hasta*) juveniles by deletion method. Aquacult Nutr 00:1–9

Mozanzadeh MT, Marammazi JG, Yaghoubi M, Agh N, Pagheh E, Gisbert E (2017) Macronutrient requirements of silvery-black porgy (*Sparidentex hasta*): a comparison with other farmed Sparid species. Fishes 2(2):5

NRC (2011) Nutrient Requirements of Fish and Shrimp. The National Academic Press, Washington, D.C.

Peres H, Oliva-Teles A (2006) Effect of the dietary essential to non-essential amino acid ratio on growth, feed utilization and nitrogen metabolism of European sea bass (*Dicentrarchus labrax*). Aquaculture 256:395–402

Rønnestad I, Kamisaka Y, Conceição L, Morais S, Tonheim S (2007) Digestive physiology of marine fish larvae: hormonal control and processing capacity for proteins, peptides and amino acids. Aquaculture 268:82–97

Rosalki S, Mcintyre NI (1999) Biochemical investigations in the management of liver disease. Oxf Text Clin Hepatol 2:503–521

Solovyev M, Gisbert E (2016) Influence of time, storage temperature and freeze/thaw cycles on the activity of digestive enzymes from gilthead sea bream (*Sparus aurata*). Fish Physiol Biochem 42:1383–1394

Tang L, Wang GX, Jiang J, Feng L, Yang L, Li SH, Kuang SY, Zhou XQ (2009) Effect of methionine on intestinal enzymes activities, microflora and humoral immune of juvenile Jian carp (*Cyprinus carpio* var. Jian). Aquacult Nutr 15:477–483

Tang L, Feng L, Sun CY, Chen GF, Jiang WD, Hu K, Liu Y, Jiang J, Li SH, Kuang SY (2013) Effect of tryptophan on growth, intestinal enzyme activities and TOR gene expression in juvenile Jian carp (*Cyprinus carpio* var. Jian): studies in vivo and in vitro. Aquaculture 412:23–33

Von der Decken A, Lied E (1993) Metabolic effects on growth and muscle of soya-bean protein feeding in cod (*Gadus morhua*). Br J Nutr 69:689–697

Wilson RP (2002) Amino Acids and Proteins. In: Hardy JEHW (ed) Fish Nutrition, 3rd edn. Academic Press, San Diego, pp 143–179

Worthington C (1991) Worthington enzyme manual related Biochemical. Freehold, New Jersey

Wu G, Wu Z, Dai Z, Yang Y, Wang W, Liu C, Wang B, Wang J, Yin Y (2013) Dietary requirements of "nutritionally non-essential amino acids" by animals and humans. Amino Acids 44:1107–1113

Wu G, Bazer F, Dai Z, Li D, Wang J, Wu Z (2014) Amino acid nutrition in animals: protein synthesis and beyond. Ann Rev Anim Biosci 2:387–417

Yaghoubi M, Mozanzadeh M, Marammazi J, Safari O, Gisbert E (2017) Effects of dietary essential amino acid deficiencies on the growth performance and humoral immune response in silvery-black porgy (*Sparidentex hasta*) juveniles. Aquacult Res 00:1–13

10

Food-deprivation-induced suppression of pituitary–testicular-axis in the tilapia *Oreochromis mossambicus*

R. P. Pikle · R. M. Jatiger · C. B. Ganesh

Abstract In the present study, the tilapia *Oreochromis mossambicus* were exposed to food deprivation for a period of 6 or 12 days and changes in the luteinizing hormone (LH) immunoreactivity in the proximal pars distalis (PPD) of the pituitary gland and the testicular activity were examined. Intensely immunoreactive LH content was noticed in the PPD of the pituitary gland in the initial controls, controls on days 6 and 12, and fasting fish on day 6, whereas the LH immunoreactivity was moderate or weak in fasting fish on day 12. In addition, although the mean gonadosomatic and hepatosomatic indices among different experimental groups did not show any statistically significant difference, the mean numbers of spermatogonia, spermatocytes, and spermatids were significantly lower in food-deprived fish on days 6 or 12 compared to those of controls. The inhibition of the spermatogenesis was accompanied by the presence of abundant spermatozoa in the lumen of seminiferous tubules of the testis in food-deprived fish, whereas the occurrence of spermatozoa was relatively infrequent in initial controls and controls. Furthermore, refeeding to food-deprived fish on day 6 onwards resulted in occurrence of few intensely stained LH secreting cells and significantly higher numbers of spermatocytes and spermatids concomitant with sparse spermatozoa in majority of tubules compared to those of food-deprived fish. These results suggest that prolonged exposure to food-deprivation causes suppression of the LH secretory activity in the pituitary gland and disruption in the spermatogenesis in *O. mossambicus*.

Keywords Starvation · Luteinizing hormone · Testis · Spermatogenesis · Fish reproduction · *Oreochromis mossambicus*

Introduction

Nutrition influences growth, reproduction, and health of fish (Jobling 2015). While availability of food is a great limiting factor, food-deprivation has diverse effects on tissue and plasma components that can interfere with various physiological functions (Mendez and Wieser 1993). Most fishes undergo natural period of food-deprivation throughout a normal life cycle (Silverstein et al. 1998; Grone et al. 2012; Jobling 2015). Several studies have examined the effect of food deprivation/starvation on reproduction in fish. For example, limitation in feeding causes reduction in the growth rate and delays the spawning time in the European seabass *Dicentrachus labrax* (Cerda et al. 1994), whereas food-deprivation due to mouthbrooding substantially suppresses ovarian development and androgen and estrogen levels in the African cichlid fish, *Astatotilapia burtoni* (Grone et al. 2012). On the other hand, exposure to starvation for a period of 40 days at 20 °C induces

R. P. Pikle · R. M. Jatiger · C. B. Ganesh (✉)
Neuroendocrinology Research Lab, Department of Studies in Zoology, Karnatak University, Dharwad 580 003, India
e-mail: ganeshkcd@gmail.com

testicular regression, but does not affect the onset of recrudescence in the estuarine gobiid fish *Gillichthys mirabilis* (De Vlaming 1971). Likewise, a significant decrease in the gonadosomatic index (GSI) is observed following food deprivation in the Atlantic cod *Gadus morhua* (Dutil et al. 2003), whereas starvation with no intermittent feeding causes a significant reduction in the testicular development and maturation in the iteroparous Arctic charr *Salvelinus alpinus* (Rice 1999). Similarly, feed restriction affects the testicular activity in a phase-dependent manner in the catfish *Clarias gariepinus* (Suchiang and Gupta 2011). Although there has been extensive studies on the relationship between nutritional status and reproductive health of fishes, majority of these are limited to seasonal breeders except for *A. burtoni* (Grone et al. 2012) in which suppression of sex steroid hormones levels is observed following starvation due to mouthbrooding. However, duration-dependent effects of food deprivation on luteinizing hormone (LH) secretion and testicular activity in non-seasonally breeding fish are not known. The tilapia *Oreochromis mossambicus* is a continuous breeder that shows spermatogenic cycle throughout the year (Dadzie 1969). Since many fish are exposed to natural period of starvation at certain times during aquacultural practices (Jobling 2015), the understanding of the effect of food-deprivation and refeeding on reproductive activity of the fish is imperative to obtain maximum benefits. The objective of the present investigation is to determine whether exposure to short- and long-term food-deprivation affects the LH secretion and the testicular activity in *O. mossambicus*, and if so, whether this effect is restored after refeeding.

Materials and methods

Experimental procedure

Adult *O. mossambicus* were collected from ponds around Dharwad (longitude 75°01′E, latitude 15°27′N) and brought to the laboratory. The fish were acclimatized to the laboratory conditions 2 months prior to the commencement of experiment. Confirmation of males was done by visual observation of genital openings. The sexual maturity was confirmed by their ability to release milt upon gentle pressing at the ventral surface near the genital opening. The fish weighing 25–35 g were maintained at a stocking density of six per aquaria (Vol. 70-l; size, 36 × 12 × 18 inches, L × W × H) under natural conditions (photoperiod, 11.33 ± 0.7; water temperature, 30.86 ± 1.02; dissolved oxygen, 7.16 ± 0.25). The water was recirculated daily to avoid algal growth and accumulation of planktons that might interfere with the nutritional status of the fish in all aquaria during the period of experimentation.

The experimental setup included six groups ($n = 6$ in each group). The first group of male fish was sacrificed on the day of commencement of experiment and served as the initial controls. The control fish in groups 2 and 3 were fed (Taiyo pet feed, Chennai, India) daily 3 g per aquaria for a period of 6 (controls 6D) and 12 (controls 12D) days, respectively. The fish in groups 4 and 5 were subjected to fasting for 6 (fasting 6D) and 12 (fasting 12D) days, respectively. The fish in sixth group (refeeding group) were food deprived for the first 6 days and then refed with the same quantity of food as in controls up to day 12. The fish in groups 2–6 were sacrificed ($n = 30$). All experimental fish were euthanized following anesthetization with 2-phenoxy ethanol (1:1500, Sigma, USA). At autopsy, the weights of the body, the testis, and the liver were recorded and GSI (100 × gonad mass/body weight) and hepatosomatic index (HSI; 100 × liver weight/body weight) were calculated.

Histology and morphometry of the testis

The testes were fixed in Bouin's fluid and embedded in paraffin wax after dehydration in ascending series of ethyl alcohol. Five micron-thick serial sections were cut and stained with hematoxylin and eosin. Depending upon the morphology and the size, different spermatogenic cell types were identified. The mean size and characteristics of each spermatogenic cell type are provided in Table 1. Measurements and quantification of different spermatogenic cell types were done using Q capture pro7.1 image analysis system consisting of BX53 microscope attached to a digital CCD color camera (Olympus, Tokyo, Japan). In each testis section, counting of different cell types (spermatogonia, spermatocytes and spermatids) was made from ten randomly chosen tubules. This procedure was done in ten discretely selected sections (each being at least 50 sections

Table 1 Morphology and size of different spermatogenic cells in the testis of *O. mossambicus*

Spermatogenic cell type	Characteristics	Cell diameter (Mean ± SE)
1. Spermatogonia	Spherical nucleus, less condensed chromatin, single nucleolus, and faintly stained cytoplasm	12.00 ± 0.38 μm
2. Spermatocytes	Spherical nucleus, deeply condensed chromatin in various meiotic stages	4.92 ± 0.20 μm
3. Spermatid	Indistinct outline with scanty cytoplasm and dense spherical nuclei	1.75 ± 0.06 μm
4. Spermatozoa	Deeply stained spherical head with the flagellum, located in lumen	1.39 ± 0.04 μm

away) to avoid repeated counting of the cells in the same tubule. The total number of each spermatogenic cell type was obtained from 100 tubules/testis/fish. The mean values obtained from each experimental fish ($n = 6$) were finally expressed as mean ± SE. The density of spermatozoa was determined as sparse or abundant based on the area occupied by the spermatozoa in the lumen.

Immunocytochemistry

The tilapia *O. mossambicus* were perfused transcardially with 20 ml of ice-cold phosphate buffered saline (PBS, pH 7.4) followed by 20-ml ice-cold 4% paraformaldehyde. The brains were dissected out and post-fixed in the same fixative for 12 h. The tissues were rinsed in PBS and cryoprotected in ice-cold 10% (2 h), 20% (2 h), and 30% (overnight) sucrose solution. Transverse and sagittal frozen sections were cut at 14 μm thickness using cryostat (CM1510S; Leica Microsystems, Wetzlar, Germany). For immunocytochemical localization of the LH secreting cells in the proximal pars distalis (PPD) region of the pituitary gland, streptavidin–biotin-peroxidase method was used as described previously (Chabbi and Ganesh 2012, 2015, 2016). Briefly, the sections were rinsed in PBS for 15 min and incubated with rabbit polyclonal human LHβ antiserum (National Hormone and Peptide Program, Harbor-UCLA Medical Centre, CA, USA) at 1:8000 dilution for the detection of the LH secreting cells in the PPD region of the pituitary gland. The sections were rinsed in PBS for 10 min and incubated for 2 h with biotinylated goat antirabbit immunoglobulin G (Sigma, USA) followed by streptavidin peroxidase (EXTRA 3; Sigma, USA) for 1 h. The sections were again rinsed in PBS and stained with chromogen 3-amino-9-ethyl carbazole to visualize the antigen–antibody complex in the form of a reddish brown precipitate. The sections were rinsed in distilled water and mounted with glycerol–gelatin. The following control procedures were included to verify the specificity of the antibodies: (a) omission of primary antibody, (b) replacement of primary antisera with 1% BSA, (c) omission of secondary antibody, and (d) preadsorption of diluted primary antibody for LH with either salmon pituitary gland LH or human pituitary gland LH (Sigma, USA) at 10^{-5} M for 24 h prior to incubation. These control procedures completely blocked the immunoreactions, confirming the specificity of the antibody.

The photographs of the sections of pituitary gland were adjusted for brightness and contrast using PHOTOSHOP CS5, version 12.0 (Adobe Systems Inc, San Jose, CA, USA). The immunoreactivity was evaluated using the Image J software version 1.45 (NIH, Bethesda, MD, USA) as described previously (Chabbi and Ganesh 2015, 2016). Briefly, the pixel intensities were measured in both non-immunoreactive background and immunoreactive content. The threshold pixel intensity of the background was subtracted from the total intensity. The threshold intensity of the immunoreactive cells was measured in five sections of each pituitary gland ($n = 6$) and expressed as integrated units in parentheses (mean ± SE) under results.

Statistical analysis

The mean values of spermatogenic cell type/tubule diameter were obtained from 100 tubules of each testis of experimental fish. The data on these morphometric parameters and GSI and HSI were computed from six fish belonging to each experimental group and expressed as mean ± SE. All parameters were analyzed by one-way analysis of variance (ANOVA) and post hoc Student–Newman–Keuls multiple comparison procedure using SigmaStat 3.5 software. Significant differences for all comparisons were considered at $P < 0.05$ level.

Results

The mean GSI and HSI did not show any significant difference among different experimental groups (Fig. 1a, b). Histological examinations of the testis in the initial control and control fish revealed the presence of several germ cell cysts that occur along the circumference of the seminiferous tubules (Fig. 2b, c). These cysts consisted of different spermatogenic cell types such as spermatogonia, spermatocytes, spermatids, and spermatozoa (Fig. 2a–c). The spermatogonia were large round cells with faint cytoplasm located close to the periphery of testicular lobules (Fig. 2b). The nucleus was large, spherical, and contained poorly condensed chromatin and a darkly stained small nucleolus in it (Fig. 2b). The spermatocytes were smaller than spermatogonia and had spherical nuclei with deeply stained condensed chromatin in various meiotic stages (Fig. 2b). The spermatids were smaller than spermatocytes and located close to the lumen of the tubule. These cells had indistinct outline with scanty cytoplasm and dense spherical nuclei (Fig. 2b). The spermatozoa were the smallest of the spermatogenic lineage with deeply stained round heads and found in the tubular lumen without any arrangement (Fig. 2a, b).

The mean number of spermatogonia, spermatocytes, and spermatids did not differ significantly among the initial controls and controls 6D and 12D groups (Fig. 3a–c). On the other hand, the seminiferous tubules in the testis of food-deprived fish showed fewer spermatogenic stages (Fig. 2d–f). There was a significant decrease in the number of spermatogonia, spermatocytes, and spermatids in fasting 6D and 12D groups compared to those of initial controls and controls (Fig. 3a–c). The testis in fasting 6D group was characterized by enlarged seminiferous tubules filled with much greater number of sperms concomitant with relative lack of germinal epithelium, spermatogonia, spermatocytes, and spermatids in majority of the tubules (Fig. 2d). The tubule border was still intact in fasting 6D group similar to that of controls 6D group (Fig. 3c). Although large seminiferous tubules were filled with spermatozoa, severe degeneration, collapsing, and necrotic changes in

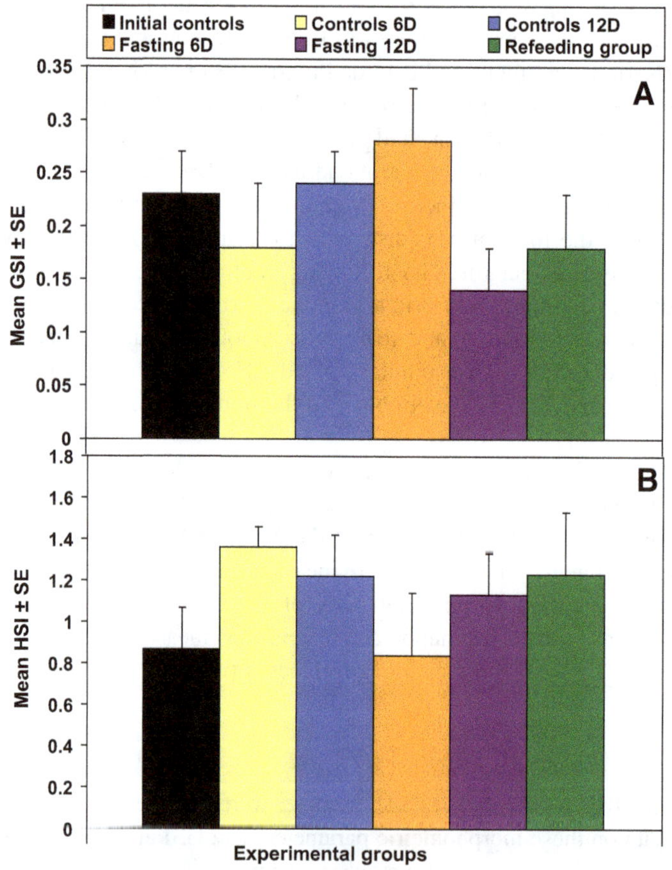

Fig. 1 **a** and **b** Effect of food-deprivation on GSI (**a**) and HSI (**b**) in *O. mossambicus*. Values are mean ± SE. One-way ANOVA followed by post hoc Student–Newman–Keuls multiple comparison test: no statistically significant differences were noticed among different experimental groups

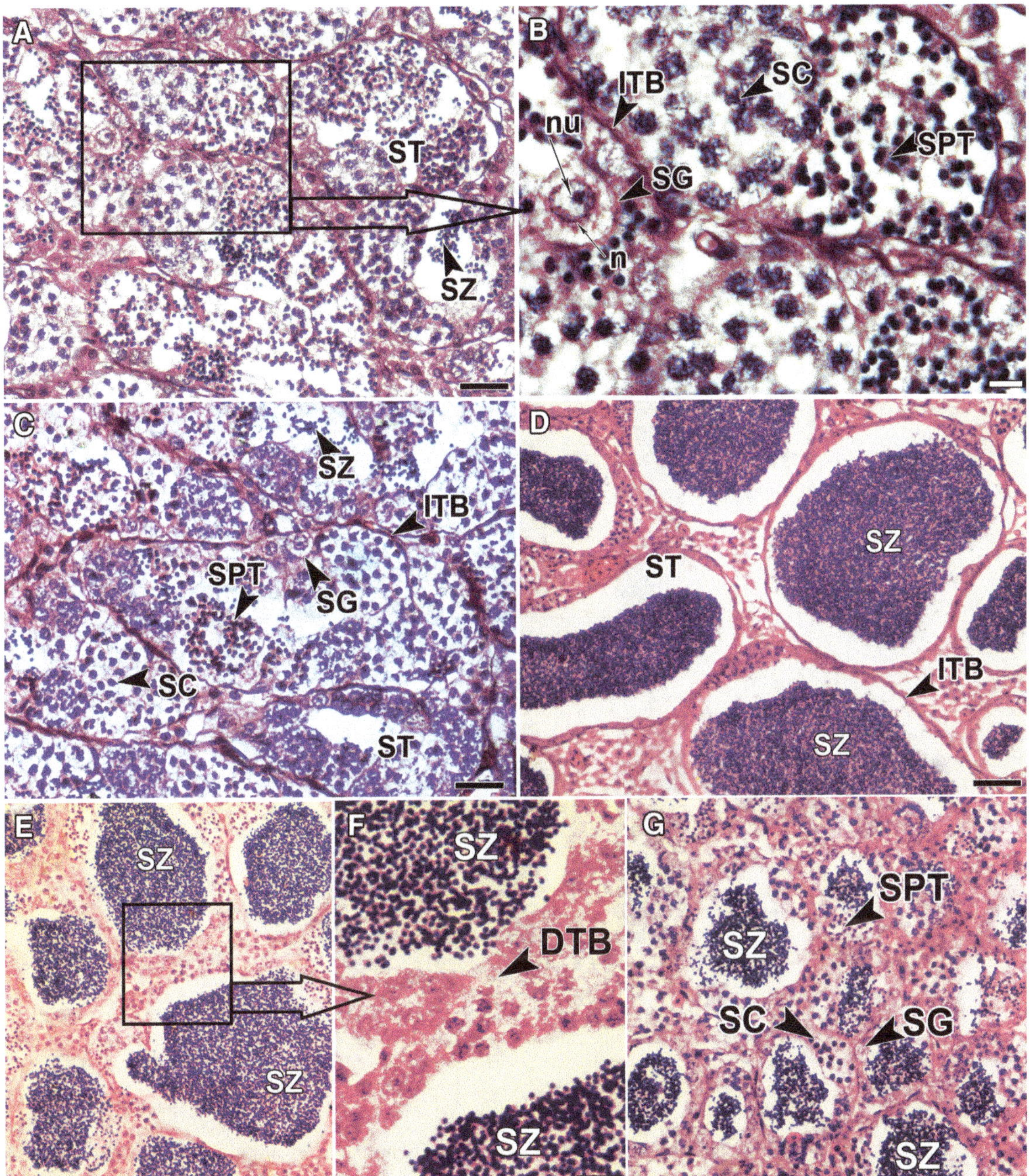

Fig. 2 Cross sections of the testis of *O. mossambicus* showing seminiferous tubules (ST) containing spermatogonia (SG), spermatocyte (SC), spermatids (SPT), and spermatozoa (SZ). Note the presence of abundant spermatozoa in fasting fish on days 6 (**d**) and 12 (**e**) compared to sparse spermatozoa and intact tubule border (ITB) in the initial controls (**a** and **b**) and controls 6D or 12D groups (**c**). These changes are partially restored in refeeding group (**g**). Haematoxylin and Eosin; *Scale bar*, **a**, **c–e**, and **g** 100 μm; **b** 10 μm; **f** 25 μm. *DTB* distorted tubule border, *cy* germ cyst, *n* nucleus, *nu* nucleolus

the germinal epithelium of the seminiferous tubules were associated with lack of spermatogenic cell types and distorted tubule border in fasting 12D group (Fig. 2e, f) compared to those of controls 12D group (Fig. 3c).

The fish in refeeding group showed tendency towards the reestablishment of the histoarchitecture of the testis as shown by the appearance of cysts along the borders of tubules and the presence of different spermatogenic cell types (Fig. 2g). Refeeding to food-deprived fish did not improve the number of spermatogonia, whereas the number of spermatocytes and spermatids increased significantly compared to those of fasting 6D or 12D groups (Fig. 3a–c). Although the mean number of spermatocytes in refeeding group was almost closer to that of controls, the mean number of spermatids was significantly lower compared to those of initial controls

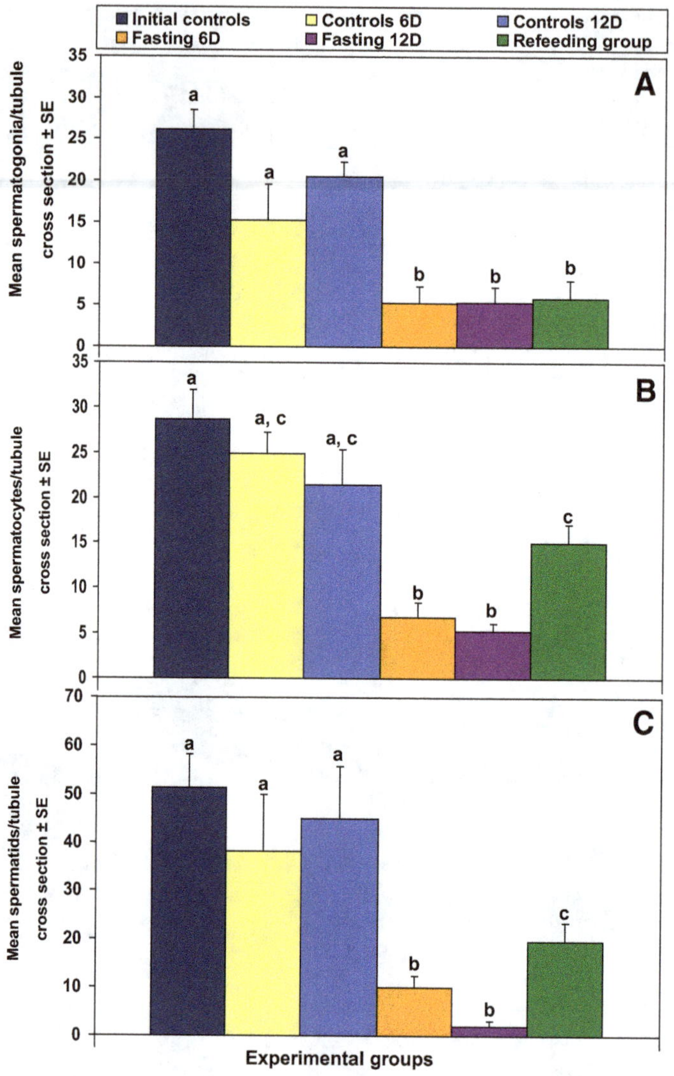

Fig. 3 Effect of food-deprivation on spermatogonia (**a**), spermatocytes (**b**) and spermatids (**c**) in *O. mossambicus*. Values are mean ± SE. One-way ANOVA followed by post hoc Student–Newman–Keuls multiple comparison test: Groups with same superscripts are not significantly different from each other, whereas *different superscripts* indicate significant ($P < 0.05$) difference

and controls (Fig. 3b, c). The mean tubule diameter did not significantly differ among different experimental groups (Fig. 4).

The LH immunostaining was intense in the initial controls (91.00 ± 0.94; Fig. 5a) and controls 6D or 12D groups (96.00 ± 1.34; Fig. 5b). Exposure of fish to fasting until day 6 also showed the LH immunoreactivity similar to that of initial controls and controls (92.20 ± 1.04; Fig. 5c), whereas the LH immunoreactivity was moderate or weak in fasting fish on day 12 (50.58 ± 1.08; Fig. 5d) compared to those of initial controls and controls. Refeeding to the fish exposed to the initial food-deprivation, resulted in some intensely stained LH immunoreactive content in the PPD region of the pituitary gland (95.00 ± 1.65; Fig. 5e). However, the extent of immunoreactive area covered in the PPD region was not similar to that of initial controls or controls.

Discussion

A tight relationship between food availability and reproduction has been observed in fish (Schneider 2004). For example, reduction in feeding rate has been shown to inhibit gonadal maturation in the female Atlantic

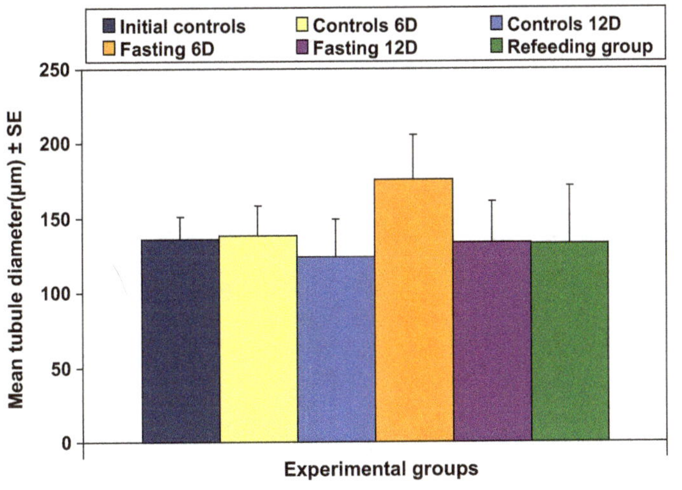

Fig. 4 Effect of food-deprivation on tubule diameter of the testis in *O. mossambicus*. Values are mean ± SE. One-way ANOVA followed by post hoc Student–Newman–Keuls multiple comparison test: No statistically significant differences were noticed among different experimental groups

salmon *Salmo salar* (Thorpe et al. 1990; Reimers et al. 1993), whereas prolonged fasting caused reduction in the weights of the body and the ovary and increased the appearance of atretic follicles in immature female Coho salmon *Oncorhynchus kisutch* (Yamamoto et al. 2011). Likewise, reduction in the plasma levels of estradiol and the size of eggs as well as the larvae were observed following 6 months of broodstock feeding at a half feed ration in the European seabass *D. labrax* (Cerda et al. 1994). Whereas exposure of adolescent turbot *Scophthalmus maximus* to low rations during vitellogenesis affected gonadal maturation (Bromley et al. 2000), the GSI was low in the Atlantic cod *G. morhua* subjected to food deprivation for a period of 84 days (Dutil et al. 2003). While the majority of these studies point to the vulnerability of the reproductive success following food-deprivation in females, limited studies have revealed inhibitory effect of food-deprivation in males. Restriction in spring time feeding for 2–3 months has been shown to affect maturation in the male Chinook salmon *O. tshawytscha* (Hopkins and Unwin 1997), whereas 3 weeks of partial and full feed restriction markedly reduced the testicular activity in the catfish *C. gariepinus* (Suchiang and Gupta 2011). Similarly, starvation with no intermittent feeding caused a significant reduction in the testicular development and maturation in the iteroparous Arctic charr *S. alpinus* (Rice 1999). In the present study, exposure of *O. mossambicus* to food-deprivation did not significantly affect the GSI as well as the HSI in both fasting 6D and 12D groups compared to those of initial controls or controls. However, the present study reveals the negative effect of food-deprivation on the spermatogenetic activity for the first time in a non-seasonally breeding fish *O. mossambicus*. A significant reduction in the mean number of spermatogonia, spermatocytes, and spermatids associated with distortion in the tubule borders and relative lack of germinal epithelium in food-deprived fish compared to those of controls or initial controls clearly suggest that exposure to food-deprivation causes detrimental effect on the testicular activity.

On the contrary, several studies have also suggested no influence of feed restriction or starvation in both male and female fish. For instance, no significant differences were noticed in egg hatchability between the fed and starved rainbow trout *S. gairdneri* (Ridelman et al. 1984), whereas the mean diameter of egg size and GSI were not affected following exposure to limited feeding regime for a period of 550 days from first feeding to sexual maturity in the *Tilapia zillii* (Coward and Bromage 1999). Similarly, reduction in feeding ratio throughout a year (cycles of 3 days fed/2 days unfed) did not affect the gonadal development in both male and female fish *Brycon amazonicus* (Carvalho 2001). In adolescent male turbot *S. maximus,* low ration dietary regime did not significantly influence the maturation (Bromley et al. 2000). Likewise, food restriction (1 g food daily/weekly) for 4 months did not affect the testis development in the convict cichlid *Cichlasoma nigrofasciatum* (Townshend and Wootton 1984). These results appear to be dependent on the experimental conditions, duration or timing of food restriction, and fish species.

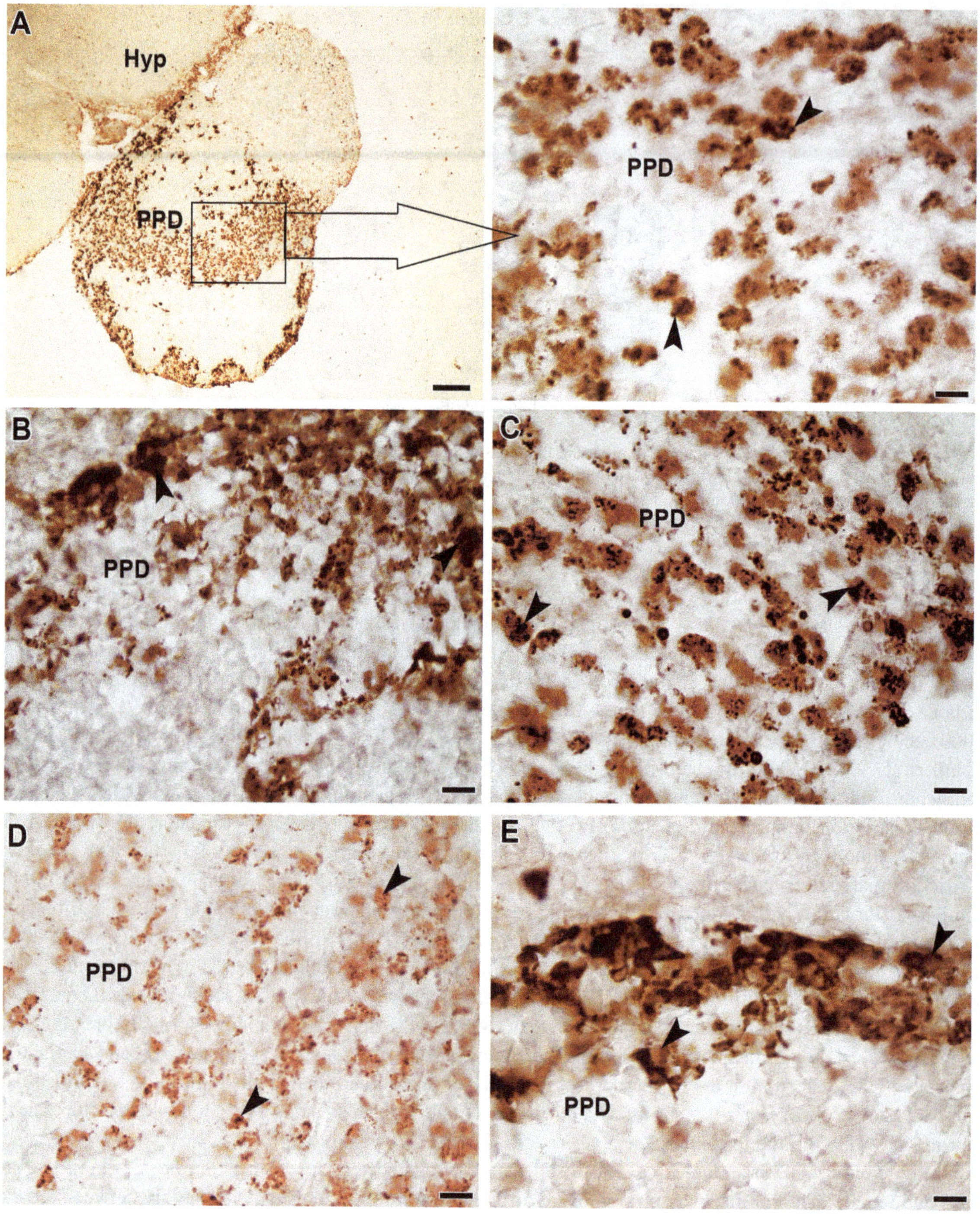

Fig. 5 Effect of food-deprivation on the LH secretory activity in the proximal distalis (PPD) of the pituitary gland in *O. mossambicus*. Note weak or moderate LH immunostaining in transverse sections of the pituitary gland in fish exposed to fasting for 12 days (**d**) compared to intense LH immunostaining in initial control (**a**), controls on days 6 or 12 (**b**), and in fish exposed to fasting on day 6 (**c**). In refeeding group (**e**), intensely labeled LH immunoreactive content is seen in the PPD region. *Hyp* hypothalamus. *Scale bar A* 100 μm; all other figures and magnified view of *A*, 25 μm

In fish, the loss of germinal cells via apoptosis is a normal process throughout the spermatogenesis (Almeida et al. 2008). Franek and Chladkova-Sramkova (1995) demonstrated a close relationship between apoptosis and nutrition in mammalian hybridoma cells. They found artificial starvation due to saline containing medium-induced apoptosis in vitro, whereas refeeding with an amino acid mixture suppressed the starvation-induced apoptosis in the cultured hybridoma cells. Escobar et al. (2014) observed an increased apoptosis in germ cells of the testis following exposure of the sea bass *D. labrux* to food restriction for a period of 9 months. The present study also reveals an inhibition of spermatogenesis in the testis of fish exposed to food-deprivation as shown by a significantly lower number of spermatogonia, spermatocytes, and a relatively lower number of spermatids compared to those of controls. Although it is plausible that the proliferation of spermatogonia from the germinal epithelium was affected, the possibility of loss of spermatogenic cells via food deprivation-induced apoptosis cannot be ruled out. Further studies are required to address this issue.

Sasayama and Takahashi (1972) noticed the presence of large amount of spermatozoa in the lumen surrounded by thin-walled testicular lobules retaining only a small number of spermatogonia in the goldfish *Carassius auratus* exposed to starvation. In the present study, the organization of germ cell cysts was severely disturbed in fasting fish on day 6 or 12, whereas the integrity of tubule border was lost in fasting fish on day 12 as compared to the presence of intact tubule border in fasting 6D and control fish groups. These changes suggest a duration-dependent detrimental effect of food deprivation on tubule organization in the testis. Furthermore, exposure of the tilapia to food-deprivation for 6 or 12 days resulted in accumulations of abundant spermatozoa compared to the presence of relatively sparse spermatozoa the lumen of the testis in initial controls and controls. Taken together, sparsely found spermatogenic cell types and germ cysts within the intact tubule border concomitant with lumens filled with spermatozoa in fasting 6D group suggest the possibility of rapid completion of spermatogenesis and maturation of spermatids under food-deprivation condition. However, similar changes accompanied by the distortion of tubule border on day 12 of fasting indicate the probability of onset of apoptosis leading to loss of spermatogenic cells and damage to germinal epithelium. Further studies involving the use of apoptotic markers would address this issue.

We may also recall that refeeding at ad libitum in the present study, improved the spermatogenesis as shown by the significantly higher numbers of spermatocytes and spermatids compared to those of food-deprived fish concomitant with occurrence of sparse spermatozoa and germ cysts in the majority of tubules. While these results confirm that the inhibition of spermatogenetic process is induced by food-deprivation in the present study, the reestablishment of germ cell cysts in tubules of the testis in refed fish might be due to the migration of spermatogonial cells into the cysts from a permanent germinal layer.

In teleosts, the testicular activity is primarily controlled by the pituitary gonadotrophic hormones (GtHs) such as GtH-1 (follicle stimulating hormone, FSH) and GtH-II (LH), which bind to specific receptors in the testis to regulate spermatogenesis and steroidogenesis (Yaron et al. 2003). Kawauchi et al. (1989) reported an increase in the number of GtH-II cells in the pituitary gland at the time of final maturation in teleosts. The expression of the GtH genes was not affected by food restriction in mature female goldfish (Sohn et al. 1998), whereas when food-restricted rats were given unrestricted access to food, the frequency of LH and growth hormone (GH) pulses and the mean levels of FSH increased significantly within 2 days (Sisk and Bronson 1986). Even at the hypothalamic level, although 1 week of food restriction did not result in alterations in the gonadotrophin-releasing hormone (GnRH) neuronal system, food restriction for 2–3 weeks was associated with increased GnRH-immunoreactive neuronal soma size and numbers as well as an increase in fibre intensity in the main fibre pathway to the median eminence in male prairie vole *Microtus ochrogaster* (Kriegsfeld et al. 2001). These results suggest a stimulatory influence of starvation on hypothalamo-pituitary–gonad (HPG) axis. Furthermore, although the regulation of GtH release is under the control of several factors in teleosts, potential influence of neuropeptide Y (NPY) is implicated (Kah et al. 1989). Stimulatory effect of NPY on LH secretion was more pronounced in fasted fish (Cerda-Reverter et al. 1999). Such a possibility cannot be excluded for the present study as exposure of *O. mossambicus* to 6-day food-deprivation did not affect the LH secretory activity, whereas the LH immunolabeling was moderate or diminished in fish exposed to food-deprivation for 12 days. The duration-dependent variation in LH immunoreactivity might be due to prolonged fasting status and/or involvement of multiple hormones in regulation of reproductive axis. For example, exposure to food-deprivation suppressed the plasma T4 and T3 levels and reduced thyroid tissue sensitivity to thyroid stimulating hormone in fish (Cerda-Reverter et al. 1996). Indeed, suppression of plasma T4 and T3 levels in catfish *C. gariepinus* maintained under partial and full feed restriction were associated with reduced spermatogenesis (Suchiang and Gupta 2011). Since the thyroid hormones have been shown to modulate the HPG axis in teleosts (Swapna and Senthilkumaran 2007), the dwindling germinal cysts and spermatogenic cell types associated with abundance of spermatozoa in fasting fish 6D group as well as decreased LH immunoreactivity in fasting fish 12D group might be due to down-regulation of hypothalamo–

pituitary–thyroid axis following starvation-induced energy crisis. However, this possibility needs to be further confirmed in the tilapia. In addition, in the present study, a tendency for reinstatement of pituitary–testicular activity following refeeding apparently substantiates the negative effects of prolonged food-deprivation along the reproductive axis in *O. mossambicus*.

In conclusion, the present study suggests that exposure to food-deprivation for prolonged duration results in deleterious effects on reproduction, which might lead to impairment in the reproductive success of *O. mossambicus*. Although, the short-term exposure to fasting, perhaps, allows the fish to complete the reproductive process under food scarcity conditions, at later stages of food-deprivation, suppression of LH secretory activity in the pituitary gland and disruption in the spermatogenesis are observed. These results have potential implications in aquaculture practices as production of quality gametes is extremely important to achieve reproductive success.

Acknowledgements The work was supported by a grant from UGC-SAP (No. 18/2015/DSA-I), New Delhi. The infrastructural facility received under research grant (No. SR/S0/AS-52/2013) from Science & Engineering Research Board, Department of Science and Technology (SERB-DST), New Delhi, Govt. of India is gratefully acknowledged.

References

Almeida FFL, Kristofferson C, Taranger GL, Schulz RW (2008) Spermatogenesis in Atlantic Cod (*Gadus morhua*): a novel model of cystic germ cell development. Biol Reprod 78:27–34

Bromley JP, Ravier C, Withames PR (2000) The influence of feeding regime on sexual maturation, fecundity and atresia in first-time spawning turbot. J Fish Biol 56:264–278

Carvalho EG (2001) Restricao alimentar e reproducao em peixes. Jaboticabal: Centro de Aquicultura da UNESP. PhD thesis

Cerda J, Carrillo M, Zanuy S, Ramos J (1994) Effect of food ration on estrogen and vitellogenin plasma levels, fecundity and larval survival in captive sea bass, *Dicentrarchus labrax*: preliminary observations. Aquat Living Resour 7:255–266

Cerda-Reverter JM, Zanuy S, Carrillo M, Kah O (1996) Development of enzyme immunoassays for 3,5,3′-triiodo-L-thyronine and L-thyroxine: time-course studies on the effect of food deprivation on plasma thyroid hormones in two marine teleosts, sea bass (*Dicentrarchus labrax* L.) and sea bream (*Sparus aurata* L.). Gen Comp Endocrinol 103:290–300

Cerda-Reverter JM, Sorbera LA, Carrillo M, Zanuy S (1999) Energetic dependence of NPY-induced LH secretion in a teleost fish (*Dicentrarchus labrax*). Am J Physiol 277:R1627–R1634

Chabbi A, Ganesh CB (2012) Stress-induced inhibition of recruitment of ovarian follicles for vitellogenic growth and interruption of spawning cycle in the fish *Oreochromis mossambicus*. Fish Physiol Biochem 38:1521–1532

Chabbi A, Ganesh CB (2015) Evidence for the Involvement of dopamine in stress-induced suppression of reproduction in the cichlid fish *Oreochromis mossambicus*. J Neuroendocrinol 27:343–356. doi:10.1111/jne.12269

Chabbi A, Ganesh CB (2016) Neuroanatomical evidence for the involvement of β-endorphin during reproductive stress response in the fish *Oreochromis mossambicus*. J Chem Neuroanat 77:161–168

Coward K, Bromage NR (1999) Spawning frequency, fecundity, egg size and ovarian histology in groups of *Tilapia zillii* maintained upon two distinct food ration sizes from first-feeding to sexual maturity. Aquat Living Resour 12:11–22

Dadzie S (1969) Spermatogenesis and the stages of maturation in the male cichlid fish *Tilapia mossambica*. J Zool 159:399–403

De Vlaming VL (1971) The effects of food deprivation and salinity changes on reproductive function in the estuarine Gobiid fish (*Gillichthys mirabilis*). Biol Bull 141:458–471

Dutil JD, Lambert Y, Chabot D (2003) Winter and spring changes in condition factor and energy reserves of wild cod compared with changes observed during food-deprivation in the laboratory. ICES J Mar Sci 60:780–786

Escobar S, Felip A, Salah M, Zanuy S, Carrillo M (2014) Long-term feeding restriction in prepubertal male sea bass (*Dicentrarchus labrax* L.) increases the number of apoptotic cells in the testis and affects the onset of puberty and certain reproductive parameters. Aquaculture 433:504–512

Franek F, Chladkova-Sramkova K (1995) Apoptosis and nutrition: involvement of amino acid transport system in repression of hybridoma cell death. Cytotechnology 18:113–117

Grone BP, Carpenter RE, Lee M, Maruska KP, Fernald RD (2012) Food deprivation explains effects of mouthbrooding on ovaries and steroid hormone, but not brain neuropeptide and receptor mRNAs in an African Cichlid fish (*Astatotilapia burtoni*). Horm Behav 62:18–26

Hopkins CL, Unwin MJ (1997) The effect of restricted spring time feeding on growth and maturation of freshwater-reared chinook salmon, *Oncorhynchus tshawytscha* (Walbaum). Aquacult Res 28:545–549

Jobling M (2015) Fish nutrition research: past, present and future. Aquacult Int 24:767–786

Kah O, Pontet A, Danger JM, Dubourg P, Pelletier G, Vaudry H, Calas A (1989) Characterization, cerebral distribution and gonadotropin release activity of neuropeptide Y (NPY) in the goldfish. Fish Physiol Biochem 7:69–76

Kawauchi H, Suzuki K, Itoh H, Swanson P, Naito N, Nagahama Y, Nozaki M, Nakai Y, Itoh S (1989) The duality of teleost gonadotropins. Fish Physiol Biochem 7:29–38. doi:10.1007/BF00004687

Kriegsfeld LJ, Ranalli NJ, Trasy AG, Nelson RJ (2001) Food restriction affects the gonadotropin releasing hormone neuronal system of male prairie voles (*Microtus ochrogaster*). J Neuroendocrinol 13:791–798

Mendez G, Wieser W (1993) Metabolic responses to food deprivation and refeeding in juveniles of *Rutilus rutilus* (Teleostei Cyprinidae). Environ Biol Fish 36:73–81

Reimers E, Kiorrefjord AG, Stavostrand SM (1993) Compensatory growth and reduced maturation in second sea winter farmed Atlantic salmon following starvation in February and March. J Fish Biol 43:805–810

Rice CD (1999) Restricted feeding, spermatogenesis and growth in Arctic Charr (*Salvelinus alpinus*)-The identification of two possible gametogenic control points. Masters thesis, Memorial University of Newfoundland, USA

Ridelman JM, Hardy RW, Brannon EL (1984) The effect of short-term starvation on ovarian development and egg viability in rainbow trout (*Salmo gairdneri*). Aquaculture 37:133–140

Sasayama Y, Takahashi H (1972) Effect of starvation and unilateral castration in male goldfish (*Carassius auratus*), and a design of bioassay for fish gonadotropin using starved goldfish. Bull Fac Fish Hokkaido Univ 22:267–279

Schneider JE (2004) Energy balance and reproduction. Physiol Behav 81:289–317

Silverstein J, Breninger TJ, Baskin DG, Plisetskaya EM (1998) Neuropeptide Y-like expression in the salmon brain increases with fasting. Gen Comp Endocrinol 110:157–165

Sisk CL, Bronson FH (1986) Effects of food restriction and restoration on gonadotropin and growth hormone secretion in immature male rats. Biol Reprod 35:554–561

Sohn YC, Yoshiura Y, Kobayashi M, Aida K (1998) Effect of sex steroids on the mRNA levels of gonadotropin I and II subunits in the goldfish *Carassius auratus*. Fish Sci 64:715–721

Suchiang P, Gupta BBP (2011) Effects of partial and full feed restriction on the plasma levels of thyroid hormones and testicular activity in the male air breathing catfish (*Clarias gariepinus*) during different phases of breeding cycle. Int J Biol 3:32–42

Swapna I, Senthilkumaran B (2007) Thyroid hormones modulate the hypothalamo–hypophyseal–gonadal axis in teleosts: molecular insights. Fish Physiol Biochem 33:335–345

Thorpe JE, Talbot C, Miles MS, Keay DS (1990) Control of maturation in culture Atlantic salmon, *Salmo salar*, in pumped seawater tanks, by restricting food intake. Aquaculture 86:315–326

Townshend TJ, Wootton RJ (1984) Effects of food supply on the reproduction of the convict cichlid, *Cichlasoma nigrofasciatum*. J Fish Biol 24:91–104

Yamamoto Y, Luckenbach JA, Middleton MA, Swanson P (2011) The spatiotemporal expression of multiple coho salmon ovarian connexin genes and their hormonal regulation in vitro during oogenesis. Reprod Biol Endocrinol 9:52. doi:10.1186/1477-7827-9-52

Yaron Z, Gur G, Melamed P, Rosenfeld H, Elizur A, Levavi-Sivan B (2003) Regulation of fish gonadotropins. Int Rev Cytol 225:131–185

11

Adenosine monophosphate degradation and inosinic acid accumulation in the shrimp *Penaeus japonicus*

Hiroko Seki · Kaori Nakazato · Naoko Hamada-Sato

Abstract Shrimp muscle contains taste component compounds, such as glutamic acid and inosinic acid (IMP). When both are present, taste intensity increases by fourfold to eightfold compared to that when only glutamic acid is present. Inosinic acid is generated via adenosine triphosphate degradation in two metabolic pathways in which adenosine monophosphate (AMP) is generated and then degraded to either IMP or adenosine (AdR). We investigated post-mortem AMP degradation pathways in the shrimp *Penaeus japonicus* by measuring the activity of enzymes extracted from shrimp muscle and by isolating bacteria from shrimp muscle and examining their role in AMP degradation. The enzyme extract degraded AMP to IMP through high AMP deaminase activity, and in addition, we identified *Shewanella* sp. and *Exiguobacterium* sp. as mediating AMP degradation to AdR. Therefore, preventing an increase in bacterial numbers during storage is important for preventing AMP from degrading to AdR.

Keywords Adenosine · Adenosine monophosphate · Bacteria · Enzyme · Inosinic acid · Shrimp

Introduction

The shrimp *Penaeus japonicus*, an economically important crustacean, inhabits the Indian Ocean, western Pacific Ocean, and Mediterranean Sea (Motoo 1990). Shrimp has substantial quantities of free amino acids (Yamanaka 2008), an appealing flavor, and high market value (Hamano 2015). The most common free amino acids in shrimp are glycine and alanine, which provide a sweet taste, and glutamic acid, which is an important taste component (Matsumoto and Yamanaka 1990; Matsumoto et al. 1991). Because flavor is improved by the simultaneous presence of glutamic acid and inosinic acid (IMP) (Zhang et al. 2008; Kobayashi et al. 2012), IMP is also important for shrimp flavor.

IMP is generated through adenosine triphosphate (ATP) degradation after an organism dies. ATP degradation produces a series of compounds: ATP → adenosine diphosphate (ADP) → adenosine monophosphate (AMP) → IMP → inosine (HxR) → hypoxanthine (Hx). ATP is generally degraded to IMP and accumulates relatively quickly in fish muscle (Srirangsan et al. 2010). AMP accumulates in substantial quantities in shrimp,

---

H. Seki · K. Nakazato · N. Hamada-Sato (✉)
Tokyo University of Marine Science and Technology, 4-5-7, Konan, Minato-ku, Tokyo 108-8477, Japan
e-mail: hsnaoko@kaiyodai.ac.jp

H. Seki
e-mail: supergirlhiro1010@gmail.com

K. Nakazato
e-mail: io91.okan@gmail.com

and it degrades to either IMP or adenosine (AdR) (Fujisawa and Yoshino 1985; Matsumoto 1996; Koseki et al. 2006; Koyama et al. 2008). AdR does not appear to affect flavor, whereas the degradation of AMP in shrimp muscle to IMP increases flavor.

ATP-related compounds are generally degraded by muscle enzymes, but IMP is degraded by bacteria (Matsumoto and Yamanaka 1991; Yokoyama et al. 1996; Hayashi and Nakata 2003; Seki and Hamada-Sato 2014). Other ATP-related compounds might be degraded not only by muscle enzymes but also by bacteria. No detailed reports on AMP degradation in shrimp or on AMP degradation by muscle or bacterial enzymes have appeared in the literature. In this study, we investigated AMP degradation in shrimp to ascertain which compounds are produced in the largest quantities. We prepared an enzyme solution from shrimp and examined its effect on AMP degradation, and we isolated and identified the bacteria that degrade AMP in shrimp. In addition, we measured the amounts of ATP-related compounds and obtained an estimate of bacterial numbers in shrimp over a few days of storage. We suggest an optimal approach to the maintenance of flavor components in stored shrimp.

Materials and methods

Sample preparation

Live shrimp (*P. japonicus*, length approximately 17 cm; weight approximately 33 g) from aquacultures in Ōita Prefecture, Japan, were transported to the laboratory, where they were stunned in sterilized ice water and killed quickly by severing the first ganglia. Killing involved either immediate loss of sensibility or induction of insensibility without discomfort or pain, followed by death, while the animal was insensible.

Shrimp muscle enzyme degradation of AMP

AMP-degrading enzyme was extracted as part of an enzyme solution. Five grams of muscle was collected from each shrimp and crushed in three times its weight of ultrapure water. The mixture was dialyzed against water for 2 days, and then, the dialysate was filtered through a No. 1 filter (Advantec Co., Ltd.; Tokyo, Japan) and then a 0.20 μm filter (Toyo Roshi Kaisha, Ltd.; Tokyo, Japan) at 10 °C to remove all bacteria from the enzyme solution. Forty samples of a 3 mL reaction mixture consisting of 0.5 mL AMP (25 mM), 2 mL buffer (50 mM maleic acid/Tris/NaOH, pH 7.5), and 0.5 mL enzyme solution were incubated at 0, 5, 10, 20, 30, 40, 50, or 60 °C for 5 days, and the reaction was stopped every day in three samples for each temperature by the addition of 2 mL 10% perchloric acid. The mixture was neutralized using 10% KOH, and the resulting precipitate was removed by centrifugation (13,697×*g*, 5 min, 5 °C). The supernatant was then transferred to a 15 mL centrifuge tube, diluted with sterilized water to obtain a final volume of 10 mL, and filtered through a 0.20 μm filter (Millex-LG, Millipore, Merck). AMP degradation compounds were measured by high-performance liquid chromatography (HPLC) with the following equipment and parameters: pump, L2130 (Hitachi, Tokyo, Japan); column, Shodex Asahipak GS-320 HQ (Showa Denko K.K., Tokyo, Japan); mobile phase, 200 mM NaH_2PO_4; flow rate, 0.6 mL/min; temperature, 30 °C; detector, Hitachi L7420; and wavelength, 260 nm. Four standard curves were established, each with 0.025, 0.050, 0.125, or 0.250 mmol/L of each compound (AMP, IMP, AdR, HxR, and Hx), respectively.

Bacterial degradation of AMP

Shrimp muscle (10 g) was homogenized with 90 mL sterilized 0.9% NaCl solution in a sterilized bag (PYXON-30, ELMEX Ltd., Tokyo, Japan) using an Exnizer-400 Stomacher (ORGANO Co., Ltd., Tokyo, Japan) (30 s, 260 rpm). A 1 mL aliquot of the supernatant was diluted with 10 mL sterilized 0.9% NaCl solution. The diluted supernatant (100 μL) was smeared on Plate Count Agar (PCA) and incubated at 35 °C for 48 h to stimulate bacterial growth. In addition, the supernatant was smeared onto PCA containing 1% NaCl and incubated at 20 °C for 168 h. Previous reports (Fujii 1985; Fukuda et al. 2012) had indicated that bacteria from seawater on fresh fish grew better on PCA containing 1% NaCl and incubated at 20 °C than on PCA without added NaCl at 35 °C. After incubation, six types of bacteria that varied in color and form and had that many colonies were selected. Each bacterial type was added to 1 mL of sterilized 0.9% NaCl solution, and the suspensions were diluted 10-, 100-, and 1000-fold; 100 μL of each diluted supernatant was then smeared on PCA containing 1% NaCl and incubated at 20 °C for 48 h to produce a pure culture. Colonies were extracted from these cultures and suspended in 1 mL of sterilized 0.9% NaCl solution. The suspensions were then diluted fivefold with 0.9% NaCl solution and used in the analyses.

AMP degradation was measured using 3.5 mL of a reaction mixture consisting of 0.5 mL AMP (Oriental Yeast Co., LTD., Tokyo, Japan) (25 mM), 2 mL buffer (50 mM maleic acid/Tris/NaOH, pH 7.5), 0.5 mL autoclaved enzyme solution, and 0.5 mL bacterial suspension. Ninety of the reaction mixtures and 30 of the control reaction mixtures (with autoclaved enzyme solution and without bacterial suspension, and without autoclaved enzyme solution or bacterial suspension) were incubated at 20 °C for 5 days, and the reaction was stopped every day in three samples of each bacterial suspension by the addition of 1.5 mL 10% perchloric acid. AMP degradation compounds were measured by HPLC using the same procedure previously outlined. To determine the bacterial count in the reaction mixture, 0.1 mL of the reaction mixture, before the addition of perchloric acid, was diluted by a factor of five with a sterilized solution of 0.9% NaCl. The diluted supernatant (100 μL) was then smeared on PCA plus 1% NaCl and incubated at 20 °C for 48 h. Bacterial counts were made daily.

Identification of AMP-degrading bacteria

Bacteria that showed AMP-degrading activity were identified by sequencing the 16S rRNA gene. Sequencing was performed by TechnoSuruga Laboratory Co., Ltd. (16S r-DNA-500 service; Shizuoka, Japan). Cells were lysed with achromopeptidase (Wako Pure Chemical Industries, Ltd., Osaka, Japan) to obtain a DNA template. PCR was conducted in a 25 μL reaction with 7.0 μL buffer (buffer 5.0 μL + dNTPs mixture 2.0 μL), 0.50 μL each of forward and reverse primers, 0.25 μL PrimeSTAR HS DNA polymerase (Takara Bio Inc., Shiga, Japan), and 0.50 μL template DNA. The primer set was 9F (5′-GAGTTTGATCCTGGCTCAG-3′) and 1406R (5′-CGGGCGGTGTGTAC-3′) (Nakagawa et al. 2002). The thermal cycle program consisted of 30 cycles of 98 °C for 10 s, 57 °C for 5 s, and 72 °C for 90 s. The BigDye Terminator v3.1 Cycle Sequencing Kit (Applied Biosystems, Foster City, CA, USA) was used for the sequencing reactions. The primer set was the 9F from PCR and 536R (5′-GTATTACCGCGGCTGCTG-3′), and the thermal cycle program was 25 cycles of 96 °C for 10 s, 50 °C for 5 s, and 96 °C for 1 min, followed by a final extension of 60 °C for 4 min. An ABI PRISM 3130xl Genetic Analyzer System (Applied Biosystems) was used with the DB-BA11.0 database (Techno Suruga Laboratory, Shizuoka, Japan) and ChromasPro 1.7 (Technelysium Pty Ltd., Tewantin, Australia). Species were identified with the bacterial identification software system DB-BA11.0 (Techno Suruga Laboratory Co., Ltd.) and international nucleotide sequence databases (GenBank/DDBJ/EMBL) using BLAST (Altschul et al. 1997). CLUSTALW (Thompson et al. 1994) was used to align sequences. Distances were calculated using the Kimura 2-parameter model (Kimura 1980), and trees were constructed with neighbor-joining (Saitou and Nei 1987) using CLUSTALW (Thompson et al. 1994). The bootstrap method (Felsenstein 1985) was used with 1000 replicates to evaluate support for the trees.

Changes in levels of ATP-related compounds and bacterial counts

Shrimp were stored at 4 °C for 7 days. Three were selected each day, and ATP-related compounds were extracted by placing 2 g of muscle from one shrimp in 5 mL of 10% perchloric acid; the mixture was shaken vigorously. The muscle was pelleted by centrifugation (11,509×*g*, 10 min, 5 °C), the supernatant was transferred to a 50 mL centrifuge tube, and 10% perchloric acid was added to obtain a volume of 25 mL. The mixture was neutralized using KOH, and the precipitate was removed by centrifugation (13,697×*g*, 5 min, 5 °C). The supernatant was then transferred to a 15 mL centrifuge tube and diluted with sterilized water to obtain a final volume of 10 mL. The solution was filtered (0.20 μm filter, Millex-LG), and ATP-related compounds were measured by HPLC using the same method outlined above. Each extract was measured daily for 7 days; extracts were stored at 4 °C over this period.

A freshly prepared bacterial suspension was used for counting bacteria in shrimp muscle. Extraction and incubation were carried out, as described above in "Bacterial degradation of AMP." For bacterial isolation and counting, 100 μL of the prepared suspension was spread on PCA and incubated at 35 °C for 48 h. In addition, the suspension was spread onto PCA containing 1% NaCl and incubated at 20 °C for 168 h. The number of bacterial colonies was counted on both plates every other day for 7 days.

Statistical analysis

Data were subjected to a one-way analysis of variance using the least significant difference method. *T* tests were used for pairwise comparisons; $P < 0.05$ was considered statistically significant. Analyses were conducted with Microsoft Excel, and figures were prepared using Microsoft Excel.

Results

Shrimp muscle enzyme degradation of AMP

Table 1 shows how 2.5 mM AMP was degraded by shrimp muscle enzyme extracts after 24 h. At 0 °C, AMP was still detected at a 2.2 mM concentration after the incubation period. In contrast, at 30 °C, the AMP concentration was 0.9 mM. The amount of remaining AMP decreased as the temperature changed from 0 to 30 °C, but it increased between 30 and 60 °C ($P < 0.05$). At 0 °C, 0.4 mM IMP was produced, whereas at 30 °C, 2.1 mM was produced. The amount of IMP increased from 0 to 30 °C and decreased from 30 to 60 °C ($P < 0.05$). AMP-degrading enzymes in shrimp muscle showed the highest activity at 30 °C (Table 1). No AdR was produced, indicating that over a 24-h period, shrimp muscle enzymes degraded AMP to IMP but did not degrade AMP to AdR. We investigated AMP-degrading enzyme activity over a period of 5 days at 20, 30, and 40 °C (Fig. 1); the temperatures with the highest levels of enzyme activity. At all temperatures, AMP was degraded to IMP within 24 h, and AMP concentration on day 0 was significantly different from that on day 1 ($P < 0.05$). AMP concentration at 20 °C was 1.1 mM on day 1 and decreased to 0.59 mM on day 5 ($P < 0.05$). AMP concentrations at 30 °C were 0.88 mM on day 1 and decreased to 0.61 mM on day 3 ($P < 0.05$), remained steady between days 3 and 4, ($P > 0.05$), and decreased to 0.51 mM on day 5 ($P < 0.05$). AMP concentration at 40 °C was between 1.1 and 1.0 mM from days 1–4 ($P > 0.05$), and it decreased 0.83 mM on day 5 ($P < 0.05$). AMP concentrations at 30 °C were significantly lower than those at 20 and 40 °C between days 1 and 5 ($P < 0.05$). IMP concentrations at 20 °C increased to 2.1 mM on day 2 ($P < 0.05$), remained steady between days 3 and 4 ($P > 0.05$), and decreased to 1.8 mM on day 5 ($P < 0.05$). IMP concentrations at 30 °C increased to 2.3 mM on day 2 ($P < 0.05$), remained steady between days 3 and 4 ($P > 0.05$), and decreased to 1.8 mM on day 5 ($P < 0.05$). IMP concentrations at 40 °C were between 1.9 and 2.0 mM from days 1–4 ($P > 0.05$) and decreased to 1.6 mM on day 5 ($P < 0.05$). IMP concentrations at 30 °C were significantly higher than those at 20 and 40 °C between days 1 and 5 ($P < 0.05$). We also found that the level of HxR increased after 2 days. Over the 5-day period, we confirmed IMP production but found no evidence of AdR production.

Table 1 Degradation of AMP by enzymes extracted from shrimp muscle

Temperature (°C)	AMP (mM)	IMP (mM)	AdR (mM)	HxR (mM)	Hx (mM)
0	2.2 ± 0.02	0.4 ± 0.01	0.1[a]	0.0[a]	0.0[a]
5	2.0 ± 0.04	0.8 ± 0.01	0.1	0.0	0.0
10	1.5 ± 0.03	1.4 ± 0.02	0.1	0.0	0.0
20	1.1 ± 0.01	1.9 ± 0.03	0.1	0.0	0.0
30	0.9 ± 0.01	2.1 ± 0.01	0.0	0.1	0.0
40	1.1 ± 0.01	1.9 ± 0.02	0.1	0.0	0.0
50	1.8 ± 0.03	1.1 ± 0.03	0.1	0.0	0.0
60	1.9 ± 0.04	0.9 ± 0.02	0.1	0.0	0.0

Data are presented as the mean ± SD of three replicates. The initial concentration of AMP was 2.5 mM. Concentrations are shown after 24 h of enzyme reaction

[a] SD of 0

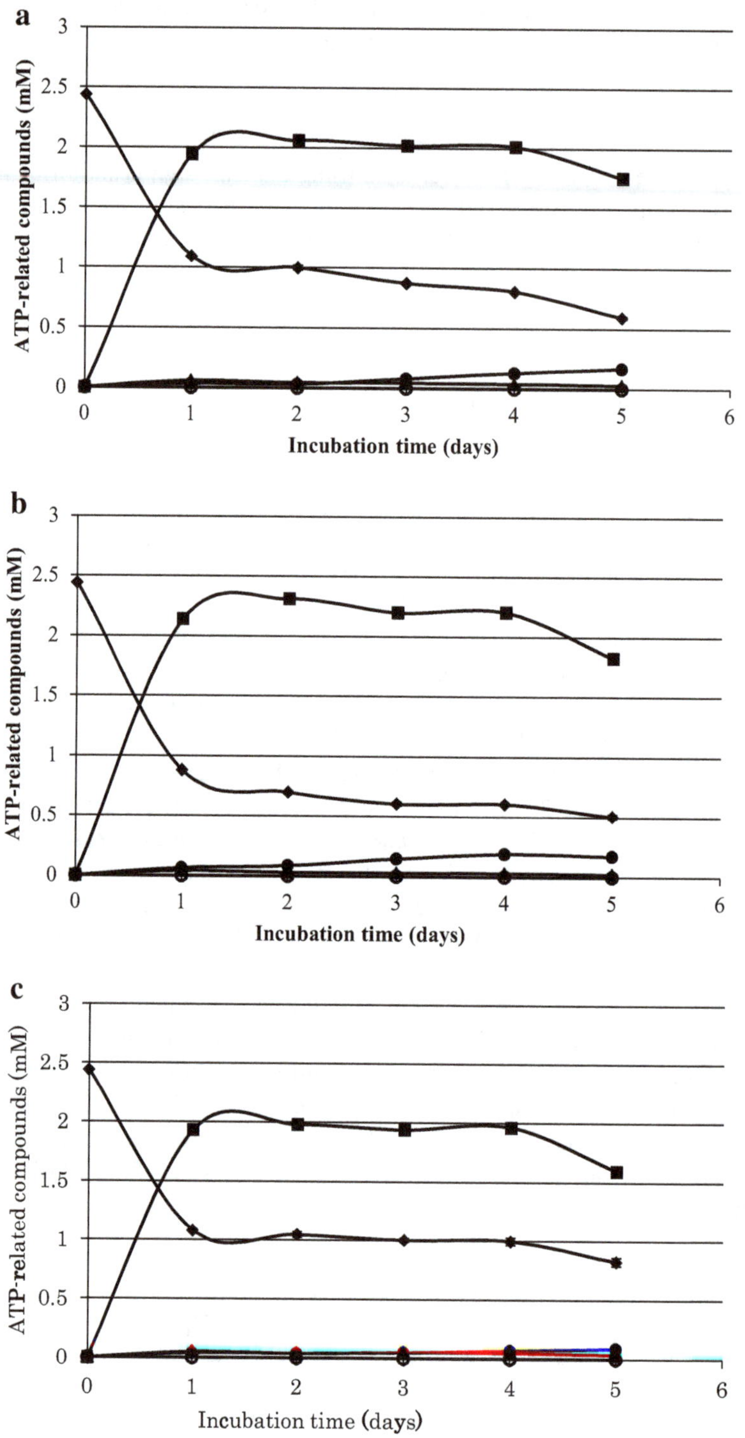

Fig. 1 Concentrations of AMP, IMP, AdR, HxR, and Hx from day 0 to day 5 during enzymatic degradation at 20 °C (**a**), 30 °C (**b**), and 40 °C (**c**) at pH 7.5. *Filled diamonds* AMP, *filled squares* IMP, *filled triangles* AdR, *filled circles* HxR, *open circles* Hx

Bacterial degradation of AMP

Few colonies grew on pure PCA, so we investigated AMP degradation using six strains (Table 2) isolated on PCA containing 1% NaCl. After 48 h of incubation, AMP, IMP, and AdR were not detected, but HxR and Hx were present in cultures with bacterial isolates 1, 5, and 6. Bacterial isolate 2 produced AdR and Hx but no

Table 2 Degradation of AMP by bacteria isolated from shrimp muscle

Bacterial isolates	AMP (mM)	IMP (mM)	AdR (mM)	HxR (mM)	Hx (mM)
No. 1	0	0	0	0.92	0.98
No. 2	0.42	0	1.40	0	0.04
No. 3	2.30	0	0	0	0
No. 4	2.30	0	0	0	0
No. 5	0	0	0	1.60	0.61
No. 6	0	0	0	0.88	1.10

The initial concentration of AMP was 2.5 mM. The concentrations after 48 h of bacterial degradation at pH 7.5 are shown
AMP was not degraded in the controls (one containing autoclaved enzyme solution without bacterial suspension, and the other without both autoclaved enzyme solution and bacterial suspension)

IMP or HxR. Our analysis indicated that bacterial isolates 1, 2, 5, and 6 could degrade AMP. In contrast, bacterial isolates 3 and 4 did not degrade AMP. Bacterial isolate 2 degraded AMP to AdR. Bacterial isolates 1, 5, and 6 degraded AMP to HxR and Hx, but it was not clear whether IMP or AdR was an intermediate. Although AdR was detected in cultures with bacterial isolate 2, the possibility that AMP was degraded to IMP could not be discounted, so the AMP degradation pathway in bacterial isolates 1, 2, 5, and 6 was investigated further. In addition, it was confirmed that AMP was not degraded in the control. We also assessed the number of bacteria in the reaction cultures over the 5-day period. In this experiment, no medium was used, but the autoclaved enzyme solution from shrimp muscle served as the nutrient source.

Bacterial isolate 1 almost completely degraded AMP by day 2. No IMP was detected, but AdR was present. The level of AdR increased to 0.74 mM at day 2, but decreased at day 3. HxR increased to 1.3 mM at day 3, but then decreased by day 5. Hx increased after day 2 and reached 2.1 mM by day 5 (Fig. 2a). Bacterial isolate 2 slowly degraded AMP to day 4. IMP was not detected, but AdR was present. The level of AdR increased to 2.1 mM at day 4 and then decreased at day 5. The level of Hx increased after day 2 to a maximum of 0.80 mM at days 3 and 4, and then decreased at day 5. HxR was not detected (Fig. 2b). Bacterial isolate 5 slowly degraded AMP to a minimum of 1.1 mM on day 5. IMP was not detected, but AdR was present at 0.19 mM on day 3. The level of HxR increased after day 4 to a maximum of 0.82 mM on day 5. Hx increased after day 2 to a maximum of 0.64 mM on day 3, then decreased slowly to day 5 (Fig. 2c). Bacterial isolate 6 almost completely degraded AMP by day 2. IMP was not detected. AdR was detected and increased after day 1 to a maximum of 0.52 mM on day 2, but decreased after day 3. HxR increased after day 1 to a maximum of 0.90 mM on day 3 and then decreased to day 5. Hx increased after day 2 to a maximum of 3.2 mM on day 4 and then decreased on day 5 (Fig. 2d). These results indicated that the bacteria degraded AMP to AdR.

Initial bacterial concentrations and growth during the experiment varied by isolate. The initial counts of isolates 1, 2, 5, and 6 were 3.3, 6.0, 1.7, and 4.2 log CFU/mL, respectively. By day 4, the isolates had counts of 11, 10, 5.3, and 7.1 log CFU/mL, respectively (Fig. 3).

Identification of AMP-degrading bacteria

Genetic analysis suggested that isolates 1 and 6 were probably *Shewanella baltica* NCTC 10735^{T} (98.6 and 98.4% homology, respectively), while isolate 5 was most similar to *Shewanella xiamenensis* S4^{T} (97.6% homology), and isolate 2 was very similar to *Exiguobacterium oxidotolerans* T-2-2^{T} (99.8% homology, Table 3). For each phylogenetic analysis, the top ten matches were included. Outgroup taxa included species of *Aeromonas* for isolates 1, 5, and 6 and *Salirhabdus* for isolate 5. These bacteria were the closest match from a different bacterial genus. Isolates 1 and 6 had lower homology rates (Table 3) and bootstrap support (74 and 65%, respectively; Fig. 4) with *S. baltica* than isolate 2 did with *E. oxidotolerans* (100% bootstrap support). Isolate 5 was very weakly supported (40%) as sister to a clade including *S. xiamenensis*, which was its closest match in the nucleotide databases. The 16S sequences for isolates 1, 2, 5, and 6 were submitted to the DNA Data Bank of Japan (Table 3).

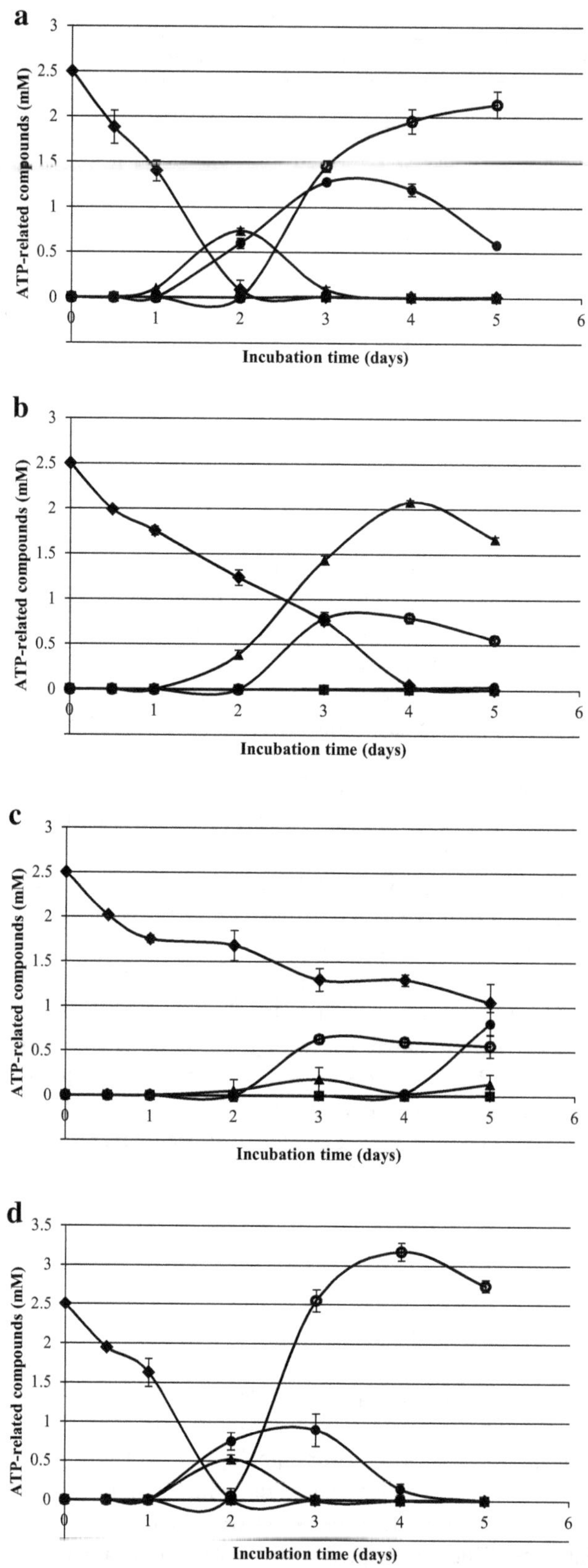

Fig. 2 Concentrations of AMP, IMP, AdR, HxR, and Hx from days 0 to 5 during bacterial degradation at 20 °C, pH 7.5. Bacterial isolate 1 (**a**), 2 (**b**), 5 (**c**), and 6 (**d**). *Filled diamonds* AMP, *filled squares* IMP, *filled triangles* AdR, *filled circles* HxR, *open circles* Hx

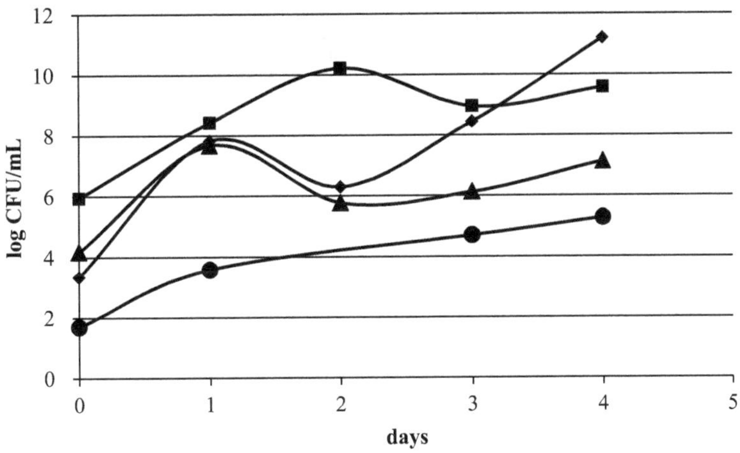

Fig. 3 Counts of AMP-degrading bacteria on Plate Count Agar containing 1% NaCl and incubated at 20 °C. *Filled diamonds* LC160132 (isolate 1), *filled squares* LC160133 (isolate 2), *filled circles* LC160134 (isolate 5), *filled triangles* LC160135 (isolate 6)

Changes in levels of ATP-related compounds and bacterial counts

Changes in the levels of ATP-related compounds in shrimp extracts were monitored for 7 days at 4 °C (Fig. 5). The initial level of ATP (4.3 μmol/g) approached zero (0.26 μmol/g) by day 2. The initial amount of ADP (6.6 μmol/g) decreased to 1.1 μmol/g on day 2. Thereafter, the level of ADP decreased slowly to 0.58 μmol/g on day 7. The initial amount of AMP was 1.2 μmol/g, which increased to 4.4 μmol/g on day 2 and then decreased to 1.5 μmol/g on day 7. The initial amount of IMP was 0.11 μmol/g, which increased to 7.1 μmol/g on day 4 and then decreased. IMP was the most abundant ATP-related compound after day 3. AdR was not detected during the 7-day experiment. HxR was not detected initially, but was present on day 2 and rose to a maximum level of 2.0 μmol/g on day 6. Hx was not detected initially, but was present on day 2 and reached a maximum level of 1.7 μmol/g on day 7. Overall, ATP degradation in shrimp increased with time, and degradation to AdR was not observed; all the AMP degraded to IMP.

Changes in bacterial numbers in shrimp extracts over the 7-day period are shown in Fig. 6. A bacterial count of <2.0 log CFU/g was found from samples incubated at 35 °C on day 7. In contrast, the bacterial count for samples incubated at 20 °C was initially 3.5 log CFU/g, and after day 2, the bacterial count increased and reached 7.2 log CFU/g by day 7. It was found that there were increases in the numbers of bacteria that can grow in the presence of NaCl.

Discussion

Shrimp muscle enzyme degradation of AMP

Shrimp muscle extracts demonstrated enzymatic activity that degraded AMP to IMP (Table 1; Fig. 1); however, AdR production from AMP degradation was detected only at a low level and was significantly less than IMP production from AMP degradation ($P < 0.05$). AMP degradation to IMP is catalyzed by AMP deaminase through excision of the amino group. When AMP is degraded to AdR, AMP-5′-nucleotidase excises the phosphate group. IMP accumulates in fish owing to an inherently high AMP deaminase activity (Yokoyama and Sakaguchi 1998). AMP is reported to degrade to both IMP and AdR in shrimp (Koseki et al. 2006; Koyama et al. 2008), presumably through both AMP deaminase and AMP-5′-nucleotidase activities. Both IMP and AdR have been detected in the pink shrimp *Pandalus hypsinotus*, and it was reported that the primary degradation pathway in this species is AMP to AdR mediated by AMP-5′-nucleotidase (Arai and Terasaki 1966). Therefore, the activity of AMP-5′-nucleotidase is higher than that of AMP deaminase. In the Japanese spiny lobster *Panulirus japonicus*, AMP was degraded to both IMP and AdR, although IMP

Table 3 Sequences used in genetic analyses to identify bacterial isolates from shrimp

Species name	Homology rate (%)[a]	Accession number[b]
Isolate 1	–	LC160132
Shewanella baltica NCTC 10735[T]	98.6	AJ000214
Shewanella aestuarii SC18[T]	98.2	JF751004
Shewanella pacifica KMM3597[T]	97.1	AF500075
Shewanella japonica KMM3299[T]	96.8	AF145921
Shewanella gaetbuli TF-27[T]	95.5	AY190533
Shewanella putrefaciens ATCC8071[T]	96.0	X82133
Shewanella frigidimarina ACAM591[T]	95.1	U85903
Shewanella vesiculosa M7[T]	94.9	AM980877
Shewanella livingstonensis LMG19866[T]	94.9	AJ300834
Aeromonas sobria ATCC 43979[T,c]	94.9	X74683
Isolate 2	–	LC160133
Exiguobacterium oxidotolerans T-2-2[T]	99.8	AB105164
Exiguobacterium artemiae 9AN[T]	96.9	AM072763
Exiguobacterium acetylicum DSM20416[T]	96.9	DQ019167
Exiguobacterium indicum HHS_31[T]	96.7	AJ846291
Exiguobacterium soli DVS3y[T]	96.3	AY864633
Exiguobacterium antarcticum DMS14480[T]	96.3	DQ019164
Exiguobacterium undae L2[T]	95.9	AJ344151
Exiguobacterium sibiricum 255-15[T]	96.7	CP001022
Exiguobacterium aurantiacum DSM6208[T]	96.9	DQ019166
Salirhabdus euzebyi CVS-14[T,c]	88.8	AM292417
Isolate 5	–	LC160134
Shewanella xiamenensis S4[T]	97.6	FJ589031
Shewanella hafniensis P010[T]	97.0	AB205566
Shewanella putrefaciens ATCC8071[T]	96.3	X82133
Shewanella oneidensis MR-1[T]	96.1	AE014299
Shewanella decolorationis CCTCC_M_203093[T]	95.0	AJ609571
Shewanella vesiculosa M7[T]	96.0	AM980877
Shewanella livingstonensis LMG19866[T]	95.7	AJ300834
Shewanella frigidimarina ACAM591[T]	94.5	U85903
Shewanella donghaensis LT17[T]	94.7	AY326275
Shewanella sairae SM2-1[T]	95.2	AB081762
Shewanella marinintestina IK-1[T]	95.2	AB081757
Shewanella schlegeliana HRKA1[T]	94.9	AB081760
Shewanella gaetbuli TF-27[T]	94.9	AY190533
Shewanella aestuarii SC18[T]	94.5	JF751004
Aeromonas molluscorum 848[T,c]	93.7	AY532690
Isolate 6	–	LC160135
Shewanella baltica NCTC 10735[T]	98.4	AJ000214
Shewanella aestuarii SC18[T]	98.0	JF751004
Shewanella pacifica KMM3597[T]	96.9	AF500075
Shewanella vesiculosa M7[T]	94.7	AM980877
Shewanella livingstonensis LMG19866[T]	94.7	AJ300834
Shewanella frigidimarina ACAM591[T]	94.9	U85903
Shewanella putrefaciens ATCC8071[T]	95.7	X82133

Table 3 continued

Species name	Homology rate (%)[a]	Accession number[b]
Shewanella gaetbuli TF-27[T]	95.3	AY190533
Aeromonas sobria ATCC 43979[T,c]	94.7	X74683

[T] Type strain

[a] Similarity determined by BLAST search

[b] Accession number in GenBank

[c] Outgroup taxon used for phylogenetic reconstruction

accumulated to a greater extent, indicating that AMP deaminase activity was higher than AMP-5′-nucleotidase activity (Shirai et al. 1996). In the shovel-nosed lobster *Ibacus ciliatus*, a higher level of AdR than IMP accumulated, indicating higher AMP-5′-nucleotidase activity than AMP deaminase activity (Shirai et al. 1996). In *Pandalus borealis*, AMP deaminase activity was high, while AMP-5′-nucleotidase activity was minimal (Fijisawa and Yoshino 1987). These reports indicate that the relative activities of AMP deaminase and AMP-5′-nucleotidase vary by species. In the present study, AdR production was significantly lower than IMP production ($P < 0.05$; Fig. 1). This indicates that AMP deaminase activity was higher than AMP-5′-nucleotidase activity.

The pathway and rate of AMP degradation depend on temperature (Koyama et al. 2008). In the vannamei shrimp *Litopenaeus vannamei* (syn. *Penaeus vannamei*), HxR and Hx accumulated at 0–5 °C, but IMP accumulated at 20–40 °C, indicating that different AMP-degrading enzymes are active at these temperature ranges (Koyama et al. 2008). Therefore, in this study, the amount of IMP and AdR that were generated by shrimp enzymes was measured at different temperatures on day 5 to investigate temperature dependency of AMP deaminase and AMP-5′-nucleotidase activity in shrimp (Fig. 7). From Fig. 7, it is clear that AMP deaminase activity was lowest at 0 °C and highest at 30 °C. Activity decreased when the temperature increased above 30 °C. In contrast, AMP-5′-nucleotidase activity was the highest at 0 °C and decreased to 30 °C, before increasing again above 30 °C. These results are similar to those reported for vannamei shrimp, in which AMP deaminase and AMP-5′-nucleotidase activities were temperature dependent (Koyama et al. 2008). AMP deaminase and AMP-5′-nucleotidase are AMP-degrading enzymes, and it is likely that AMP is degraded to IMP by AMP deaminase and further to AdR by AMP-5′-nucleotidase. Figure 7a shows AMP deaminase activity, and Fig. 7b shows AMP-5′-nucleotidase activity. As both enzymes are different, the substrates are different as well. Overall, we concluded that AMP deaminase and AMP-5′-nucleotidase are active in shrimp at certain temperatures. In addition, as AMP deaminase activity is higher than AMP-5′-nucleotidase activity, almost all AMP is degraded to IMP in shrimp.

Bacterial degradation of AMP

We demonstrated that *S. baltica*, *E. oxidotolerans*, and *S. xiamenensis* closely matched the isolated bacteria that degraded AMP in shrimp. *Shewanella baltica* has been isolated from the oyster *Crassostrea gigas* and from seawater in Delaware Bay (Richards et al. 2008), plaice *Pleuronectes platessa*, cod *Gadus morhua*, and flounder *Platichthys flesus* from the Baltic Sea (Vogel et al. 2005), and tuna *Thunnini* sp., swordfish *Xiphias gladius* (Serio et al. 2014), and vannamei shrimp *Penaeus vannamei* (Mace et al. 2014) from markets. *Exiguobacterium oxidotolerans* has been isolated from a drain in a fish processing plant in Hokkaido (Yumoto et al. 2004) and from sea mud (Chen et al. 2010). *Shewanella xiamenensis* has been isolated from sea floor sediment (Ng et al. 2014; Huang et al. 2010). These three species of bacteria are present on the sea floor where shrimp live, so it was not surprising to encounter them in our samples.

In this study, we investigated whether the bacteria from shrimp degrade AMP. We showed that bacterial isolates 1, 2, and 6 from shrimp could degrade AMP to AdR in muscle (Table 2; Fig. 2). Isolates 1, 2, and 6 degraded AMP not to IMP but to AdR. On the other hand, the amount of AdR in isolate 5 increased slightly at day 3, after which IMP was no longer detected. The amount of Hx in isolate 5 increased after day 2, indicating that isolate 5 degrades AMP to Hx immediately and has an enzyme that shows very strong activity with AMP. The bacterial counts in the reaction mixture increased with time (Fig. 3), which indicated that AMP

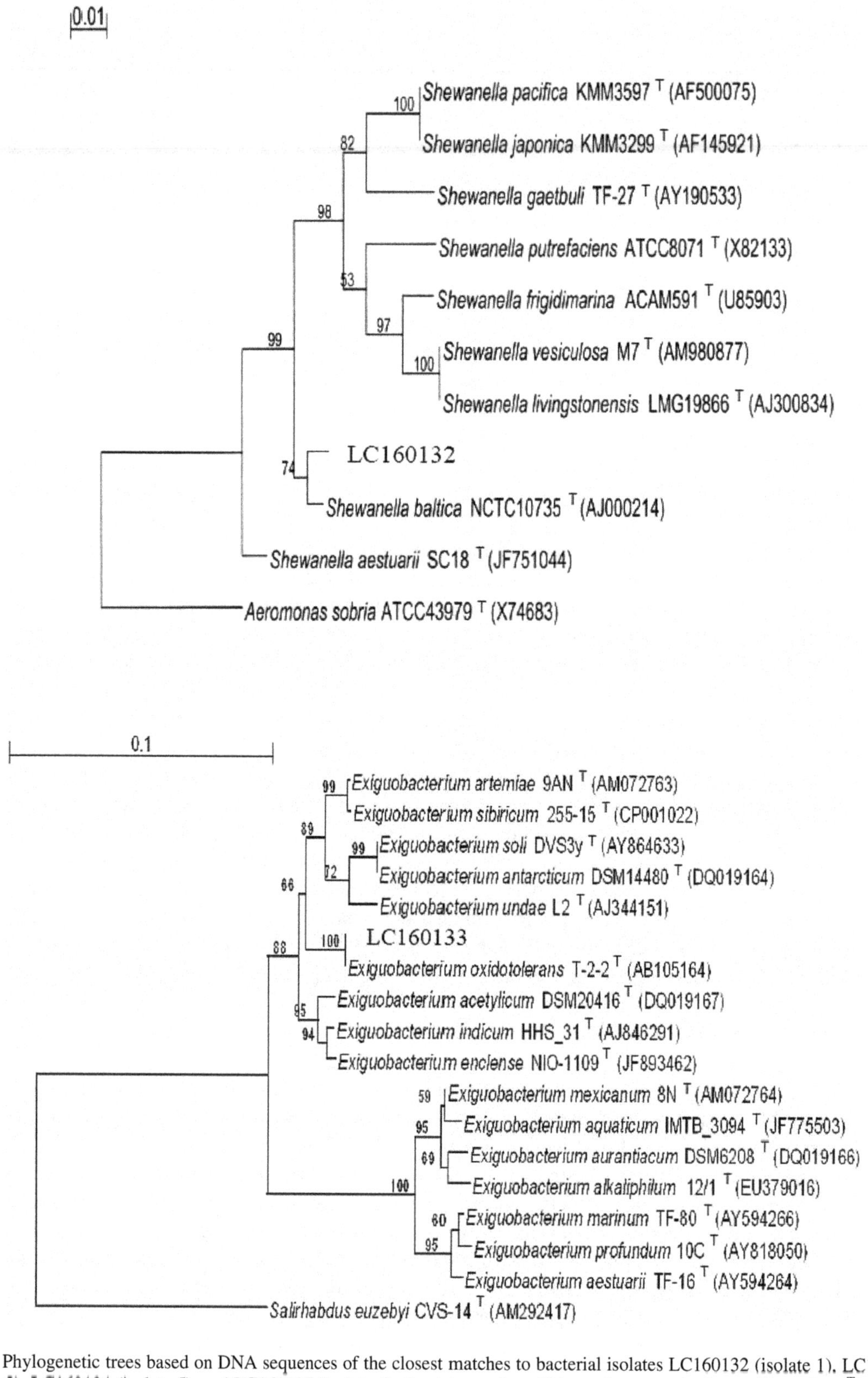

Fig. 4 Phylogenetic trees based on DNA sequences of the closest matches to bacterial isolates LC160132 (isolate 1), LC160133 (isolate 2), LC160134 (isolate 5), and LC160135 (isolate 6). Bootstrap values (%) are shown on the internal branches. [T]indicates type strain

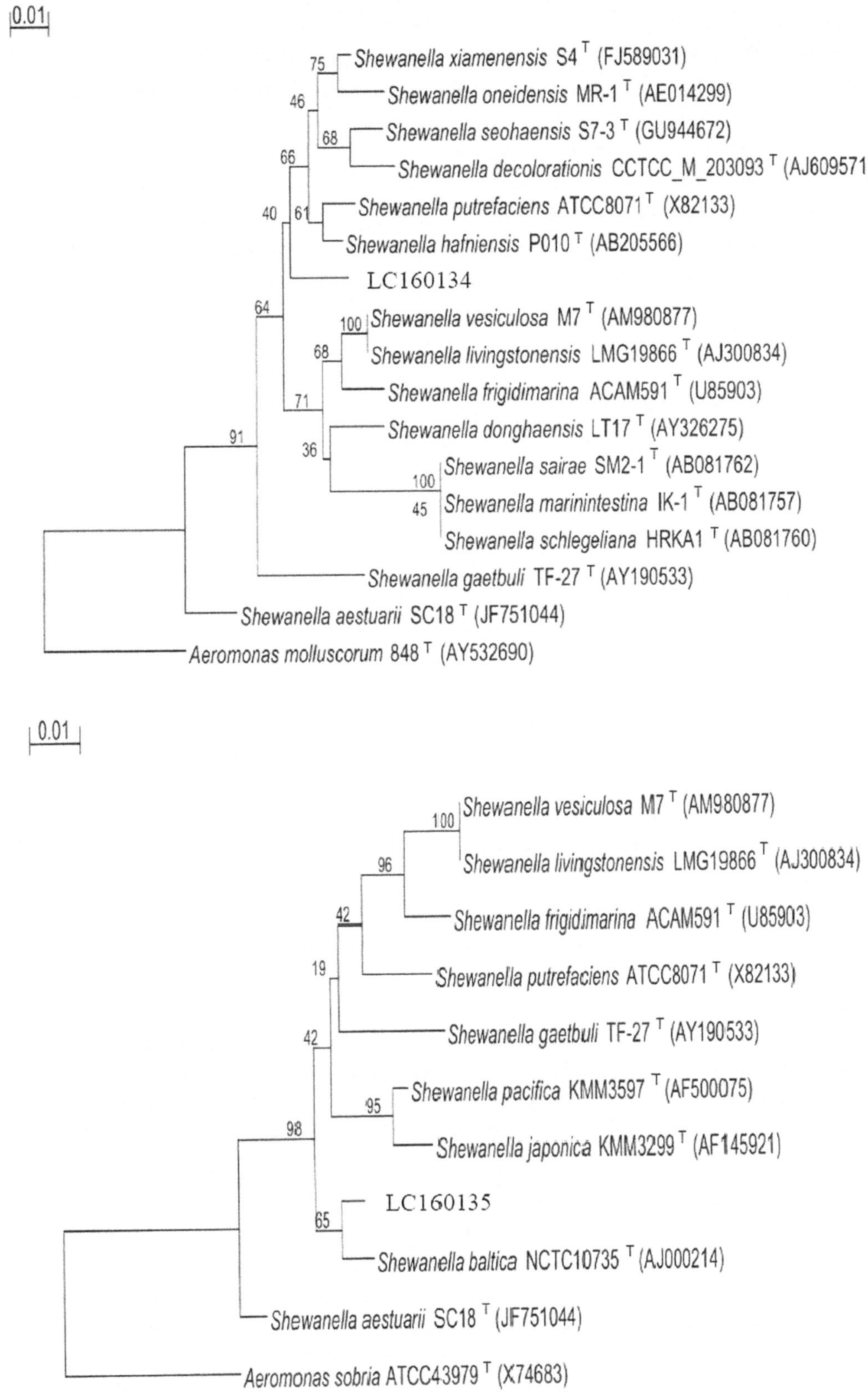

Fig. 4 continued

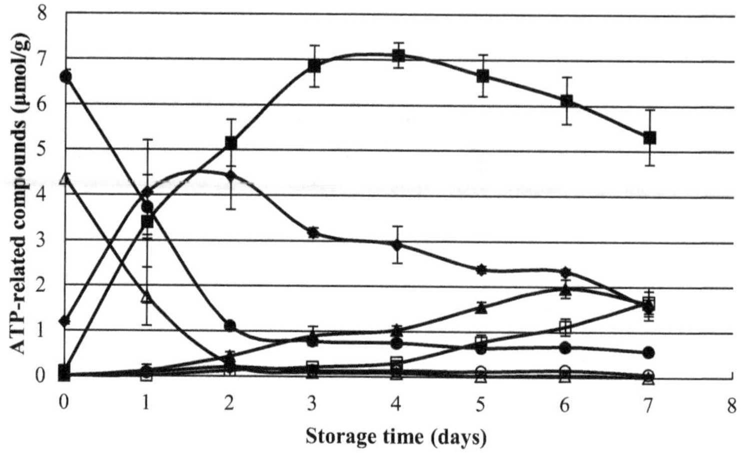

Fig. 5 Levels of ATP-related compounds at different times in shrimp stored at 4 °C. *Bars* denote the standard deviation ($n = 9$). *Open triangles* ATP, *filled circles* ADP, *filled diamonds* AMP, *filled squares* IMP, *open circles* AdR, *filled triangles* HxR, *open squares* Hx

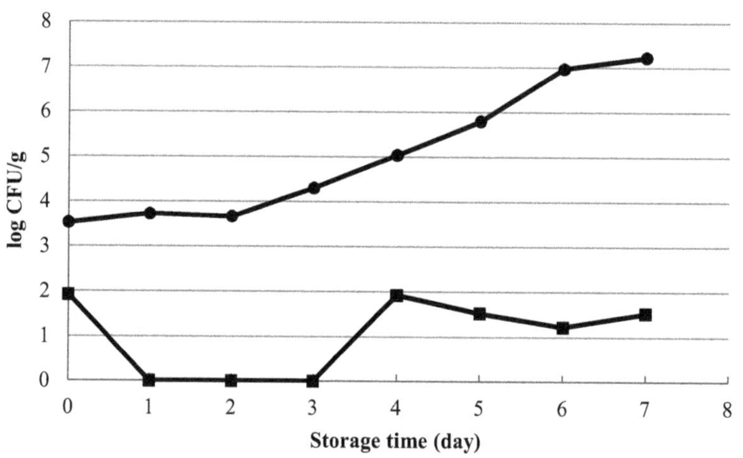

Fig. 6 Counts of bacteria on Plate Count Agar at 35 °C (*filled squares*) and on Plate Count Agar containing 1% NaCl at 20 °C (*filled circles*)

degradation was dependent on the bacteria, because only the bacteria isolated from shrimp degraded AMP. When AMP is degraded to AdR, the phosphate group is removed by a phosphatase, such as AMP-5′-nucleotidase. This enzyme removes phosphate from a nucleotide composed of a base + carbohydrate + phosphate. In this study, there is a possibility that the isolated bacteria from shrimp yielded phosphatases that removed the phosphate group from AMP.

Alkaline phosphatase has been found in marine bacteria isolated from the sea urchin *Strongylocentrotus intermedius* (Nedashkovskaya et al. 2005a), green alga *Ulva fenestrata* (Nedashkovskaya et al. 2005b), and soft coral *Paragorgia arborea* (Nedashkovskaya et al. 2005c). In addition, an alkaline phosphatase that was active at low temperatures was refined from a species of *Shewanella* (Ishida et al. 1998; Tsuruta et al. 1998). The *Shewanella* spp. identified in this study may also have alkaline phosphatase. However, the alkaline phosphatase from the *Shewanella* sp. studied by Tsuruta and Aizono (1999) has been shown to remove phosphoric acid only from phosphotyrosine, not from ATP, ADP, or AMP. The species of *Shewanella* identified in our study had different activity, but as there are many different species within the genus, multiple phosphatases are likely. At present, the type(s) of phosphatase(s) in the three species identified in this study are not known, although several studies (Ishida et al. 1998; Murakawa et al. 2002; Yasuda et al. 2002; Yamaguchi et al. 2006; Tsuruta et al. 2010) have examined phosphatase production in *Shewanella* spp. The species of bacteria identified in our study probably produce a phosphatase that degrades AMP to AdR in shrimp. A

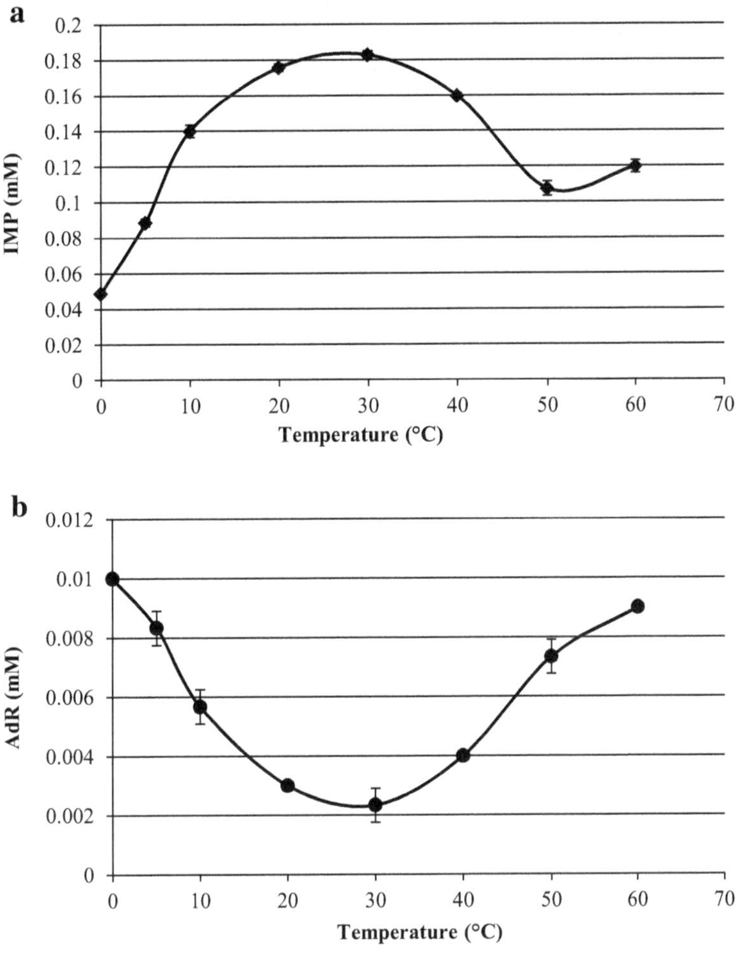

Fig. 7 Effects of temperature on the amount of IMP (**a**, *filled diamonds*) and AdR (**b**, *filled circles*) generated by shrimp enzymes from muscle by day 5 of incubation. *Bars* denote the standard deviation (n = 3). IMP is produced from degradation by AMP deaminase, and AdR is produced from degradation by AMP-5′-nucleotidase

deaminase is required to degrade AMP to IMP, and although a glucosamine-6-phosphate deaminase has been extracted from *Vibrio cholerae* non-O1 (Fujishima et al. 1996, 1997), it is unlikely that this deaminase affects AMP degradation, because IMP was not detected (Table 2; Fig. 2).

Changes in levels of ATP-related compounds and bacterial counts

We found that ATP and ADP were degraded relatively quickly and that AMP and IMP accumulated in the shrimp (Fig. 5). Shrimp and lobsters fall into two groups with respect to AMP degradation: in one group, AMP degrades to IMP, and in the other, AMP degrades to AdR (Koseki et al. 2006; Koyama et al. 2008). AdR was not detected in this study, so *P. japonicus* belongs to the first group. In the vannamei shrimp *P. vannamei*, the level of IMP accumulation depended on temperature (Koyama et al. 2008). IMP has been detected in shrimp held at −1 to 0 °C (Matsumoto and Yamanaka 1990) or at 2 °C (Nakamura and Ishikawa 1986). In this study, IMP was detected in shrimp extracts at 5 °C, indicating that AMP was degraded to IMP, because AMP deaminase activity was higher than AMP-5′-nucleotidase activity.

We found that the bacteria from the shrimp degraded AMP to AdR (Table 2; Fig. 2) and that the bacterial numbers increased after 2 days of incubation on agar plates with 1% NaCl at 20 °C (Fig. 6). However, AdR was not detected in shrimp tissue stored for up to 7 days at 4 °C (Fig. 5). These apparently contradictory results are due to the low level of enzyme produced by bacteria at 4 °C in comparison with the high level of active enzyme in shrimp muscle at the same temperature. Alternatively, it is possible that the types of bacteria that can grow in the presence of NaCl but do not degrade AMP increased in shrimp muscle.

Our results indicate that suppressing the growth of bacteria during storage is necessary to prevent AMP degradation to AdR, which negatively affects shrimp flavor. Bacteria are not generally present in muscle; however, after death, bacteria may invade the muscle from organs and the shrimp's external environment. Therefore, it is important to remove organs and wash the surface of the shrimp as soon as possible.

Conclusions

In this study, we investigated AMP degradation in shrimp using enzymes extracted and bacteria isolated from shrimp muscle. An enzyme with high AMP deaminase activity in shrimp degraded AMP to IMP. In contrast, bacterial isolates 1, 2, and 6 degraded AMP to AdR with a phosphatase. Isolate 5 produced an enzyme with high activity that immediately degraded AMP to Hx. The amount of enzyme that was produced from bacteria was less than that in shrimp muscle, so no AdR but considerable IMP was confirmed when shrimp was stored for 7 days at 5 °C, demonstrating that refrigeration is important for preserving the taste component in shrimp and preventing bacteria from degrading IMP.

Acknowledgements The gene identification analysis was supported by TechnoSuruga Laboratory Co., Ltd. (Shizuoka, Japan).

References

Altschul SF, Madden TF, Schäffer AA, Zhang J, Zhang Z, Miller W, Lipman DJ (1997) Gapped BLAST and PSI-BLAST: a new generation of protein database search programs. Nucleic Acids Res 25:3389–3402

Arai K, Terasaki M (1966) Nucleotides in the muscles of marine invertebrates. Nippon Suisan Gakk 32:174–179

Chen S, Hong Y, Shao Z, Liu Z (2010) A cold-active β-glucosidase (Bgl1C) from a sea bacteria *Exiguobacterium oxidotolerans* A011. World J Microbiol Biotechnol 26:1427–1435

Felsenstein J (1985) Confidence limits on phylogenies: an approach using the bootstrap. Evolution 39:783–791

Fijisawa K, Yoshino M (1987) Activities of adenylate-degrading enzymes in muscles from vertebrates and invertebrates. Comp Biochem Physiol B Biochem Mol Biol 86:109–112

Fujii T (1985) Comparative studies on methods for determination of bacterial counts in seafoods. I. Comparisons of media, incubation temperatures and plating methods. Bull Tokai Reg Fish Res Lab 118:71–79

Fujisawa K, Yoshino M (1985) Distribution of AMP deaminase and adenosine deaminase in muscle of mammals, fish, molluscs and crustaceans. J Jpn Soc Nutr Food Sci 38:322–326

Fujishima S, Yamano N, Wang J (1996) Degradation of chitin by marine bacteria *Vibrio* sp. Chitin Chitosan Res 2:120–121

Fujishima S, Yamano N, Maruyama A, Higashihara T (1997) *N*-Acetylglucosamine metabolism of marine bacteria. Chitin Chitosan Res 3:164–165

Fukuda T, Furushita M, Shiba T (2012) Viable counts of bacteria determined on food fishes by the standard-agar-plate method incubated at 35 °C and nutrient-agar-plate method incubated at 20 °C. J Natl Fish Univ 60:183–188

Hamano K (2015) Strategies to improve aquaculture and disease controlling agent of prawn in Japan. The report of Setouchi, vol 21, pp 10–12. http://feis.fra.affrc.go.jp/event/h26seika_happyoukai/houkoku/youshi3_hamano.pdf. Accessed May 2015

Hayashi M, Nakata K (2003) Effect of contaminating bacteria on the inosinic acid content of chicken meat. J Home Econ 54:441–448

Huang J, Sun B, Zhang X (2010) *Shewanella xiamenensis* sp. nov., isolated from coastal sea sediment. Int J Syst Evol Microbiol 60:1585–1589

Ishida Y, Tsuruta H, Tsuneta ST, Uno T, Aizono Y, Watanabe K (1998) Characteristics of psychrophilic alkaline phosphatase. Biosci Biotechnol Biochem 62:2246–2250

Kimura M (1980) A simple method for estimating evolutionary rates of base substitutions through comparative studies of nucleotide sequences. J Mol Evol 16:111–120

Kobayashi H, Kadowaki M, Fujimura S (2012) Regulation of muscular free glutamate content by the nutrition of animal diet. Proc Jpn Soc Anim Nutr Metab 56:1–12

Koseki S, Kitakami S, Kato N, Arai K (2006) Rigor mortis of fish and shellfish and evaluation of freshness of their muscles as K value. J School Mar Sci Technol Tokai Uni 4:31–46

Koyama N, Matsukawa M, Shimada M, Sato R (2008) The change of ATP and its related compounds of the *Penaeus vannamei* muscle and effect of the sense of taste. Nippon Suisan Gakk 74:1068–1074

Mace S, Cardinal M, Jaffres E, Cornet J, Lalanne V, Chevalier F, Serot T, Pilet MF, Dousset X, Joffraud JJ (2014) Evaluation of the spoilage potential of bacteria isolated from spoiled cooked whole tropical shrimp (*Penaeus vannamei*) stored under modified atmosphere packaging. Food Microbiol 40:9–17

Matsumoto M (1996) Freshness and extractive components of marine product invertebrate animals. J Cook Sci Jpn 29:306–313

Matsumoto M, Yamanaka H (1990) Changes in contents of glycolytic metabolites and free amino acids in the muscle of kuruma prawn during storage. Nippon Suisan Gakk 56:1515–1520

Matsumoto M, Yamanaka H (1991) Influences of antibiotics chloramphenicol on post-mortem biochemical changes in the muscle of Kuruma Prawn during storage. Nippon Suisan Gakk 57:2291–2297

Matsumoto M, Yamanaka H, Hatae K (1991) Effect of "aria" treatments on the biochemical changes in the Kuruma Prawn muscle. Nippon Suisan Gakk 57:1383–1387

Motoo H (1990) Resource and use of a shrimp. The biology of tiger shrimps. J Food Sci 145:24–30

Murakawa T, Yamagata H, Tsuruta H, Aizono Y (2002) Cloning of cold-active alkaline phosphatase gene of a psychrophile, *Shewanella* sp., and expression of the recombinant enzyme. Biosci Biotechnol Biochem 66:754–761

Nakagawa Y, Sakane T, Suzuki M, Hatano K (2002) Phylogenetic structure of the genera Flexibacter, Flexithrix, and Microscilla deduced from 16S rRNA sequence analysis. J Gen Appl Microbiol 48:155–165

Nakamura K, Ishikawa S (1986) Changes in freshness of kuruma prawn muscle during chill-storage. Bull Tokai Reg Fish Res Lab 120:69–72

Nedashkovskaya OI, Frolova GM, Mikhailov VV, Kim SB, Bae KS, Lysenko AM, Lee DH, Kim IS (2005a) *Gramella echinicola* gen. nov., sp. nov., a novel halophilic bacterium of the family Flavobacteriaceae isolated from the sea urchin *Strongylocentrotus intermedius*. Int J Syst Evol Microbiol 55:391–394

Nedashkovskaya OI, Frolova GM, Mikhailov VV, Kim SB, Bae KS, Lysenko AM (2005b) *Bizionia paragorgiae* gen. nov., sp. nov., a novel member of the family Flavobacteriaceae isolated from the soft coral *Paragorgia arborea*. Int J Syst Evol Microbiol 55:375–378

Nedashkovskaya OI, Shevchenko LS, Frolova GM, Mikhailov VV, Kim SB, Lee DH, Lee KH, Bae KS, Lysenko AM (2005c) *Roseivirga ehrenbergii* gen. nov., sp. nov., a novel marine bacterium of the phylum 'Bacteroidetes', isolated from the green alga *Ulva fenestrata*. Int J Syst Evol Microbiol 55:231–234

Ng IS, Chen T, Lin R, Zhang X, Ni C, Sun D (2014) Decolorization of textile azo dye and Congo red by an isolated strain of the dissimilatory manganese-reducing bacterium *Shewanella xiamenensis* BC01. Appl Microbiol Biotechnol 98:2297–2308

Richards GP, Watson MA, Crane EJ III, Burt IG, Bushek D (2008) *Shewanella* and *Photobacterium* spp. in oysters and seawater from the Delaware Bay. Appl Environ Microbiol 74:3323–3327

Saitou N, Nei M (1987) The neighbor-joining method: a new method for reconstructing phylogenetic trees. Mol Biol Evol 4:406–425

Seki H, Hamada-Sato N (2014) Identification of bacteria that contribute to IMP degradation in horse mackerel. J Food Process Technol 5:363

Serio A, Fusella GC, Chaves LC, Sacchetti G, Paparella A (2014) A survey on bacteria isolated as hydrogen sulfide-producers from marine fish. Food Control 39:111–118

Shirai T, Hirakawa Y, Koshikawa Y, Toraishi H, Suzuki T, Hirano T, Terayama M (1996) Taste components of Japanese spiny and shovel-nosed lobsters. Fish Sci 62:283–287

Srirangsan P, Hamada-Sato N, Kawai K, Watanabe M, Suzuki T (2010) Improvement of fish freshness determination method by the application of amorphous freeze-dried enzymes. J Agric Food Chem 58:12456–12461

Thompson JD, Higgins DG, Gibson TJ (1994) CLUSTAL W: improving the sensitivity of progressive multiple sequence alignment through sequence weighting, positions-specific gap penalties and weight matrix choice. Nucleic Acids Res 22:4673–4680

Tsuruta H, Aizono Y (1999) Enzymatical properties of psychrophilic phosphatase I. J Biochem 125:690–695

Tsuruta H, Tsuneta ST, Ishida Y, Uno T, Aizono Y, Watanabe K (1998) Purification and some characteristics of phosphatase of a psychrophile. J Biochem 123:219–225

Tsuruta H, Mikami B, Higashi T, Aizono Y (2010) Crystal structure of cold-active alkaline phosphatase from the psychrophile *Shewanella* sp. Biosci Biotechnol Biochem 74:69–74

Vogel BF, Gram L, Venkateswaran K, Satomi M (2005) Identification of *Shewanella baltica* as the most important H_2S-producing species during iced storage of Danish marine fish. Appl Environ Microbiol 71:6689–6697

Yamaguchi H, Tsuruta H, Yamagata H, Aizono Y (2006) Enzymatic characteristics of cold-active alkaline phosphatase. Mem Grad Sch Sci Technol Kobe Univ 24:A23–A31

Yamanaka H (2008) Fish brands 11) live prawn. Food Sci 51:31–34

Yasuda J, Nakasine K, Kato C, Horikoshi K (2002) Cloning and sequencing of the phoA gene encoding alkaline phosphatase from a deep-sea piezophilic bacterium, *Shewanella violacea* strain DSS12. JAMSTEC Rep Res Dev 45:69–74

Yokoyama Y, Sakaguchi M (1998) ATP metabolism in muscle after death of fish/shellfish and related items. Jpn Soc Comp Physiol Biochem 15:193–200

Yokoyama Y, Sakaguchi M, Azuma Y, Kawai F, Kanamori M (1996) Postmortem changes of ATP and its related compounds in oyster tissues in the presence of antibiotic chloramphenicol. Fish Sci 62:312–316

Yumoto I, Hishinuma-Narisawa M, Hirota K, Shingyo T, Takebe F, Nodasaka Y, Matsuyama H, Hara I (2004) *Exiguobacterium oxidotolerans* sp. nov., a novel alkaliphile exhibiting high catalase activity. Int J Syst Evol Microbiol 54:2013–2017

Zhang F, Xu H, Pronin A, Liu H, Tachdjian C, Li X, Klebansky B, Fine RM (2008) Molecular mechanism for the umami taste synergism. Proc Natl Acad Sci USA 105:20930–20934

12

Chironomidae larvae associated with *Eichhornia azurea* leaf detritus: decomposition community structure and colonization dynamics

Carolina Vieira da Silva · Raoul Henry

Abstract Chironomidae stand out among aquatic insects associated with organic matter decomposition due to their abundance, cosmopolitan distribution, the varied forms in which they use detritus, as well as the feeding plasticity of their larvae. The objective of this study was to investigate the structure and composition of the Chironomidae community (Diptera) in the decomposition of *Eichhornia azurea* leaves in a marginal lake. The working hypothesis is that the taxonomic and functional density and richness of Chironomidae increase over time during the degradation of *E. azurea* leaves. Decomposition was analyzed in leaves kept in 2-mm mesh litter bags and collected at set successive sampling intervals. Significant differences were found in Chironomidae density and composition in the time scale. The density of individuals increased significantly during the experiment, in contrast to the taxonomic and functional richness. Subfamily Chironominae was the most representative in terms of density and taxonomic richness. The densities of taxa *Labrundinia* sp., *Tanytarsus* sp., *Dicrotendipes* sp., *Endotribelos* sp. *Chironomus* sp. and *Ablabesmyia* sp. were mainly responsible for intragroup similarity in the groups formed at each sampling time. In contrast, the taxonomic composition varied throughout the experiment with characteristic ecologic succession and dynamic stabilization of the colonizing community towards the end of the experiment. In conclusion, the Chironomidae community structure presented an increase in larva density during the experimental period associated with changes in taxon composition; however, the total taxonomic richness variation was low and functional richness did not vary.

Keywords Macroinvertebrates · Aquatic insects · Macrophytes · Ecologic succession · Lentic ecosystem

Introduction

Decomposition is a key process in the metabolism of aquatic ecosystems (Gimenes et al. 2010) and is often the focus of various ecological studies (Graça and Canhoto 2006). When considering that plants serve as the base for aquatic food webs, the degradation of plant material is ideal for analyses of aquatic biodiversity. Aquatic plants have a high growth potential and contribute to the production of allochthonous material, in addition to serving as an important source of organic matter in the detritus food chain. The production of detritus is more intense in tropical lentic ecosystems, which are characterized by high temperatures and high availability of nutrients (Bianchini 2007). A large amount of detritus is consumed by larvae of the family Chironomidae

C. V. da Silva (✉) · R. Henry
Departamento de Zoologia, Instituto de Biociências, Universidade Estadual Paulista "Julio de Mesquita Filho" (UNESP), Botucatu, SP, Brazil
e-mail: cvscarol@hotmail.com

(Henriques-Oliveira et al. 2003; Silveira et al. 2013; Saito and Fonseca-Gessner 2014), which consists mostly of omnivorous species (Berg 1995; Galizzi et al. 2012).

Chironomidae larvae can be divided into six functional trophic groups based on their feeding habits: gathering collectors, filtering collectors, scrapers, shredders and predators. However, most Chironomids are not restricted to a single feeding habit, and taxa of the same genus may vary in feeding "preferences" according to resource availability, nutritional value, size and type (Berg 1995). Shredders are suggested to consume conditioned plant organic matter, which is organic matter colonized and degraded by microorganisms during decomposition (Graça et al. 2001). However, relationships between invertebrates and plant detritus are considered complex due to the physical and chemical alterations that occur during the decomposition of organic matter (Gonçalves et al. 2004).

The mechanisms to colonize detritus are determined by the compositions of microorganisms and other fauna. Microorganisms colonize decomposing organic matter and increase in biomass through cell division. In contrast, the biomass of macroinvertebrates associated with plant detritus increases mainly through immigration of individuals to the new substrate for colonization (Hieber and Gessner 2002). The colonization of decomposing plant matter by macroinvertebrates is related to their feeding habits, which may include the ingestion of detritus or predation on smaller detritivorous invertebrates (Lancaster and Downes 2013).

The relationship between decomposing macrophytes and associated organisms requires further investigation to improve the understanding of the ecological processes that drive the metabolism of aquatic ecosystems, such as nutrient cycling, intra- and interspecific interactions, organic matter degradation, food chain, etc. Therefore, this study examined changes in the structure and composition of the Chironomid community in decomposing *E. azurea* leaf detritus and its role in the degradation of organic material.

The objective of this study was to examine changes in the structure and composition of the Chironomidae community in decomposing *E. azurea* leaf detritus and its role in the decomposition of organic matter. Exponential increases in density and taxonomic and functional richness during the decomposition process and a tendency towards stabilization at the end of the experiment are expected. As to taxonomic richness, the number of Chironomidae genera is presumed to increase significantly, mostly in the beginning of the decomposition process, when *E. azurea* leaves make up a "new substrate" available for colonization.

Materials and methods

Study area

This study was conducted on a marginal lake at the mouth of the Paranapanema River into Jurumirim Reservoir (Fig. 1) in the municipality of Angatuba, state of São Paulo, Brazil. This region is characterized by a high diversity of aquatic ecosystems, consisting of marginal lakes with distinct connections to the river. The hydrologic connectivity, as well as the river bed outflow and lateral flooding, is controlled by the Jurumirim Reservoir located downstream (Henry 2003, 2005).

Experimental design and sampling interval

The *E. azurea* leaf decomposition experiment was performed between April and July 2013, which corresponds to the end of the rainy season and beginning of the dry season, respectively. Litter bags of 2-mm mesh and measuring 20 × 25 cm were used to hold 20.0 g of *E. azurea* leaves. The leaves were previously dried at room temperature for 3 weeks before being incubated at 30 cm under the water surface. The litter bags were tied to floaters located at four sites, where seven litter bags were placed at each site on Barbosa Lake (Fig. 1). Four litter bags, one from each sampling site, were collected at days 1, 3, 7, 14, 35, 56 and 70, totaling 28 litter bags in whole study.

The litter bags were stored in plastic bags upon sampling and kept in thermo boxes with ice until analyses in the laboratory. For analysis, the leaf detritus was carefully washed with running water over a stack of sieves organized by decreasing mesh size (0.25, 0.125, 0.105 and 0.053 mm) for retention of the associated invertebrate fauna. The material collected from each sieve was stored in a plastic jar with a 70% alcohol solution.

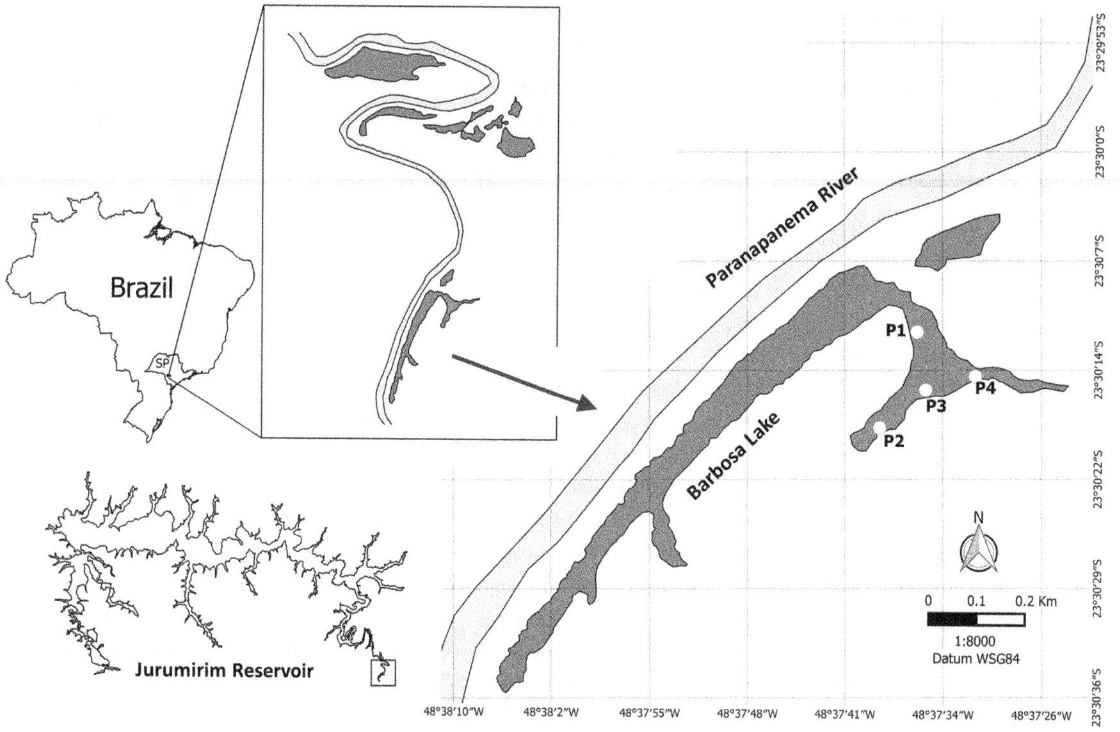

Fig. 1 Study area: Barbosa Lake, marginal to and at the mouth zone of the Paranapanema River into Jurumirim Reservoir (São Paulo, Brazil)

The invertebrates were sorted under a stereoscopic microscope and identified through specialized dichotomous key (Trivinho-Strixino 2011).

After the invertebrates were collected, three sets of five 1.5-cm diameter disks were taken from five *E. azurea* leaves from each litter bag, labeled and stored frozen at − 20 °C. They were used for determination of dry mass, organic matter, microbial and fungal biomass. Ash-free dry mass (AFDM) and organic matter content were determined, using a muffle furnace at 550 °C for 4 h (Taylor and Chauvet 2014; Martins et al. 2017). Carbon content, were calculated based on the rate of 47% of the organic matter (Wetzel 1975). Total microbial biomass, in $gAFDM^{-1}$, was estimated from the ATP content found in detritus (in nmoles), determined by enzymatic activity bioluminescence, following Abelho (2005). Fungal biomass, in μg $gAFDM^{-1}$, was determined by ergosterol content following Gessner (2005).

Abiotic and biotic variables

Water variables were determined on the water surface within macrophyte stands at each of the four sampling sites. Dissolved oxygen was determined using Winkler's method. Water pH and electrical conductivity (corrected to 25 °C) were measured using laboratory bench meters. Suspended solids were determined by gravimetry, water temperature with mercury thermometer, water transparency with a Secchi disk, and photosynthetic pigments were extracted in acetone (90%) for 12 h and the concentration determined with a spectrophotometer. Suspended solids were determined according to Teixeira and Kutner (1962) and all the others according to Golterman et al. (1978). Daily precipitation data during the study period (from April 26 to July 05, 2013) were obtained from the pluviometric station at Departamento de Água and Energia elétrica in the Municipality of Angatuba.

The functional trophic group of each Chironomid genus was determined following Coffman and Ferrington (1996), Henriques-Oliveira et al. (2003) and Galizzi et al. (2012). The functional trophic groups were defined following methods described in Gonçalves et al. (2003), Janke and Trivinho-Strixino (2007), Zilli et al. (2008), Silva et al. (2008), Ligeiro et al. (2010), Silva et al. (2010), and Silva et al. (2011).

Analysis of the decomposition process

The decomposition coefficient (k) was determined by fitting the percent mass loss to the exponential model (Olson 1963).

The variation of the ATP content was estimated using a first-order mathematical model that considered two processes simultaneously, microorganism growth and ATP concentration decrease (Levenspiel 1974). Fitting was performed by non-linear regressions using the Levenberg–Marquardt iterative algorithm (Press et al. 2007).

$$ATP_t = ATP_{max} \times \mu \times \left(\frac{e^{-\mu t}}{k-\mu} + \frac{e^{-kt}}{\mu - k}\right),$$

where ATP = ATP concentration, μ = coefficient of increase in ATP concentration (day^{-1}), k = coefficient of decrease in ATP concentration (day^{-1}), t = time (day).

The variation of the concentration of ergosterol was evaluated using the exponential growth model (Eq. 1). The duplication time (T_d) was estimated with Eq. (2). Fitting was performed with non-linear regressions using the iterative Levenberg–Marquardt algorithm (Press et al. 2007).

$$N_t = N_0 \times e^{\mu \times t}, \qquad (1)$$

where N_t = value of the biotic variable in time; N_0 = initial value of the biotic variable; μ = growth coefficient (day^{-1}); t = time (day).

$$t_d = \frac{\ln 2}{\mu}, \qquad (2)$$

where t_d = duplication time in days, ln 2 = Neper logarithm of 2, μ = growth coefficient of the biotic variable.

Quantitative analysis of invertebrates

Density (ind.100 g DM^{-1}) was calculated from the total abundance of Chironomidae in the remaining ash-free dry mass of *E. azurea* leaf detritus from each of the seven sampling periods. The mean daily colonization rate (x) of macrophyte detritus by Chironomidae, expressed in terms of abundance (daily variation rate of individual abundance day^{-1}) and richness (daily variation rate of taxon richness day^{-1}), in the successive sampling intervals was estimated by Chironomidae abundance and richness over time using $x_A = (A_2 - A_1)/(t_2 - t_1)$ (A = abundance and t = time) and $x_R = (R_2 - R_1)/(t_2 - t_1)$ (R = richness and t = time).

Statistical analysis

Statistically significant differences ($p < 0.05$) in variables between sampling days were examined by variance analysis using Sigma Plot 11 software. Multiple data comparisons were carried out through Tukey test. "One-way" ANOVA was used when the data were normally distributed and the Kruskal–Wallis test when normality was not observed.

Chironomidae associated with *E. azurea* leaf detritus and their respective biotic were related to Barbosa Lake environmental variables, by means of simple correlation matrix/Spearman correlation coefficient, using Statistica 7 (Statsoft 2009) software. Chironomidae density values recorded at each sampling time were used. The similarity in Chironomidae density and composition (presence or absence) between leaf detritus samples on successive sampling days was tested by Non-Metric Multidimensional Scaling analysis (NMDS) using the Bray–Curtis dissimilarity coefficient. The main taxa responsible for groupings shaped by NMDS were identified by SIMPER analysis and the differences in Chironomidae density and composition by ANOSIM analysis. All analyses were carried out with Primer v6 (Clarke and Gorley 2006) software using square root-transformed data (Clarke and Warwick 2001).

Statistical tests were carried out using log $(x + 1)$ transformed data, except pH values and NMDS, SIMPER, and ANOSIM analyses.

Results

Abiotic variables

All values of the abiotic variables are showed in Table 1. The highest water temperature was recorded in the beginning of the experiment, which then decreased until day 56 and showed increase on day 70. The mean dissolved oxygen concentration varied little between successive sampling days, while the percent saturation showed a higher degree of variation. Higher pH, suspended solids and total nitrogen values were observed on day 56. Conductivity increased after day 7 and peaked on day 35. The lowest water transparency value was recorded during the period with the greatest depth. The total phosphorus concentration in the water column was the highest on day 70.

Biotic variables

Values of the biotic variables are showed in Table 2. Higher values of total pigments were observed at the end of the experiment (days 35 to 70), with a peak on day 56 (222.6 μg L^{-1}). The remaining *E. azurea* leaf biomass showed a loss of 1.81% after the first seven days. The decomposition rates at the subsequent sampling periods were low as well, demonstrating a small loss of mass during the entire experiment. The percent mass loss was the highest at day 35 (4.4% of the initial value), followed by mass increases at days 56 and 70. The ash content of the remaining dry biomass was the highest at day 35 (4.8%) and decreased thereafter until the end of the experiment. In contrast, the carbon content was the lowest on day 35 (44.9%) and then increased until the end of the experiment. The nitrogen and phosphorus concentrations of the remaining dry biomass showed little variation and both concentrations were low at the end of the experiment. Ergosterol was detected

Table 1 Variation of mean values and standard deviations ($N = 4$) of Barbosa Lake environmental variables between April and July 2013

Variables/days	Initial	1st	3rd	7th	14th	35th	56th	70th
Water temperature (°C)	28 ± 0	26 ± 1	26 ± 1	26 ± 1	21 ± 1	19 ± 0	19 ± 0	21 ± 1
Dissolved oxygen concentration (mg L^{-1})	7 ± 2	8 ± 2	8 ± 1	10 ± 2	9 ± 3	8 ± 2	7 ± 2	7 ± 1
Oxygen saturation (%)	101 ± 22	110 ± 29	110 ± 20	124 ± 20	105 ± 38	96 ± 23	78 ± 23	82 ± 8
Water pH	6 ± 0	6 ± 0	6 ± 0	6 ± 0	6 ± 0	7 ± 0	8 ± 1	6 ± 0
Organic suspended matter (mg L^{-1})	5 ± 2	10 ± 5	13 ± 7	11 ± 4	12 ± 7	25 ± 13	45 ± 32	20 ± 22
Inorganic suspended matter (mg L^{-1})	3 ± 1	4 ± 2	3 ± 1	3 ± 2	4 ± 2	3 ± 2	7 ± 6	12 ± 8
Water electrical conductivity (μS cm^{-1})	33 ± 1	33 ± 1	33 ± 1	32 ± 2	36 ± 1	38 ± 4	35 ± 1	34 ± 2
Water transparency (m)	1.1 ± 0	1.5 ± 0	1.5 ± 0	1.4 ± 0	1.2 ± 0	1.6 ± 0	1.5 ± 0	0.7 ± 0
Water depth (m)	1.8 ± 0	1.8 ± 0	1.8 ± 0	1.8 ± 0	1.7 ± 0	1.6 ± 0	1.8 ± 0	2.3 ± 1
Total nitrogen concentration (μg L^{-1})	490 ± 141	507 ± 160	458 ± ±27	487 ± 36	592 ± 110	792 ± 78	975 ± 364	752 ± 297
Total phosphorus concentration (μg L^{-1})	22 ± 3	37 ± 16	25 ± 9	34 ± 15	26 ± 7	46 ± 23	46 ± 46	57 ± 24

Table 2 Variation in means and standard deviation ($N = 4$) values of total pigments and the variables determined from the remaining biomass of *E. azurea* leaves during decomposition on Barbosa Lake between April and July 2013

Variables/days	1st	3rd	7th	14th	35th	56th	70th
Total pigments (μg L^{-1})	58.1 ± ±47.9	60.6 ± 39.6	68.8 ± 33.6	43.5 ± 22.6	192.3 ± 111.5	222.6 ± 190.7	102.3 ± 124.2
Remaining biomass (%)	98.5 ± 0.2	98.3 ± 0.1	98.2 ± 0.2	97.1 ± 1.1	95.6 ± 0.6	97.7 ± 0.5	98.1 ± 0.6
Ash content (%)	2.0 ± 0.1	2.2 ± 0.1	2.3 ± 0.2	3.4 ± 1.0	4.8 ± 0.6	2.9 ± 0.5	2.5 ± 0.6
Carbon content (%)	46.3 ± 0.1	46.2 ± 0.1	46.1 ± 0.1	45.6 ± 0.5	44.9 ± 0.3	45.9 ± 0.2	46.1 ± 0.3
Nitrogen content (mg $g^{-1)}$	0.5 ± 0.0	0.7 ± 0.2	1.0 ± 0.6	1.1 ± 0.7	1.0 ± 0.8	0.7 ± 0.1	0.8 ± 0.2
Phosphorus content (mg g^{-1})	0.2 ± 0.1	0.1 ± 0.0	0.1 ± 0.0	0.2 ± 0.0	0.1 ± 0.1	0.1 ± 0.0	0.1 ± 0.0
Ergosterol biomass (μg g^{-1} AFDM)	0.0 ± 0.0	0.0 ± 0.0	0.0 ± 0.0	0.0 ± 0.0	19.7 ± 23.8	25.9 ± 10.1	58.8 ± 19.4
ATP biomass (nmoles g^{-1} AFDM)	145.5 ± 105.7	45.3 ± 25.3	62.2 ± 35.2	39.1 ± 7.0	34.9 ± 8.9	25.8 ± 14.7	27.6 ± 9.4

on day 35 and its concentration gradually increased thereafter. In contrast, the microbial community biomass was high on day 1 and decreased until day 7, and showed little variation until day 70.

The remaining decay rate of *E. crassipes* was extremely low, while the half-life was very high (Table 3). Excluding data on days 56 and 70, it was obtained $k = 0.00148$ day^{-1}, $r^2 = 0.52$ and $t_{1/2} = 468$ days.

The density of Chironomidae increased over time from 72 ± 69 ind.100 g DM^{-1} at days 1 to 3,049 ± 915 ind.100 g DM^{-1} at day 70. The subfamily Chironominae showed the highest density and taxonomic richness (Fig. 2a, b). The mean rate of colonization by the Chironomids (richness; taxa day^{-1}) was highest at day 1, but remained low for the rest of the experiment (Fig. 2c). The mean taxonomic richness increased over time, with the highest richness observed on day 70 (Fig. 2d). Three peaks were observed at days 1, 7 and 70 for the abundance of invertebrate fauna (ind day^{-1}), with 11 ind day^{-1}, 8 ind day^{-1} and 9 ind day^{-1}, respectively (Fig. 2e).

Analyses of abundance over time showed that *Endotribelos* sp. (Chironominae subfamily) was the dominant genus at most of the sampling periods, with the greatest relative abundance shown at day 70 (38.5%). *Corynoneura* sp. (Orthocladiinae subfamily) and *Larsia* sp (Tanypodinae subfamily) showed the lowest relative abundance (Fig. 3).

The subfamily Chironominae had the greatest number of taxa, which showed a peak in richness on day 56 (Table 4). The richness of the functional trophic groups showed no variation over time, but six functional categories were observed throughout the entire experimental period with the exception of *Stenochironomus* sp. (subgenus *Petalopholius*), which was observed from day 35 onwards. This particular genus is a miner of submersed aquatic macrophyte leaves (Trivinho-Strixino 2011).

The functional taxonomic richness of Chironomidae showed no variation throughout the experimental period. On the other hand, variation was shown for the relative abundance of trophic groups (Fig. 4). Only *Polypedilum* sp., the most representative taxon at the beginning of the experiment, was classified in the three functional groups of collectors, shredders and scrapers. As shown in Table 4, *Ablabesmyia* sp. was a collector only in its first larval instars. Nevertheless, developmental stages were not considered for the richness of the functional trophic groups. Therefore, *Ablabesmyia* sp. was included in the collector and predator functional groups, together with *Parachironomus* sp., and its relative abundance did not vary significantly from days 3 to 70. On the first day of colonization, no larvae were observed from either *Ablabesmyia* sp. or *Parachironomus* sp. From day 7, the collectors and shredders were the most representative groups at all sampling periods (Fig. 4). *Endotribelos* sp. and *Chironomus* sp. composed a large fraction of the two groups, especially towards the end of the experiment (Fig. 3).

Statistically significant differences were shown for most environmental variables over time, with the exception of the dissolved oxygen (Kruskal–Wallis: $H = 7.795$; $p = 0.351$), oxygen saturation (Kruskal–Wallis: $H = 10.358$; $p = 0.169$) suspended inorganic matter (Kruskal–Wallis: $H = 9.091$; $p = 0.246$) and phosphorus (ANOVA: $F = 1.406$; $p = 0.249$). For the ecological attributes of the invertebrate community and variables determined for the *E. azurea* leaf detritus, only the total nitrogen in detritus showed no significant changes over time (Kruskal–Wallis: $H = 10.974$; $p = 0.089$).

The density (ANOVA: $F = 41.159$, $p = <0.001$) and richness (ANOVA: $F = 14.540$, $p = <0.001$) of the Chironomid subfamilies associated with *E. azurea* leaf detritus were statistically different between sampling periods (Table 5).

NMDS and ANOSIM analyses revealed significant differences in the density (Fig. 5a) (ANOSIM: $p = 0.001$; R global $= 0.6$) and taxon composition (Fig. 5b) (ANOSIM: $p = 0.001$; R global $= 0.5$) of

Table 3 Values of decay rate (k), growth (μ) and coefficient of determination (r^2) recorded for half-life ($t_{1/2}$) and duplication (t_d) times of the models used in the analysis of the decomposition process (1), of the ATP content (2), the ergosterol concentration (3) in the remaining biomass of the *E. azurea* leaves at successive sampling times on Barbosa Lake from April to July 2013

Models	K	μ	R^2	$T_{1/2}$	T_d
(1) Remaining leaf biomass	0.00051 day^{-1}	–	0.92	1.359 days	–
(2) Microorganism biomass (ATP)	0.01808 day^{-1}	0.61029 day^{-1}	0.86	7 days	3 days
(3) Fungal biomass (ergosterol)	–	0.0261 day^{-1}	0.84	–	26 days

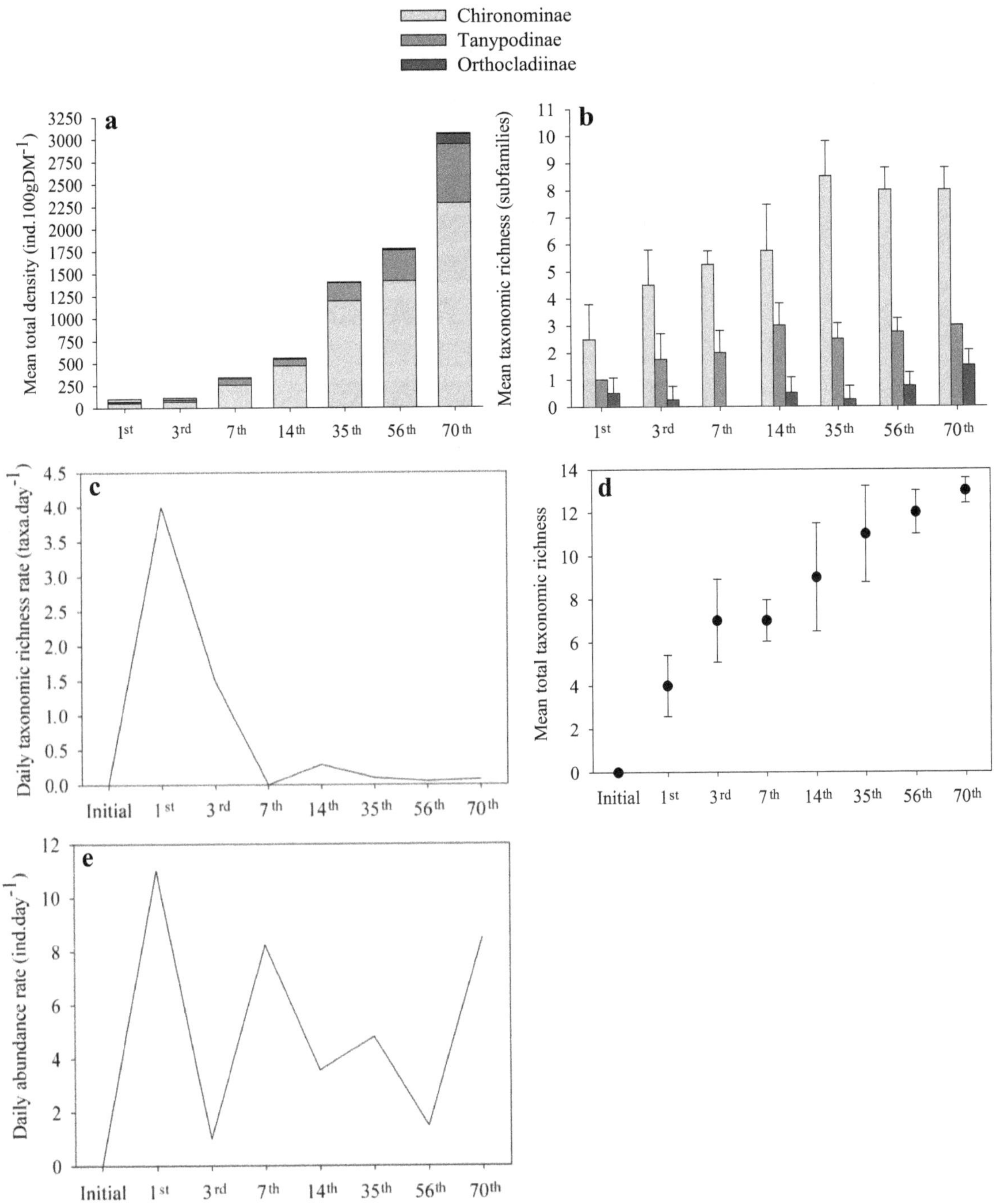

Fig. 2 Variation in mean values ($N = 4$) of total density (ind.100 g DM^{-1}) of Chironomidae (**a**) subfamilies, taxonomic richness per Chironomidae subfamily (**b**), daily variation rate of taxon colonization (richness) (taxa day^{-1}) (**c**), total taxonomic richness (**d**) and daily mean variation of the number of colonizing individuals (abundance) of Chironomidae fauna (ind day^{-1}) (**e**) associated with leaf detritus of *E. azurea* on Barbosa Lake in successive sampling times between April and July 2013

Chironomidae sampled at the seven periods, which suggests a change in the Chironomid community structure and composition over time during the experimental period. However, from day 35 on, the sample data were similar, indicating that the community was stable (Fig. 5a, b).

In the beginning of the experiment, *Labrundinia* sp. (SIMPER: 46.8%), *Tanytarsus* sp. (SIMPER: 21.3%) and *Dicrotendipes* sp. (SIMPER: 19.1%) were the main taxa responsible for the similarity within the groups at each sampling period. From day 7 on, *Endotribelos* sp. (27.7%) contributed the most to the NMDS result. At

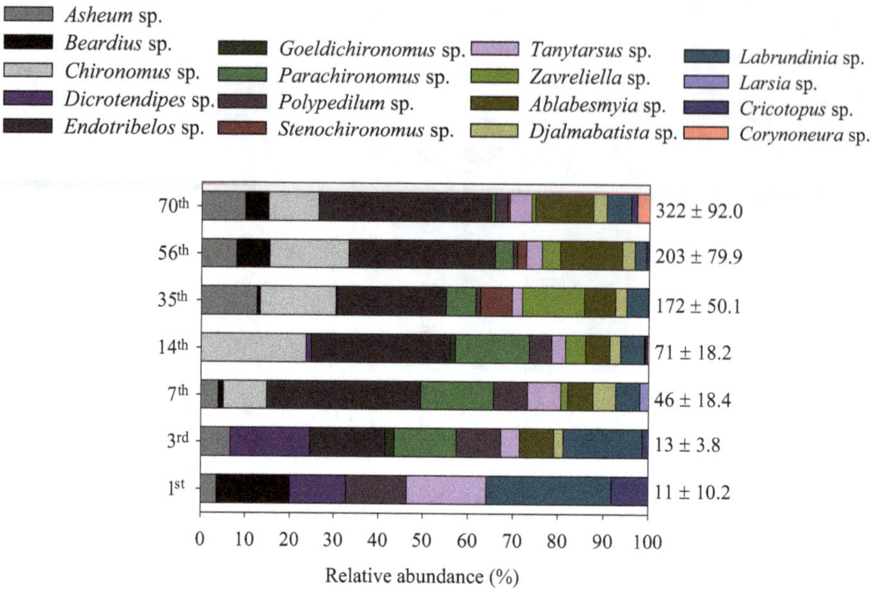

Fig. 3 Variation of mean values ($N = 4$) of relative abundance (%), total abundance (right column) and standard deviations (±) of Chironomidae genera associated with leaf detritus of *E. azurea* on Barbosa Lake in successive sampling times between April and July 2013

Table 4 Taxon occurrence, functional trophic groups, subfamily richness (Chironominae, Tanypodinae and Orthocladiinae) and total richness of the Chironomidae community associated with *E. azurea* leaf detritus on Barbosa Lake in successive sampling times between April and July 2013

Taxa	Functional trophic groups	Sampling day						
		1st	3rd	7th	14th	35th	56th	70th
Chironominae								
Asheum sp.	GC	*	*	*		*	*	*
Beardius sp.	GC	*		*		*	*	*
Chironomus sp.	GC; FC—few spp.; HS (miner)			*	*	*	*	*
Dicrotendipes sp.	GC; FC; S?	*	*		*	*	*	
Endotribelos sp.	GC; HS—facultative		*	*	*	*	*	*
Goeldichironomus sp.	GC		*		*			
Parachironomus sp.	P (engulfer); GC; PA		*	*	*	*	*	*
Polypedilum sp.	HS (miner); GC; FC?; P (engulfer)	*	*	*	*	*	*	*
Stenochironomus sp.	GC; HS (miner); FX					*	*	*
Tanytarsus sp.	GC; FC; S—few spp.	*	*	*	*	*	*	*
Zavreliella sp.	GC			*	*	*	*	*
Subfamily richness		5	7	8	8	10	10	9
Tanypodinae								
Ablabesmyia sp.	P (engulfer/piercer); GC—1st instars		*	*	*	*	*	*
Djalmabatista sp.	P (engulfer)		*	*	*	*	*	*
Labrundinia sp.	P (engulfer/piercer)	*	*	*	*	*	*	*
Larsia sp.	P (engulfer)			*	*			
Subfamily richness		1	3	4	4	3	3	3
Orthocladiinae								
Cricotopus sp.	HS (miner/shredder); GC	*	*		*		*	*
Corynoneura sp.	GC				*	*	*	*
Subfamily richness		1	1	0	2	1	2	2
Total richness		7	11	12	14	14	15	14

Chironomidae functional groups, *GC* gathering collector; *FC* filtering collector; *HS* herbivorous shredder; *XS* xylophage shredder; *P* predator; *PA* parasite; *S* scraper

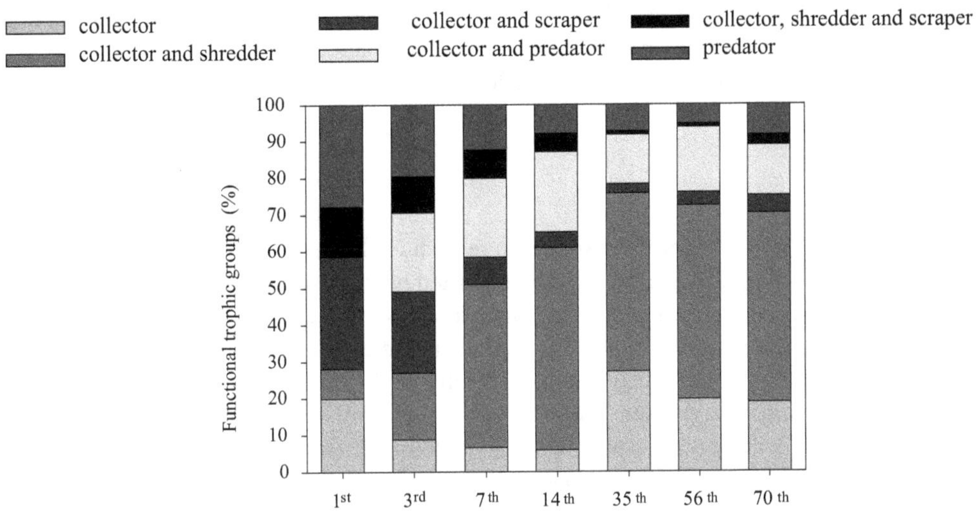

Fig. 4 Variation of mean values ($N = 4$) of relative abundance (%) of functional trophic groups of the Chironomidae community associated with *E. azurea* leaf detritus on Barbosa Lake at successive sampling times between April and July 2013

Table 5 p values obtained in statistical analyses (ANOVA "one way"/Kruskal–Wallis) and differences indicated by Tukey test for Chironomidae community density and richness and the respective subfamily densities (ind.100 g DM^{-1}) associated with leaf detritus of *E. azurea* on Barbosa Lake at successive sampling times between April and July 2013

Variables analyzed	p	Tukey test
Total Density (ind.100 g DM^{-1})	**< 0.001**	(70th ≠ 1st, 3rd, 7th, 14th); (56th ≠ 1st, 3rd, 7th, 14th); (35th ≠ 1st, 3rd, 7th); (14th ≠ 1st, 3rd, 70th); (7th ≠ 1st, 3rd, 35th, 70th)
Total Taxonomic Richness	**< 0.001**	(70th ≠ 1st, 3rd, 7th); (56th ≠ 1st, 3rd); (35th ≠ 1st, 3rd); (14th ≠ 1st); (7th ≠ 1st, 70th)
Chironominae	**< 0.001**	(70th ≠ 1st, 3rd, 7th, 14th); (56th ≠ 1st, 3rd, 7th); (35th ≠ 1st, 3rd, 7th); (14th ≠ 1st, 3rd, 70th); (7th ≠ 1st, 3rd, 35th, 56th, 70th)
Tanypodinae	**< 0.001**	(70th ≠ 1st, 3rd, 7th, 14th); (56th ≠ 1st, 3rd, 7th, 14th); (35th ≠ 1st, 3rd); (14th ≠ 1st, 56th, 70th)
Orthocladiinae	**0.001**	(70th ≠ 1st, 3rd, 7th, 14th, 35th)

Significant differences ($p < 0.05$) in bold

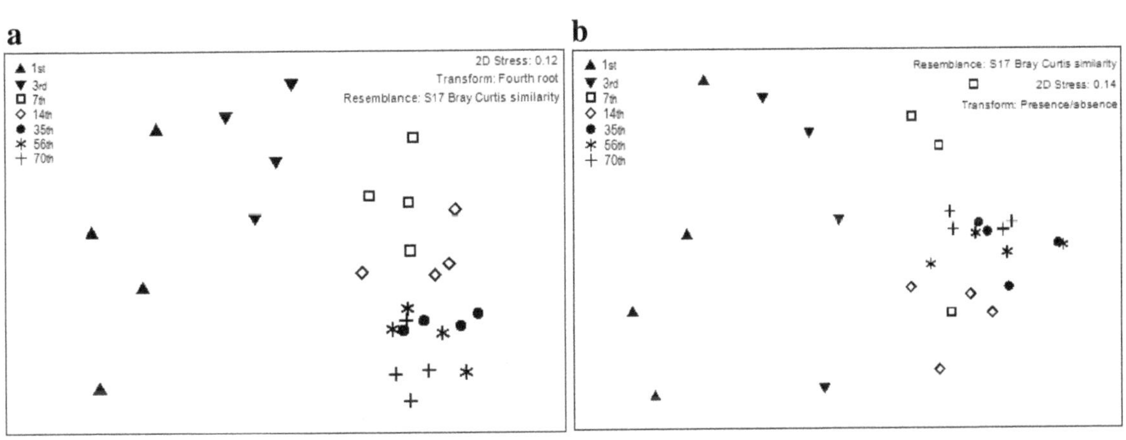

Fig. 5 Non-metric multidimensional scaling analysis (NMDS) of taxon density (**a**) and composition (presence or absence) (**b**) of the Chironomidae community associated with *E. azurea* leaf detritus on Barbosa Lake at successive sampling times between April and July 2013

days 35, 56 and 70, *Chironomus* sp., *Ablabesmyia* sp. and *Endotribelos* sp. contributed the most to the NMDS result. The ergosterol concentration and remaining detrital biomass of *E. azurea* had the greatest significant correlations with the various invertebrate taxa (Table 6).

Discussion

The procedures adopted in this experiment sought to reproduce the decomposition of aquatic macrophytes in natural conditions as close as possible; so leaves were dried naturally at room temperature before incubation on Barbosa Lake. However, the methodological procedures used may have interfered with the dynamics of decomposition of *E. azurea*, as it was slower than in other studies (Cunha-Santino et al. 2010; Martins et al. 2011; Bianchini et al. 2014). Since aquatic plants generally begin to decompose at the senescence stage, part of the labile compounds may have been lost even before incubation in water (Pagioro and Thomaz 2006). Additionally, factors like decreasing water temperature through the experiment, occasional inhibition of microorganism growth by leachate, predominance of an oligotrophic environment and low abrasion due to the environment lentic regime may have contributed to a low decomposition rate of *E. azurea* (Battle and Mihuc 2000; Grattan and Suberkropp 2001; Sangiorgio et al. 2004; Carvalho et al. 2005; Sangiorgio et al. 2008; Song

Table 6 Density of Chironomidae taxa associated with *E. azuera* leaf detritus on Barbosa Lake with significant Spearman correlations ($p \leq 0.05$) with abiotic and biotic variables in successive sampling times between April and July 2013

Taxa	Variables									
	ERG	ATP	BIO	WT	TN	pH	*K*	DP	%*A*	%CA
Total density	+ 0.804	− 0.450	− 0.896	− 0.762	+ 0.707	+ 0.499	+ 0.490	− 0.446	+ 0.485	− 0.433
Total Richness	+ 0.780	− 0.432	− 0.728	− 0.714	+ 0.595	+ 0.448	+ 0.533		+ 0.507	− 0.481
Chironominae										
Asheum sp.	+ 0.813	− 0.454	− 0.727	− 0.596	+ 0.462			− 0.465		
Beardius sp.	+ 0.719		− 0.588	− 0.412	+ 0.447	+ 0.423		− 0.473		
Chironomus sp.	+ 0.679		− 0.838	− 0.782	+ 0.756	+ 0.637	+ 0.552	− 0.493	+ 0.492	− 0.458
Dicrotendipes sp.			+ 0.392			− 0.444				
Endotribelos sp.	+ 0.814	− 0.497	− 0.921	− 0.764	+ 0.701	+ 0.486	+ 0.494	− 0.494	+ 0.497	− 0.444
Goeldichironomus sp.			+ 0.374							
Parachironomus sp.				− 0.615	+ 0.504	+ 0.568	+ 0.403		+ 0.719	− 0.715
Polypedilum sp.	+ 0.380									
Stenochironomus sp.	+ 0.555		− 0.522	− 0.722	+ 0.554	+ 0.587	+ 0.585			
Tanytarsus sp.	+ 0.782	− 0.491	− 0.739	− 0.574	+ 0.567					
Zavreliella sp.	+ 0.397		− 0.594	0.776	+ 0.750	+ 0.736	+ 0.642		+ 0.547	− 0.545
Tanypodinae										
Ablabesmyia sp.	+ 0.832	− 0.548	− 0.863	− 0.724	+ 0.659	+ 0.514	+ 0.422	− 0.495	+ 0.404	
Djalmabatista sp.	+ 0.766		− 0.688	− 0.530	+ 0.471	+ 0.384	+ 0.497	− 0.530		
Labrundinia sp.	+ 0.562		− 0.507	− 0.421						
Larsia sp.										
Orthocladiinae										
Cricotopus sp.	+ 0.386									
Corynoneura sp.	+ 0.616		− 0.553							

ERG ergosterol, *ATP* microorganism biomass, *BIO* detritus biomass, *WT* water temperature, *SOM* suspended organic matter, *TN* total nitrogen, *T* transparency, *DN* detritus nitrogen, *DP* detritus phosphorus, *% A* ash percentage in detritus, *%CA* carbon percentage in detritus, − negative, + positive

−/+ signals = negative and positive correlations, and the *R* values

et al. 2013). However, colonization of the remaining biomass by Chironomidae increased with a significant increase in the total density over time as decomposition progressed.

The feeding behavior of Chironomid larvae is also influenced by local environmental characteristics, changes in substrate, and input of allochthonous organic matter (Silva et al. 2008). Size and stage of development may also determine variations in larval feeding habits (Berg 1995). For example, Shimabukuro and Henry (in press) described the diet of *Chironomus sancticaroli* larvae as a high ingestion of coarse particulate organic matter and algae in later development stages. The consumption of coarse particulate organic matter was also reported for *Chironomus paragigas*, even in young individuals, due to their relatively large size in all stages of development (Shimabukuro and Henry in press).

The functional richness of the Chironomid community in this study showed no variation throughout the experiment, as five of the trophic groups were observed at all sampling periods (Table 2). In contrast, prior research suggested that the functional complexity of Chironomid feeding habits increases during the decomposition of macrophytes (Gonçalves et al. 2003; Silva et al. 2010). Collectors were the most representative functional group in this study. However, the same Chironomidae taxon has more than one feeding habit, such as *Polypedilum* sp. (Table 2), which can be a collector, a shredder or a predator, making classification under a single and specific functional trophic group difficult. Therefore, care must be taken with generalizations of feeding habits of closely related taxa as this approach may result in misrepresentations of the Chironomid trophic ecology (Berg 1995).

Most Chironomids are omnivorous (Berg 1995; Galizzi et al. 2012). Analysis of the Chironomid trophic structure with stable isotopes showed a broad variety of feeding behaviors between taxa, suggesting that Chironomids do not belong to specific groups in the aquatic trophic web (Henriques-Oliveira et al. 2003; Reuss et al. 2013; Silveira et al. 2013; Saito and Fonseca-Gessner 2014). Reuss et al. (2013) also found that the stomach contents of Tanypodinae (*e.g., Djalmabatista* sp., *Labrundinia* sp. and *Larsia* sp.) contained detritus, macroalgae and invertebrates, despite being commonly classified as predators. This shows that the feeding habits of Chironomidae are still relatively unknown and that further investigation of their larval feeding behavior is necessary to understand their complex trophic ecology. In an analysis of the digestive tube of Chironomidae, Shimabukuro and Henry (in press) identified algae (Diatomacea, *Closterium* sp., *Peridinium* sp., Zygnemaphyceae and *Spirogyra* sp.), animal origin resources (Rotifera, Ostracoda, Protozoa, Cladocera and animal fragments), and fine and coarse particulate organic matter. The Chironomidae studied in Shimabukuro and Henry (in press) were sampled in two lentic environments of the same location as the present study.

The collectors and shredders were the most abundant functional groups, comprised of *Chironomus* sp., *Endotribelos* sp., *Stenochironomus* sp. and *Cricotopus* sp. The genus *Endotribelos* sp. had the greatest density from day 7 to the end of the experiment. According to Coffman and Ferrington (1996), this genus is classified as both a collector and a facultative herbivorous shredder. The capacity of Chironomids to adapt their feeding strategy to their habitat (Reuss et al. 2014) affirms that *Endotribelos* sp. individuals observed in the present study may have fed on both fine and coarse particulate organic matter from *E. azurea* leaf fragmentation detritus. Galizzi and Marchese (2007) reported that *Endotribelos* sp. are also shredders and accelerate the decomposition process, as they are observed in high abundances forming miners in detritus leaf mesophyll. Herbivory in aquatic macroinvertebrates and its effects on the macrophyte biomass are well documented (Lodge 1991; Nachtrieb et al. 2011). The biomass of fungus associated with *E. azurea* leaf detritus had the greatest number of positive correlations with Chironomidae genera in this study, which affirms that the degradation of plant organic matter leads to a reduction of the polyphenol content and the simultaneous colonization by bacteria and fungi, which in turn improves the palatability and nutritional value of detritus (Newman 1991; Stripari and Henry 2002).

The presence of all functional Chironomid groups at all decomposition stages may be related to a generalist strategy to obtain food, since most taxa in this group are omnivorous (Berg 1995; Galizzi et al. 2012) and consume a large amount of detritus (Henriques-Oliveira et al. 2003; Silveira et al. 2013; Saito and Fonseca-Gessner 2014). In macrophyte decomposition experiments, detritus is noted as the most abundant food resource, which accounts for the high density of Chironomids observed in decomposition studies.

The taxonomic richness of Chironomidae observed during the experimental period increased over time with the exception of day 70. Richness was greatest on day 14, which was twofold of that observed on day 1. Richness showed no variation at the other sampling times, except on day 56. Taxon composition differed significantly over time, with smaller variations towards the end of the experiment, suggesting that the number of individuals reached equilibrium to keep the community stable in terms of number of taxa throughout most of the experiment, despite the variation in the taxonomic composition.

Conclusion

The decomposition of *E. azurea* leaves was slow when compared to previous research. For the colonization of the remaining plant biomass, however, the total density of Chironomids increased significantly as decomposition progressed with time, and no tendency towards stability at the final sampling periods was observed; so one of our working hypotheses was partially confirmed. The taxonomic and functional richness of Chironomidae followed a distinct pattern, but taxon composition varied during the experiment in a characteristic ecological succession process, with a tendency towards stability at the final sampling periods (days 35, 56 and 70), confirming another initial working hypothesis just for taxonomic richness, since the richness of the functional trophic groups showed no variation over time. On the other hand, variation was shown for the relative abundance of trophic groups, where from day 7 on, the collectors and shredders were the most representative groups. The varied forms of use of detritus by Chironomidae (e.g, as a food resource or as a feeding site), as well as their feeding plasticity, may have contributed to the present results.

Acknowledgements The authors are grateful to Fundação do Instituto de Biociências—FUNDIBIO for financially supporting this research, to Hamilton Antonio Rodrigues, Joaquim Nunes da Costa and Lucio Miguel de Oliveira for helping in the field, to Laerte José da Silva for the English language revision, to reviewers and editor for their useful comments and valuable suggestions, and to Jorge Laço Portinho and Adriano Fabbro Gandini for helping with some statistical analyses and in the mapping of the study area, respectively. The first author is also grateful to Coordenação de Aperfeiçoamento de Pessoal de Nível Superior—CAPES for the scholarship received.

References

Abelho M (2005) Extraction and quantification of ATP as a measure of microbial biomass. In: Graça MAS, Bärlocher F, Gessner MO (eds) Methods to study litter decomposition: a practical guide. Springer, Dordrecht, pp 223–229

Battle JM, Mihuc TB (2000) Decomposition dynamics of aquatic macrophytes in the lower Atchafalaya, a large floodplain river. Hydrobiologia 418:123–136

Berg HB (1995) Larval food and feeding behaviour. In: Armitage PD, Cranston PS, Pinder LCV (eds) The Chironomidae: the biology and ecology of non-biting midges. Chapman and Hall, London, pp 136–168

Bianchini I Jr (2007) A decomposição da vegetação e o consumo de oxigênio nos reservatórios: implicações ecológicas. In: Henry R (ed) Ecologia de reservatórios: estrutura, função e aspectos sociais, 2nd edn. FUNDIBIO, Botucatu, pp 627–650

Bianchini I Jr, Cunha-Santino MB, Ribeiro JU, Penteado DGB (2014) Implication of anaerobic and aerobic decomposition of *Eichhornia azurea* (Sw.) Kunth. on the carbon cycling in a subtropical reservoir. Braz J Biol 74:100–110

Carvalho P, Thomaz SM, Bini LM (2005) Effects of temperature on decomposition of a potential nuisance species: the submerged aquatic macrophytes *Egeria najas* Planchon (Hydrocharitaceae). Braz J Biol 65:51–60

Clarke KR, Gorley RN (2006) PRIMER v6: user manual/tutorial. PRIMER-E, Plymouth

Clarke KR, Warwick RM (2001) Change in marine communities: an approach to statistical analysis and interpretation, 2nd edn. PRIMER-E, Plymouth

Coffman WP, Ferrington LC Jr (1996) Chironomidae. In: Merrit RW, Cummins KW (eds) An introduction to the aquatic insects of North America, 3rd edn. Kendall Hunt, Dubuque, pp 635–754

Cunha-Santino MB, Bianchini I Jr, Okawa MH (2010) The fate of *Eichhornia azurea* (Sw.) Kunth. detritus within a tropical reservoir. Acta Limnol Bras 22:109–121

Galizzi MC, Zilli F, Marchese M (2012) Diet and functional feeding groups of Chironomidae (Diptera) in the Middle Paraná River floodplain (Argentina). Iheringia 102:117–121

Gessner MO (2005) Ergosterol as a measure of fungal biomass. In: Graça MAS, Bärlocher F, Gessner MO (eds) Methods to study litter decomposition: a practical guide. Springer, Dordrecht, pp 189–195

Gimenes KZ, Cunha-Santino MB, Bianchini I Jr (2010) Decomposição de matéria orgânica alóctone e autóctone em ecossistemas aquáticos. Oecol Aust 14:1075–1112

Golterman KL, Clymo RS, Ohmstad MAM (1978) Methods for physical and chemical analysis of freshwaters, 2nd edn. Scientific Publications, Oxford

Gonçalves JF Jr, Esteves FA, Callisto M (2003) Chironomids colonization on *Nymphaea ampla* L. detritus during a degradative ecological succession experiment in a Brazilian coastal lagoon. Acta Limnol Bras 15:21–27

Gonçalves JF Jr, Santos AM, Esteves FA (2004) The influence of the chemical composition of *Typha domingensis* and *Nymphaea ampla* detritus on invertebrate colonization during decomposition in a Brazilian coastal lagoon. Hydrobiologia 527:125–137

Graça MAS, Canhoto C (2006) Leaf litter processing in low order streams. Limnetica 25:1–10

Graça MAS, Cressa C, Gessner MO, Feio MJ, Callies KA, Barrios C (2001) Food quality, feeding preferences, survival and growth of shredders from temperate and tropical streams. Freshw Biol 46:947–957

Grattan RM, Suberkropp K (2001) Effects of nutrient enrichment on yellow poplar leaf decomposition and fungal activity in streams. J N Am Benthol Soc 20:33–43

Henriques-Oliveira AL, Nessimian JL, Dorvillé LFM (2003) Feeding habits of Chironomid larvae (Insecta: Diptera) from a stream in the Floresta da Tijuca, Rio de Janeiro, Brazil. Braz J Biol 63:269–281

Henry R (2003) Os ecótonos nas interfaces dos ecossistemas aquáticos: conceitos, tipos, processos e importância. Estudo de aplicação em lagoas marginais ao rio Paranapanema na zona de sua desembocadura na represa de Jurumirim. In: Henry R (ed) Ecótonos nas interfaces dos ecossistemas aquáticos. FUNDIBIO, Rima, São Carlos, pp 1–28

Henry R (2005) The connectivity of the Paranapanema river with two lateral lakes in its mouth zone into the Jurumirim reservoir. Acta Limnol Bras 17:57–69

Hieber M, Gessner MO (2002) Contribution of stream detrivores, fungi and bacteria to leaf breakdown based on biomass estimates. Ecology 83:1026–1038

Janke H, Trivinho-Strixino S (2007) Colonization of leaf litter by aquatic macroinvertebrates: a study in a low order tropical stream. Acta Limnol Bras 19:109–115

Lancaster J, Downes BJ (2013) Aquatic entomology, 1st edn. Oxford University Press, Oxford

Levenspiel O (1974) Engenharia das reações químicas, 1st edn. Edgard Blücher, São Paulo

Ligeiro R, Moretti MS, Gonçalves JF Jr, Callisto M (2010) What is more important for invertebrate colonization in a stream with low-quality litter inputs: exposure time or leaf species? Hydrobiologia 654:125–136

Lodge DM (1991) Herbivory on freshwater macrophytes. Aquat Bot 41:195–224

Martins RT, Silveira LS, Alves RG (2011) Colonization by oligochaetes (Annelida: Clitellata) in decomposing leaves of *Eichhornia azurea* (SW.) Kunth (Pontederiaceae) in a neotropical lentic system. Annales de Limnologie. Int J Limnol 47:339–346

Martins RT, Silveira LS, Lopes MP, Alves RG (2017) Invertebrates, fungal biomass, and leaf breakdown in pools and riffles of neotropical streams. J Insect Sci 17:1–11

Nachtrieb JG, Grodowitz MJ, Smart RM (2011) Impact of invertebrates on three aquatic macrophytes: American pondweed, Illinois pondweed, and Mexican water lily. J Aquat Plant Manag 49:32–36

Newman RM (1991) Herbivory and detritivory on freshwater macrophytes by invertebrates: a review. J N Am Benthol Soc 10:89–114

Olson JS (1963) Energy storage and the balance of producers and decomposers in ecological systems. Ecology 44:322–331

Pagioro TA, Thomaz SM (2006) Effects of flooding regime upon the decomposition of *Eichhornia azurea* (Sw.) Kunth measured on a tropical, flow-regulated floodplain (Paraná River, Brazil). River Res Appl 22:791–801

Press WH, Teukolsky SA, Vetterling WT, Flannery BP (2007) Numerical recipes, the art of scientific computing. Cambridge University Press, New York

Reuss NS, Hamerlík L, Velle G, Michelsen A, Pedersen O, Brodersen KP (2013) Stable isotopes reveal that chironomids occupy several trophic levels within West Greenland lakes: implications for food web studies. Limnol Oceanogr 58:1023–1034

Reuss NS, Hamerlík L, Velle G, Michelsen A, Pedersen O, Brodersen KP (2014) Microhabitat influence on chironomid community structure and stable isotope signatures in West Greenland lakes. Hydrobiologia 730:59–77

Saito VS, Fonseca-Gessner AA (2014) Taxonomic composition and feeding habitats of Chironomidae in Cerrado streams (Southeast Brazil): impacts of land use changes. Acta Limnol Bras 26:35–46

Sangiorgio F, Pinna M, Basset A (2004) Inter- and intra-habitat variability of plant detritus decomposition in a transitional environment (Lake Alimini, Adriatic Sea). Chem Ecol 20:353–366

Sangiorgio F, Dragan S, Rosati I, Teodorof L, Staras M, Georgescu L, Basset A (2008) Decomposition of reed swamp detritus in the Danube Delta: a case study of four eutrophic systems. Transit Waters Bull 2:26–27

Shimabukuro EM, Henry R (in press) Macrophytes in tropical shallow lakes: an important food item to benthic entomofauna or an underused resource?

Silva FL, Ruiz SS, Bochini GL, Moreira DC (2008) Functional feeding habits of Chironomidae larvae (Insecta, Diptera) in a lotic system from Midwestern region of São Paulo State, Brazil. Pan-Am J Aquat Sci 3:135–141

Silva JS, Silveira WT, Albertoni EF, Palma-Silva C (2010) Diversity of Chironomidae (Diptera) in decomposing *Nymphoides indica* (L.) Kuntze in two subtropical lakes with different trophic conditions. Pan-Am J Aquat Sci 5:557–571

Silva FL, Oliveira HRN, Escarpinati SC, Fonseca-Gessner AA, Paula MC (2011) Colonization of leaf litter of two aquatic macrophytes, *Mayaca fluviatilis* Aublet and *Salvinia auriculata* Aublet by aquatic macroinvertebrates in a tropical reservoir. Rev Ambiente Água—Interdiscip J Appl Sci 6:30–39

Silveira LS, Martins RT, Silveira GA, Grazul RM, Lobo DP, Alves RG (2013) Colonization by Chironomidae larvae in decomposition leaves of *Eichhornia azurea* in a lentic system in southeastern Brazil. J Insect Sci 13:1–13

Song N, Yan ZS, Cai HY, Jiang HL (2013) Effect of temperature on submerged macrophyte litter decomposition within sediments from a large shallow and subtropical freshwater lake. Hydrobiologia 714:131–144

Stripari NL, Henry R (2002) The invertebrate colonization during decomposition of *Eichhornia azurea* Kunth in a lateral lake in the mouth zone of Paranapanema River into Jurumirim Reservoir (São Paulo, Brazil). Braz J Biol 62:293–310

Taylor BR, Chauvet EE (2014) Relative influence of shredders and fungi on leaf litter decomposition along a river altitudinal gradient. Hydrobiologia 721:239–250

Teixeira C, Kutner MB (1962) Plankton studies in a mangrove environment. I—first assessment of standing stock and ecological factors. Bol do Inst Oceanogr 12:101–124

Trivinho-Strixino S (2011) Larvas de Chironomidae: guia de identificação. Departamento de Hidrobiologia/Laboratório de Entomologia Aquática/UFSCar, São Carlos

Wetzel RG (1975) Limnology. Saunders, Philadelphia

Zilli FL, Montalto L, Marchese MR (2008) Benthic invertebrate assemblages and functional feeding groups in the Paraná River floodplain (Argentina). Limnologica 38:159–171

13

Effect of dietary soybean meal and multienzyme on intestine histology of beluga sturgeon (*Huso huso*)

M. Sohrabnezhad · M. Sudagar · M. Mazandarani

Abstract This study aimed to evaluate the effect of soybean meal and multienzyme supplementation on intestinal histopathology of beluga sturgeon (*Huso huso*). Nine diets were formulated (40% crude protein, 13% crude fat and 4150 kcal/kg gross energy) in which fish meal was replaced by soybean meal at 0, 40 and 80% supplemented or not with 1 or 2 g/kg of multienzyme Kemin. Fish (initial weight of 820 g) were fed on experimental diets four times daily at 1% body weight for 12 weeks. From histological analyses, intact villus with normal mucosa was found in the FM group, representing the presence of intact epithelium. Supplementation of enzyme at both levels and soybean meal (alone) in the diet resulted in histopathological alterations in the intestine including mild immune cells infiltration, villus blunting, villus fusion, mononuclear immune cells infiltration as well as necrosis. Soybean meal decreased the length of villuses in the proximal and distal intestines ($P < 0.05$). Enzyme was not able to ameliorate the derangements induced by 80% replacement level of fish meal with soybean meal (treatments SB80E1% and SB80E2%). However, co-administrating of the enzyme at both levels with 40% soybean meal (treatments SB40E1% and SB40E2%) was able to inhibit the soybean meal-induced histopathological impairments.

Keywords Fish meal · Soybean meal · Multienzyme · Histopathological changes · Intestine · Beluga sturgeon

Introduction

Due to increasing demand, limitation in global supplies in the near future and the dramatic increase in fish meal (FM) price, efforts to replace FM by plant protein sources as alternatives to FM have been increasing in aqua feeds (Gatlin et al. 2007). Among plant protein sources, defatted soybean meal (SBM) due to its high protein content, relatively well-balanced amino acid profiles and steady supply, has been the most cost-effective alternative for FM in diets of many fish species (Lim et al. 2011; Magalhães et al. 2016; Lin and Lou 2011). Several studies have reported that even when heat-treated, standard (solvent-extracted) and full-fat SBM-containing feeds supplemented with limiting amino acids, can lead to decreased growth and energy digestibility in farmed species (Merrifield and Olsen 2011; Krogdahl et al. 2003). However, considerable success in partially or totally replacing FM with SBM in diets for fish species has been achieved (El-Saidy and Gaber 2002; Yaghoubi et al. 2016; Zhou et al. 2011).

It is well demonstrated that different fishes exhibited varied sensitivity to SBM in their diet. In several aquatic species replacement of FM with vegetable protein sources such as SBM resulted in lower weight gain

M. Sohrabnezhad · M. Sudagar (✉) · M. Mazandarani
Department of Fisheries, Gorgan University of Agricultural Sciences and Natural Resource, Gorgan, Iran
e mail: sudagar_m@yahoo.com

Table 1 Composition and energy of experimental diets

	Dietary treatments								
	FM	FME1%	FME2%	SB40E0%	SB40E1%	SB40E2%	SB80E0%	SB80E1%	SB80E2%
Fish meal	50	50	50	30	30	30	10	10	10
Soybean meal	0	0	0	20	20	20	40	40	40
Wheat gluten	2.51	2.51	2.51	12.16	12.16	12.16	22.03	22.03	22.03
Wheat meal	24.93	24.93	24.93	15.13	15.13	15.13	3.9	3.9	3.9
Soybean oil	4.14	4.14	4.14	4.67	4.67	4.67	5.35	5.35	5.35
Fish oil	4.14	4.14	4.14	4.67	4.67	4.67	5.35	5.35	5.35
Molasses	3	3	3	3	3	3	3	3	3
Mineral mix	2	2	2	2	2	2	2	2	2
Vitamin mix	2	2	2	2	2	2	2	2	2
Dicalcium phosphate	2	2	2	2	2	2	2	2	2
Antioxidant	0.02	0.02	0.02	0.02	0.02	0.02	0.02	0.02	0.02
Mold inhibitor	0.25	0.25	0.25	0.25	0.25	0.25	0.25	0.25	0.25
Filler	0.91	0.91	0.91	0	0	0	0	0	0
Binder	3	3	3	3	3	3	3	3	3
Methionine	0.5	0.5	0.5	0.5	0.5	0.5	0.5	0.5	0.5
Lysin	0.5	0.5	0.5	0.5	0.5	0.5	0.5	0.5	0.5
Choline chloride	0.1	0.1	0.1	0.1	0.1	0.1	0.1	0.1	0.1
Kemin	0	1	2	0	1	2	0	1	2
Crude protein	40	40	40	40	40	40	40	40	40
Crude lipid	13.27	13.27	13.27	13	13	13	13	13	13
Gross energy(Kcal/Kg)	4150	4150	4150	4174	4174	4174	4179	4179	4179

(Francis et al. 2001) and histopathological damages (Aslaksen et al. 2007; Kraugerud et al. 2007; Krogdahl et al. 2003), the magnitude of which depends on the source and level of replacement (Merrifield and Olsen 2011). Thus, the change from FM to soybean meal generally causes several metabolic and health challenges for the farmed aquatics (Heikkinen et al. 2006). Hence, when using plant derived materials, especially those from soybean at high replacement levels it is necessary to consider the impacts on intestine histology as the gastrointestinal (GI) tract can be one of the important infection routes for some pathogens in fish (Merrifield and Olsen 2011; Gatlin et al. 2007).

There is growing interest to culture beluga sturgeon in Iran due to its large size, fast growth rate and tasty flesh (Ghomi et al. 2011). The aim of the present study was to evaluate the effect of soybean meal and multienzyme supplementation on intestinal histopathology of beluga sturgeon (*Huso huso*) and whether enzyme supplementation can reduce the negative impact of SBM diet on histopathological changes of beluga intestine.

Fig. 1 Histomorphological features of the proximal intestine of beluga sturgeon fed SBM diet. Cross section from proximal ▶ intestine; **a** FM (fish meal only), **b** FME1% (fish meal + 1% enzyme), **c** FME2% (fish meal + 2% enzyme), **d** SB40E0% (40% soybean meal + 0% enzyme), **e** SB40E1% (40% soybean meal + 1% enzyme), **f** SB40E2% (40% soybean meal + 2% enzyme), **f** SB80E0% (80% soybean meal + 0% enzyme), **g** SB80E1% (80% soybean meal + 1% enzyme), **h** SB80E2% (80% soybean meal + 2% enzyme). See intact villus (IV) with normal mucosa in the control group, which are representing intact epithelium (IE). Intact villus (IV), blunting of villuses (BV), fusion of villus (FV), necrosis in the villuses (N), mild immune cell infiltration in the villus (MIM), mononuclear immune cells (MoIM), severe immune cells infiltration (SIM) morphonuclear immune cells PIM) (H&E staining, ×40, ×100 and ×400 magnifications; all images are captured under 2.4 optical zoon provided by camera)

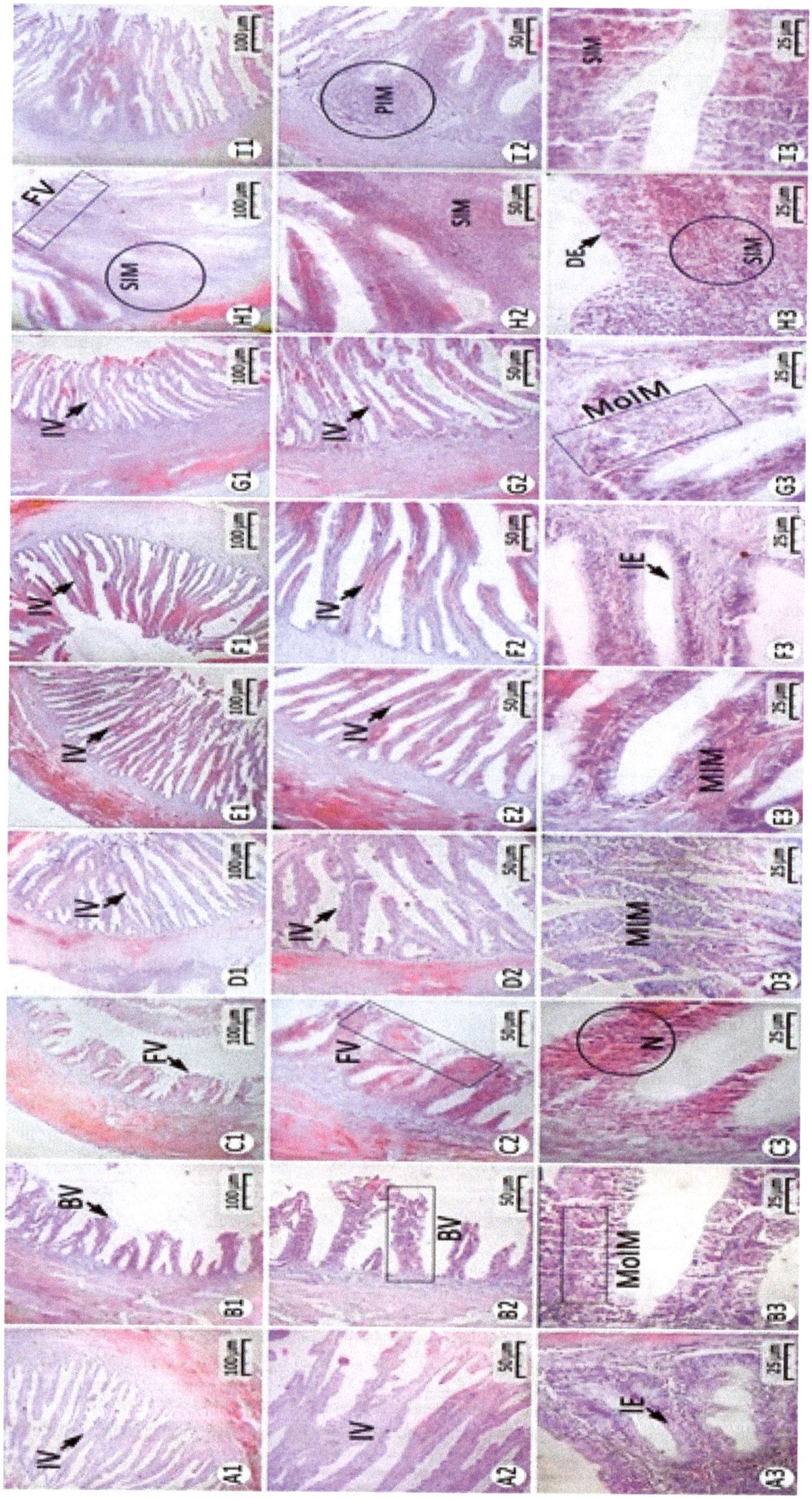
A1
IV
100 µm
B1
BV
100 µm
C1
FV
100 µm
D1
IV
100 µm
E1
IV
100 µm
F1
IV
100 µm
G1
IV
100 µm
H1
FV
SIM
100 µm
I1
100 µm
A2
IV
50 µm
B2
BV
50 µm
C2
FV
50 µm
D2
IV
50 µm
E2
IV
50 µm
F2
IV
50 µm
G2
IV
50 µm
H2
SIM
50 µm
I2
PIM
50 µm
A3
IE
25 µm
B3
MoIM
25 µm
C3
N
25 µm
D3
MIM
25 µm
E3
MIM
25 µm
F3
IE
25 µm
G3
MoIM
25 µm
H3
DE
SIM
25 µm
I3
SIM
25 µm

Fig. 2 Histomorphological features of the mid intestine of beluga sturgeon fed SBM diet. Cross section from mid intestine; **a** FM (fish meal only), **b** FME1% (fish meal + 1% enzyme), **c** FME2% (fish meal + 2% enzyme), **d** SB40E0% (40% soybean meal + 0% enzyme), **e** SB40E1% (40% soybean meal + 1% enzyme), **f** SB40E2% (40% soybean meal + 2% enzyme), **f** SB80E0% (80% soybean meal + 0% enzyme), **g** SB80E1% (80% soybean meal + 1% enzyme), **h** SB80E2% (80% soybean meal + 2% enzyme). Intact villus (IV), blunting of villuses (BV), fusion of villus (FV), necrosis in the villuses (N), mild immune cell infiltration in the villus (MIM), mononuclear immune cells (MoIM), severe immune cells infiltration (SIM) morphonuclear immune cells PIM) (H&E staining, ×40, ×100 and ×400 magnifications; all images are captured under 2.4 optical zoon provided by camera)

Materials and methods

Experimental diets

Nine experimental diets were formulated to be isonitrogenic (40% crude protein), isofat (13% crude fat) and isoenergetic (4150 kcal/kg gross energy) in a 3 × 3 factorial design in which FM was replaced by SBM at 0, 40 and 80% supplemented or not with 1 or 2 g/kg of multienzyme Kemin (phytase, lipase, xylanase, endo-1,3(4)-beta-glucanase, cellulose, alpha amylase, protease) (Kemzyme-Kemin industries, Belgium). Diet formulation has been performed suing UFFDA software (no. 6). All dietary ingredients were mixed through a commercial food mixer and pelleted using a pellet mill (Garma Electric Co., Amol, Iran) through a 3 mm die. Pellets were sieved to remove particles and stored at room temperature until use. Experimental diet compositions are indicated in Table 1.

Feeding trials

Ninety fish with initial weight of 820.2 g were obtained from Saee sturgeon farm (Sari, Iran). The trial was performed in nine fiberglass tanks (2 × 1.8 m) of 1.2 m^3 water capacity and supplied with a continuous flow of 5 L/min. Each tank was divided into two parts using a net in longitudinal direction to obtain two replication in the same tank. The same diet/treatment was supplied for each divide tank ($n = 5$ fish per each divided part). Fishes were then acclimatized for 10 days to the experimental conditions and fed with commercial Coppens diet containing 46% protein and 15% fat (Coppense Co., The Netherland) twice daily (06:00 and 18:00 h) during this period. After adaptation period, fish were fed on experimental diets by hand four times a day (06:00, 12:00, 18:00 and 24:00 h) at 1% body weight for 12 weeks. During the trial, water temperature was 17–21 °C; pH was approximately 7.6, dissolved oxygen was above 6.0 mg/l and photoperiod controlled to 12-h light and 12-h dark.

Sample collection and histological analysis

At the end of the feeding period, two fishes were removed from each divided part of the tank ($n = 4$ fish per each tank) and anesthetized using clove powder (0.3 g/l) for 3 min and then killed to remove intestines for histological analysis. During feeding trial, mortality did not occur. The obtained intestines were fixed with 4% formalin for further histological analysis. Histological measurements were performed according to Gava et al. (2015). Formalin-fixed intestinal samples were sectioned at three different parts (proximal, middle and distal sections) and routinely processed for histology including dehydration in graded ethanol and equilibration in xylene. Previously fixed intestinal samples were paraffin embedded and cut about 6 μm thick by rotary microtome (Microm, Germany). The sections (6 μm) were stained with haematoxylin and eosın (HE). After that, two glass slides of proximal, middle and distal sections from both fish were prepared and analyzed under light microscope by multiple magnifications (40×, 100× and 400×). The scale bar identification and magnification proof was performed using Measure software, version 1.1.0. Images are captured using on-board camera (Zeiss, Cyber-Shot, Japan).

Statistical analysis

All data were subjected to analysis by General Linear Model (GLM) with a 3 × 3 factorial design using SPSS 16.0 software package for Windows (SPSS, Illinois, USA). Significantly differences among the means were

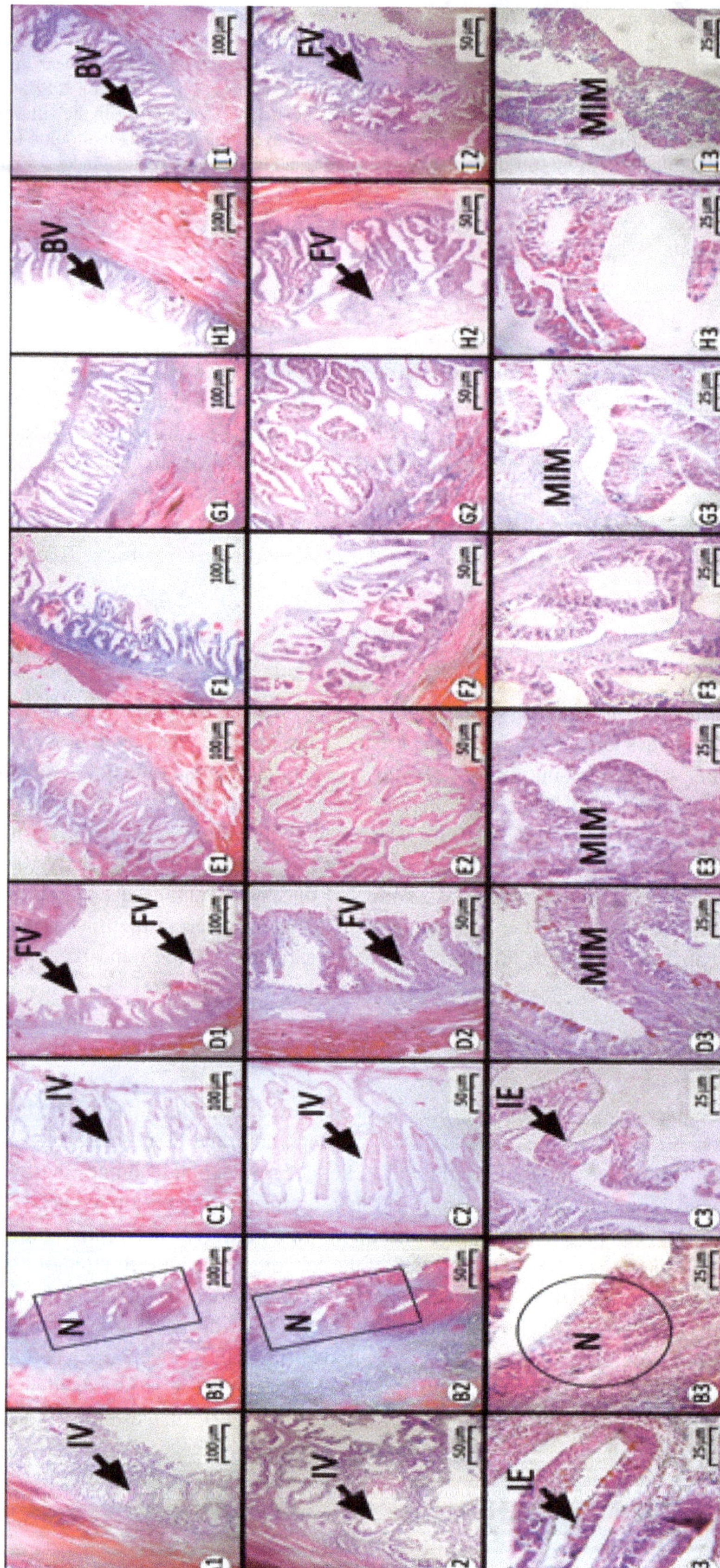
A1
A2
A3
B1
B2
B3
C1
C2
C3
D1
D2
D3
E1
E2
E3
F1
F2
F3
G1
G2
G3
H1
H2
H3
I1
I2
I3
IV
N
IE
FV
MIM
BV
100 µm
50 µm
25 µm

Fig. 3 Histomorphological features of the distal intestine of beluga sturgeon fed SBM diet. Cross section from distal intestine; ▶ **a** FM (fish meal only), **b** FME1% (fish meal + 1% enzyme), **c** FME2% (fish meal + 2% enzyme), **d** SB40E0% (40% soybean meal + 0% enzyme), **e** SB40E1% (40% soybean meal + 1% enzyme), **f** SB40E2% (40% soybean meal + 2% enzyme), **f** SB80E0% (80% soybean meal + 0% enzyme), **g** SB80E1% (80% soybean meal + 1% enzyme), **h** SB80E2% (80% soybean meal + 2% enzyme) (H&E staining, ×40, ×100 and ×400 magnifications; all images are captured under 2.4 optical zoom provided by camera)

examined using the Duncan's Multiple Range Test (DMRT) at a 95% significant difference ($P < 0.05$). All data are presented as Mean ± SD.

Results

Histopathological changes in response to types of diets were observed in the proximal part of beluga sturgeon intestine (Fig. 1). Intact villus (IV) with normal mucosa in the treatment 1 (FM) without the addition of soybean meal or multienzyme was found, representing the presence of intact epithelium (IE). Blunting of villuses (BV) in cross section of intestine from treatment 2 (FME1%) can be seen which is associated with mononuclear immune cells infiltration (MoIM) in higher magnification. The histopathological damages are developed in cross section from group treatment 3 (FME2%), which is presented with fusion of villus (FV) and sever necrosis in the villuses (N). The cross sections from treatment 4 (SB40) and treatment 5 (SB40E1%) is representing intact villus (IV), suggesting recovery condition. Mild immune cell infiltration in the villus (MIM) in cross section from treatment SB40 group can be seen similar to that cross section from SB40E1% group. The severity of damages are increased in cross sections from treatments SB80E0%, SB80E1% and SB80E2% which are representing intensive mononuclear immune cells (MoIM) in treatment SB80E0% to severe immune cells infiltration (SIM) in treatment SB80E1% that turned into poly morphonuclear immune cells (PIM) in mucosa of the cross sections from treatment SB80E2%.

The cross section of mid intestine from control group is representing intact villus, normal mucosa and submucosa layers as well as intact epithelium (Fig. 2). Necrosis (N) in cross section of fish intestine from treatment FME1% can be observed. Mild immune cells infiltration in cross sections from treatments SB40 and SB40E1% was occurred, which are enhanced in treatments SB80E0%, SB80E1% and SB80E2%. The cross sections from treatment SB80E1% and SB80E2% are representing FV and BV of villus (Fig. 2). Similar histopathological changes induced by soybean meal diet were observed in the distal part of intestine (Fig. 3).

Villus length, widths and mucosal thickness were measured in different parts of beluga intestine (Table 2). In proximal part, the length of villuses in the control was significantly higher than other treatments (1654 μm). Fish fed the SB80E0% diet had the shortest villi length (401.5 μm) ($P < 0.05$). Addition of enzyme resulted in higher villus length in treatments SB80E1% and SB80E2% (701 and 834 μm, respectively) ($P < 0.05$). The length of villuses in the mid intestine was shorted than those in the proximal and distal intestines and was not influenced by dietary treatments. Villus length in distal intestine of FM group was similar to that of proximal intestine (1604.7 μm). Similar to proximal intestine, enzyme supplementation at both levels decreased the length of villuses in treatments FME1% and FME2%.

Discussion

As reported in several studies, soybean meal is one of the most commonly used protein sources in aqua feeds due to its high protein content and favorable amino acid profile (Merrifield and Olsen 2011). However, even when heat-treated and supplemented with limiting amino acids, full-fat as well as defatted soybean meal rich feeds can lead to decreased growth, lower feed intake and reduced energy and lipid digestibility in fish species. It is known that such effects are mediated by anti-nutritional factors present in soybean meal such as saponins.

Results of the light microscopic analyses in this study showed that supplementation of multienzyme kemin (at both levels) and soybean meal (alone) in the diet of beluga sturgeon resulted in histopathological alterations in the intestine including mild immune cells infiltration, villus blunting, villus fusion, mononuclear immune cells infiltration as well as necrosis segmentally in some villuses. The observed results are in agreement with previous findings on several other fish species fed with soybean meal in their diet. Aslaksen

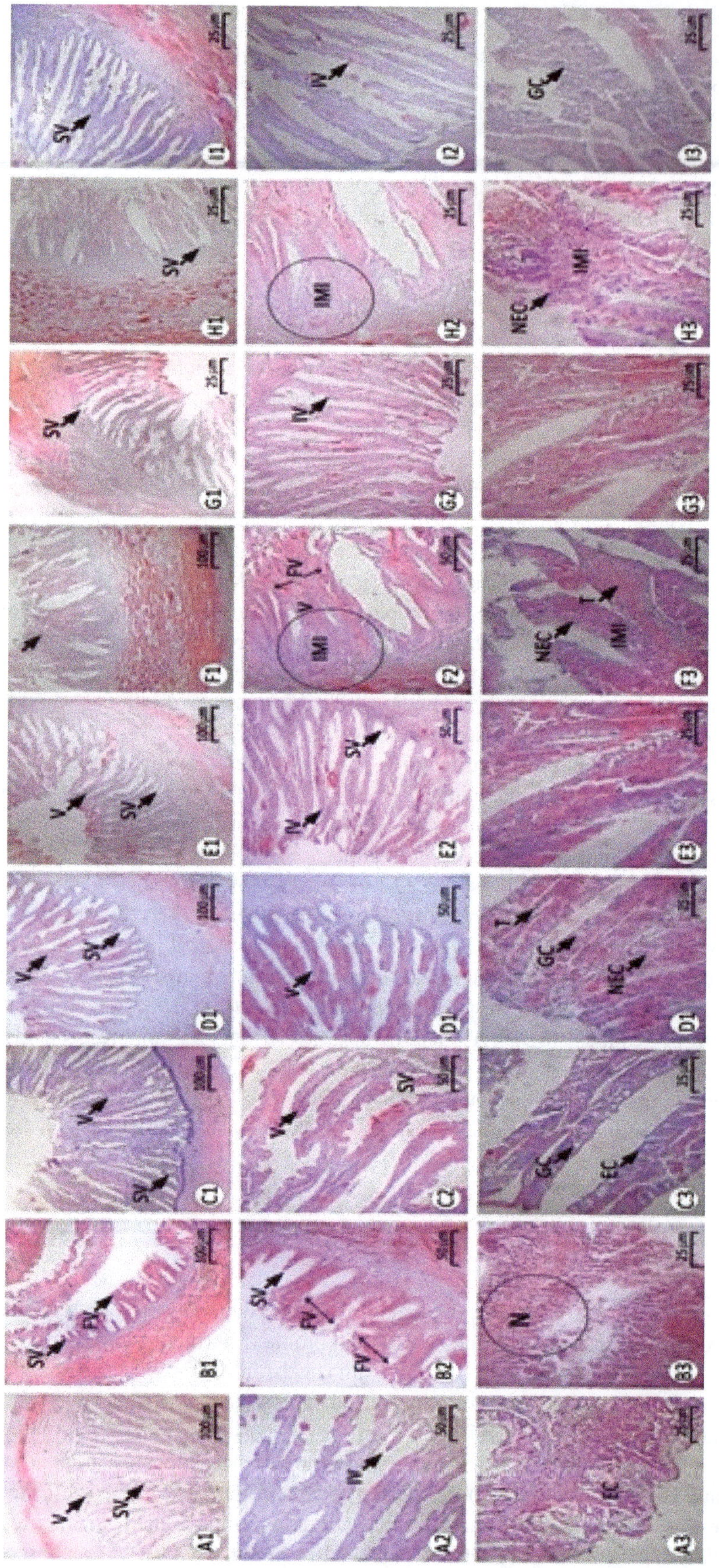

Table 2 Villus length and villus widths (μM) of different sections of beluga intestine fed SBM diet for 12 weeks

Measurements	Dietary treatments								
	FM	FME1%	FME2%	SB40E0%	SB40E1%	SB40E2%	SB80E0%	SB80E1%	SB80E2%
Villus length (μM)									
Proximal	1654.0 ± 70.7^{a}	$1203 \pm 154.9^{a,b,c}$	$1403 \pm 208.5^{b,c}$	$701 \pm 70.7^{b,c}$	$1203 \pm 70.7^{a,b}$	$1102 \pm 0.00^{a,b}$	401.5 ± 183.5^{c}	$701.0 \pm 146.7^{b,c}$	$834 \pm 70.7^{b,c}$
Middle	902.8 ± 212.6^{b}	701.5 ± 100.4^{b}	1553.5 ± 71.4^{a}	$102.7 \pm 308.8^{a,b}$	701.7 ± 141.7^{b}	902.2 ± 0.00^{b}	751.8 ± 212.6^{b}	802 ± 106^{b}	701.8 ± 70.8^{b}
Distal	1604.7 ± 708.8^{a}	1002.6 ± 160.7^{b}	1152.8 ± 212.6^{b}	$1303.2 \pm 267^{a,b}$	$902.1 \pm 70.8^{b,c}$	1103 ± 354.4^{b}	$902.2 \pm 141.7^{b,c}$	751.8 ± 354.4^{c}	601.5 ± 141.7^{c}
Villus widths (μM)									
Proximal	200.5 ± 0.00	275.6 ± 35.4	200.5 ± 70.8	225.5 ± 106.3	205.5 ± 0.00	225.5 ± 35.4	250.6 ± 70.8	225.5 ± 35.4	150.3 ± 70.8
Middle	300.7 ± 0.00	300.7 ± 70.8	325.8 ± 177.2	250.6 ± 70.8	325.8 ± 35.4	250.6 ± 70.8	250.5 ± 142	300.7 ± 141.7	250.6 ± 0.00
Distal	275.6 ± 177.2	300.7 ± 141.7	300.7 ± 70.8	451 ± 70.8	340.4 ± 14.7	325.8 ± 35.4	350.8 ± 70.8	325.8 ± 35.4	350.8 ± 70.8

Small superscript letters in the same row indicate siginificant differences ($P < 0.05$)

FM fish meal only, *FME1%* fish meal + 1% enzyme, *FME2%* fish meal + 2% enzyme, *SB40E0%* 40% soybean meal + 0% enzyme, *SB40E1%* 40% soybean meal + 1% enzyme, *SB40E2%* 40% soybean meal + 2% enzyme, *SB80E0%* 80% soybean meal + 0% enzyme, *SB80E1%* 80% soybean meal + 1% enzyme, *SB80E2%* 80% soybean meal + 2% enzyme

et al. (2007) reported morphological changes in the distal intestine of salmonids caused by dietary inclusion of full-fat as well as solvent-extracted SBM, with and without hulls. These morphological changes (enteritis) have been described as shortening of the simple and complex mucosal folds with a widening of the central stroma (lamina propria) and submucosa, shortened microvilli of the brush border membrane. In their study on Atlantic salmon, Kraugerud et al. (2007) observed morphological changes in the distal intestine of fish fed 10% dietary defatted soybean meal for 4 weeks. These changes included shortening and fusion of the simple mucosal folds, widening of the lamina propria with increased immune cells infiltration (cellularity), leucocytic cellular infiltration of the submucosa and lamina. Merrifield et al. (2009) observed the effect of replacing 50% of fish meal with soybean meal on the intestinal epithelium of rainbow trout using electron microscopy. After 16 weeks, soybean meal products caused an inflammatory response in the distal intestine. Fish fed the soybean meal enriched diet displayed missing, damaged, deformed, shorter and thicker microvilli. Additionally, enterocytes appeared to be malformed and irregular. Shorter (distal intestine) and less densely packed (proximal intestine) microvilli on the enterocyte surfaces of fish fed the soybean meal diet than fish fed the fish meal diet was found. These changes could partially explain the effects on growth parameters and feed utilization.

Comparing the damages delivered from enzyme and soybean meal (alone) with each other revealed that soybean meal exerted a higher histopathological changes. However, co-administrating of the multienzyme (at both levels) with 40% soybean meal (treatments SB40E1% and SB40E2%) was able to inhibit the soybean meal-induced histopathological impairments. Moreover, histological analyses exhibited that, enzyme (at both levels) was not able to ameliorate the derangements induced by 80% replacement level of fish meal with soybean meal (treatments SB80E1% and SB80E2%) including polymorphonuclear immune cells infiltration, necrosis, FV and BV in both intestine segments.

In conclusion, results of the present study indicated that beluga was sensitive to dietary soybean meal especially at high inclusion levels. However, co-administrating of the enzyme at both levels with 40% soybean meal (SB40E1% and SB40E2%) was able to inhibit the soybean meal-induced histopathological impairments.

References

Aslaksen MA, Kraugerud OF, Penn M, Svihus B, Denstadli V, Jørgensen HY, Hillestad M, Krogdahl A, Storebakken T (2007) Screening of nutrient digestibilities and intestinal pathologies in Atlantic salmon, *Salmo salar*, fed diets with legumes, oilseeds, or cereals. Aquaculture 272:541–555

El-Saidy DMSD, Gaber MMA (2002) Complete replacement of fish meal by soybean meal with dietary L-lysine supplementation for Nile tilapia *Oreochromis niloticus* (L.) fingerlings. J World Aquac Soc 33:297–306

Francis G, Makkar H, Becker K (2001) Antinutritional factors present in plant derived alternate fish feed ingredients and their effects in fish. Aquaculture 199:197–227

Gatlin DM, Barrows FT, Brown P, Dabrowski K, Gaylord G, Hardy R, Herman E, Hu G, Krogdahl A, Nelson R, Overturf K, Rust M, Sealy W, Skonberg D, Souza EJ, Stone D, Wilson R, Wurtele E (2007) Expanding the utilization of sustainable plant products in aquafeeds: a review. Aquac Res 38:551–579

Gava MS, Moraes LB, Carvalho D, Chitolina GZ, Fallavena LCB, Moraes HLS, Herpich J, Salle CTP (2015) Determining the best sectioning method and intestinal segment for morphometric analysis in broilers. Revista Brasileira de Ciência Avícola. doi:10.1590/1516-635x1702145-150

Ghomi MR, Shahriari R, Langroudi HF, Nikoo M, Von Elert E (2011) Effects of exogenous dietary enzyme on growth, body composition, and fatty acid profiles of cultured great sturgeon *Huso huso* fingerlings. Aquac Int. doi:10.1007/s10499-011-9453-9

Heikkinen J, Vielma J, Kemiläinen O, Tiirola M, Eskelinen P, Kiuru T, Navia-Paldanius D, Wright A (2006) Effects of soybean meal based diet on growth performance, gut histopathology and intestinal microbiota of juvenile rainbow trout (*Oncorhynchus mykiss*). Aquaculture 261:259–268

Kraugerud OF, Penn M, Storebakken T, Refstie S, Krogdahl A, Svihus B (2007) Nutrient digestibilities, gut function and morphology in Atlantic salmon (*Salmo salar*) fed diets where non-starch polysaccharides from soy replace cellulose. Aquaculture 273:96–107

Krogdahl A, Bakke-McKellep AM, Baeverfjord G (2003) Effects of graded levels of standard soybean meal on intestinal structure, mucosal enzyme activities, and pancreatic response in Atlantic salmon (*Salmo salar* L.). Aquac Nutr 9:361–371

Lim SJ, Kim SS, Ko GY, Song JW, Oh DH, Kim JD, Kim JU, Lee KJ (2011) Fish meal replacement by soybean meal in diets for Tiger puffer, *Takifugu rubripes*. Aquaculture 313:165–170

Lin S, Lou L (2011) Effects of different levels of soybean meal inclusion in replacement for fish meal on growth, digestive enzymes and transaminase activities in practical diets for juvenile tilapia, *Oreochromis niloticus* × *O. aureus*. Anim Feed Sci Technol 168:80–87

Magalhães R, Lopes T, Martins N, Díaz-Rosales P, Couto A, Pousão-Ferreira P, Oliva-Teles A, Peres H (2016) Carbohydrases supplementation increased nutrient utilization in white seabream (*Diplodus sargus*) juveniles fed high soybean meal diets. Aquaculture 463:43–50

Merrifield DL, Olsen RE (2011) Dietary effect of soybean (*Glycine max*) products on gut histology and microbiota of fish. In: El-Shemy H (ed) Soybean and nutrition. InTech, Rijeka, Croatia

Merrifield DL, Bradley G, Baker RTM, Dimitroglou A, Davies SJ (2009) Soybean meal alters autochthonous microbial populations, microvilli morphology and compromises intestinal enterocyte integrity of rainbow trout (*Oncorhynchus mykiss* Walbaum). J Fish Dis 32:755–766

Yaghoubi M, Mozanzadeh MT, Marammazi JG, Safari O, Gisbert E (2016) Dietary replacement of fish meal by soy products (soybean meal and isolated soy protein) in silvery-black porgy juveniles (*Sparidentex hasta*). Aquaculture 464:50–59

Zhou F, Song W, Shao Q, Peng X, Xiao J, Hua Y, Owari BN (2011) Partial replacement of fish meal by fermented soybean meal in diets for Black Sea bream, *Acanthopagrus schlegelii*, juveniles. J World Aquac Soc 42:184–197

14

Remarkably high ingestion ratio of acidic food in juvenile marble goby, *Oxyeleotris marmorata*

Chui-Fen Teoh · Leong-Seng Lim · Gunzo Kawamura

Abstract Behavioral test was conducted to determine the level of food pH which was preferable by the juvenile marble goby, *Oxyeleotris marmorata* (50 individuals or replicates; total length 6.6–7.0 cm) using agar gel pellets. Eight pH levels of agar gel pellets were prepared (pH 2.4, 3.0, 3.2, 3.4, 4.1, 4.9, 5.1, and the pure agar gel pellet without pH modification—pH 5.9). The ingestion ratio for each pH treatment of agar gel pellet was calculated, and the binary data (ingested or rejected) was analyzed using binomial test. Negative relationship was found between the ingestion ratio and pH level of the agar gel pellets. The highest ingestion ratio was found in the agar gel pellet with pH 2.4 (ingestion ratio, 94%), following by pH 3 (85%), pH 3.2 (3% agar gel powder; 65%), pH 3.2 (2% agar gel powder; 58%), pH 3.4 (24%), pH 4.1 (6%), pH 4.9 (6%), pH 5.1 (6%), and pH 5.9 (2%). The ingestion ratio of pH 2.4 agar gel pellet was significantly higher ($P < 0.05$) than that in the other treatments except that of pH 3.0. These results confirmed the taste preference of *O. marmorata* for acidic foods, and the most preferred was the one with pH 2.4–3.0.

Keywords Taste preference · Feeding · Dietary pH · Behavior · Weaning · *Oxyeleotris marmoratus*

Abbreviations

FT pellet	For-training pellet
PAG pellet	Pure agar gel pellet
TS pellet	Test substance pellet
HCl	Hydrochloric acid
NaCl	Sodium chloride
CaCl	Calcium chloride
CI	Confidence interval

Introduction

Marble goby, *Oxyeleotris marmorata* is a freshwater fish species with high commercial value, especially in Southeast Asia (Asia-Pacific Fishery Commission 2014). It is also a popular candidate for aquaculture in many Southeast Asian countries including Malaysia, Thailand, and Vietnam (Cheah et al. 1994; Lin and

C.-F. Teoh · L.-S. Lim (✉) · G. Kawamura
Borneo Marine Research Institute, Universiti Malaysia Sabah, Jalan UMS, 88400 Kota Kinabalu, Sabah, Malaysia
e-mail: leongsen@ums.edu.my

Kaewpaitoon 2000; Luong et al. 2005). However, it was reported that this fish generally rejects formulated feeds (Rojtinnakorn et al. 2012) and this results in its poor growth and survival. In addition to that, information on the nutritional requirement of *O. marmorata* is still limited up-to-date (Cheah et al. 1994; Lin and Kaewpaitoon 2000; Yong et al. 2015), and there is still no compounded feed commercially available for this fish. For these reasons, there is a need to find way to improve the acceptability of formulated feeds to *O. marmorata*. In the previous studies, amino acids mixture and nucleotides were reported as the suitable feeding stimulant through behavioural assays (Lim et al. 2015, 2016, 2017) and dietary supplementation of these taste substances can be practiced to improve the weaning of *O. marmorata* (Lai et al. in press). However, amino acid mixture and nucleotides are expensive and supplementation of these taste substances in the practical diet for *O. marmorata* may not be cost-effective.

According to Kasumyan and Døving (2003), fish has sensation to sour taste and its preference for food acidity is species-specific. This fact provides us the rationale to determine the possibility of improving the acceptability of formulated feeds to *O. marmorata* by manipulating the feed's pH to the fish preferable level, which is also a simple and cheaper way (by means of adding in acid) than the dietary supplementation of feeding stimulant mentioned earlier. However, information on the taste preference of *O. marmorata* for feeds with different pH levels and the level of pH in the feed most preferred by *O. marmorata* is unknown. Therefore, the present study was conducted to determine the pH level of food which is preferred by the *O. marmorata* juveniles.

Methods

Preparation of agar gel pellets

In this study, behavioral test was conducted and agar gel pellet was used as the food medium in the test. The types and composition of agar gel pellet prepared for the test are shown in Table 1. Two types of agar gel pellets were prepared: (1) the For-Training (FT) pellet that contained the extract of the commercial pellet, size 3.1 mm (Otohime, EP3 type, Nisshin Co., Tokyo, Japan), and (2) the Test Substance (TS) pellets which contained distilled water and different amounts of diluted or non-diluted hydrochloric acid—HCl (37%, 16 M, MERCK, NJ, USA). The pure agar gel (PAG) pellet that contained only distilled water (without HCl) was used as the negative control in the present study. In general, all agar gel pellets were prepared by dissolving

Table 1 Types of agar gel pellet prepared for the behavioral test in the present study with its ingredients and compositions

	Distilled water	Otohime extract	Agar gel powder[a]	Red food dye[b]	HCl[c] (molarity)
Treatments					
For-training (FT) pellet—pH 5.6	N/A	50 ml	1.0 g	0.05 g	N/A
Test-substance (TS) pellets					
(1) Pure agar gel (PAG) pellet—pH 5.9	50 ml	N/A	1.0 g	0.05 g	N/A
(2) 2% agar powder—pH 5.1	50 ml	N/A	1.0 g	0.05 g	10 μl (0.1 M)
(3) 2% agar powder—pH 4.9	50 ml	N/A	1.0 g	0.05 g	10 μl (1 M)
(4) 2% agar powder—pH 4.1	50 ml	N/A	1.0 g	0.05 g	10 μl (3 M)
(5) 2% agar powder—pH 3.4	50 ml	N/A	1.0 g	0.05 g	10 μl (16 M)
(6) 2% agar powder—pH 3.2	50 ml	N/A	1.0 g	0.05 g	18 μl (12 M)
(7) 3% agar powder—pH 3.2	50 ml	N/A	1.5 g	0.05 g	15 μl (12 M)
(8) 3% agar powder—pH 3.0	50 ml	N/A	1.5 g	0.05 g	34 μl (12 M)
(9) 3% agar powder—pH 2.4	50 ml	N/A	1.5 g	0.05 g	100 μl (12 M)

[a] Mermaid, Bangkok, Thailand

[b] Ponceau 4R, Mibo, Sabah, Malaysia

[c] Hydrochloric acid (37%, 16 M, MERCK, NJ, USA)

the agar gel powder (Mermaid, Bangkok, Thailand) (2% of the water volume) and red food dye (Ponceau 4R, Mibo, Sabah, Malaysia) (0.1%) into 50 ml distilled water. The mixture was heated on a hot plate until it boiled. Subsequently, the mixture was poured into a glass petri dish for cooling and hardening. The hardened mixture were then cut into uniform pellet size (approximately 2 mm × 2 mm × 2 mm) and stored in a refrigerator (4 °C) until further use (Lim et al. 2015, 2016, 2017).

For preparing the FT pellet, the distilled water was replaced by the commercial pellet (Otohime EP3) extract. The extract was obtained by soaking 30 g of the pellet (Otohime EP3) in 100 ml distilled water for approximately 15 min with occasional hand-stirring. When the water color turned dark brown, the homogenate was then filtered through a 60 μm mesh net to obtain the aqueous extract. The extract was prepared only prior to the preparation of the FT pellet to maintain its freshness.

To prepare the TS pellets with different pH levels, various amounts of HCl which were preliminary determined (see Table 1) were added into the mixture (distilled water + agar gel powder + red food dye) after the mixture was boiled. In total, eight types of TS pellet with pH levels of 2.4, 3.0, 3.2, 3.4, 4.1, 4.9, 5.1, and 5.9 (the PAG) were prepared. The pH levels of the TS pellets were determined following the method by Sathe (1999). In brief, a few pieces of agar gel pellet from each treatment were dropped in 100 ml distilled water, crushed with spatula, and the mixture was stirred frequently for about 3 min then the measurement was done using a pH meter (Oakton EcoTestr® pH 2, Eutech Instruments, Singapore). As the TS pellets with the extremely low pH levels (pH 2.4 and 3.0) could not harden naturally like the others, 3% of agar gel powder (instead of 2%) was used to prepare these TS pellets. To confirm that such act will not cause any bias to the results, the TS pellet with pH 3.2 was prepared with 2 and 3% agar gel powder and fed to the fish for comparison.

Experimental fish

Fifty Otohime pellet-trained *O. marmorata* juveniles (total length = 6.0–7.7 cm) was obtained and transferred from the fish hatchery to the wet laboratory of Borneo Marine Research Institute (BMRI), Universiti Malaysia Sabah (UMS). Each individual of fish was stocked in 7-L acrylic-made aquaria supplied with 4-L water and aeration, and this was done in replicate. Before the behavioral test started, all the fish were conditioned to accept agar gel pellet using the FT pellet. The fish conditioning protocol was the same as described by Lim et al. (2016, 2017). Each individual of fish was conditioned once daily at 04:00 p.m. with three pieces of FT pellet. If the fish ingested the FT pellets, they were rewarded with the commercial pellet (Otohime EP3) at the end of the daily conditioning session. The fish were starved on that day if none of the FT pellet given were ingested, and the conditioning continued on the next day. The fish were considered well-conditioned and ready for the behavioral test when it accepted all three FT pellets continuously for 3 days. In the present study, all fish were successfully conditioned to accept the FT pellets within 16 days and no mortality occurred. All experimental fish were cared and handled following the Researcher's Guideline on Code of Practice for the Care and Use of Animals for Scientific Purposes, implemented by the Universiti Malaysia Sabah (UMS).

Behavioural assays

The procedures of the behavioral test used in the present study were adopted from Lim et al. (2016, 2017). The test was started at 04:00 p.m. daily. Each fish was first fed with a FT pellet to stimulate its desire to feed, followed by a TS pellet, and then another FT pellet. The final piece of FT pellet was served to remind the fish that agar gel pellet was palatable, in case the TS pellet given was a deterrent to them. During the test, feeding responses (ingestion or rejection of the TS pellet given) were examined through naked eye and manually recorded. Each treatment of TS pellet was given only once to each fish, and only one treatment of TS pellet was tested daily. At the end of the daily test, each fish was fed with three commercial pellets (Otohime EP3) as the reward to compensate its nutritional requirement. Feces and the uneaten feed in all aquariums were syphoned out and about 20% of water was exchanged on the next day morning. The cleaning and water exchanging work was done daily at least 4 h before the behavioral test started to prevent the fish from stress. The behavioral test was conducted under indoor condition. During the test, the water temperature was at about 28 °C. The water pH before and after the test was at about 8.0–8.1.

Data collection and analysis

The total number of the ingested TS pellet from each pH level treatment was counted. Then, the ingestion ratio [total number of the ingested pellet/(total number of the given pellet, which was 50) $\times$ 100%] was calculated and expressed as the representative data. As binary data (ingested or rejected) was collected in the present study, the one-tailed binomial test was used to determine the significance of difference among the treatments. The 95% confidence interval (CI) of the total ingestion of each pH level of TS pellet was calculated and the ranges in the values were compared to each other. Significant difference ($P < 0.05$) was assumed if the range of the CI values between two treatments was not overlapped and vice versa (Sauro and Lewis 2012).

Results and discussion

In the present study, the ingestion ratio of TS pellets by the *O. marmorata* juveniles was clearly affected by the TS pellet's pH level. The ingestion ratio increased when the pH level decreased, suggesting a negative relationship between the ingestion ratio and the pH level of TS pellet. Figure 1 shows the ingestion ratio of the TS pellets by the juvenile *O. marmorata*. The highest ingestion ratio was found in the TS pellet with pH 2.4 (ingestion ratio 94%), following by pH 3 (85%), pH 3.2 (3% agar gel powder; 65%), pH 3.2 (2% agar gel powder; 58%), pH 3.4 (24%), pH 4.1 (6%), pH 4.9 (6%), pH 5.1 (6%), and pH 5.9 (2%). These results evidenced that the *O. marmorata* indeed has taste preference for food pH and they preferred the acidic food. In addition, no significant difference ($P > 0.05$) was observed between the ingestion ratio of TS pellets at pH 3.2 with 2 and 3% of agar gel powder. This result confirmed that the high ingestion ratio of the TS pellets with pH 2.4 and 3.0 was not due to the pellet's texture. In fact, many fish species were also reported to prefer the acidic food, including *Tilapia zilli* (Adams et al. 1988), European grayling *Thymallus thymallus* (Kasumyan 1997), tench *Tinca tinca* (Kasumyan and Prokopova 2001) and most of the salmonids and poecilids (Kasumyan and Døving 2003). On the other hand, acidified foods have also been reported to evoke negative response in many fish species including the puffer *Fugu pardalis* (Hidaka et al. 1978), acipenserids and many cyprinids (see Kasumyan and Døving 2003). Erteken and Nezaki (2002) also reported that the favorable food pH for the Black Sea turbot *Psetta maxima* was in the range of 7.2–7.4. Therefore, results of the present study were in agreement with the findings of Kasumyan and Døving (2003) that fish can sense sour taste and their taste preference for pH of the food is species-specific.

According to Kasumyan and Døving (2003), the classical taste substances that contained Cl^- ion such as sodium chloride (NaCl) and calcium chloride (CaCl) could act as the feeding stimulant for some fish species. In the present study, HCl was used to adjust the pH level of the TS pellets. When the amount of HCl added to

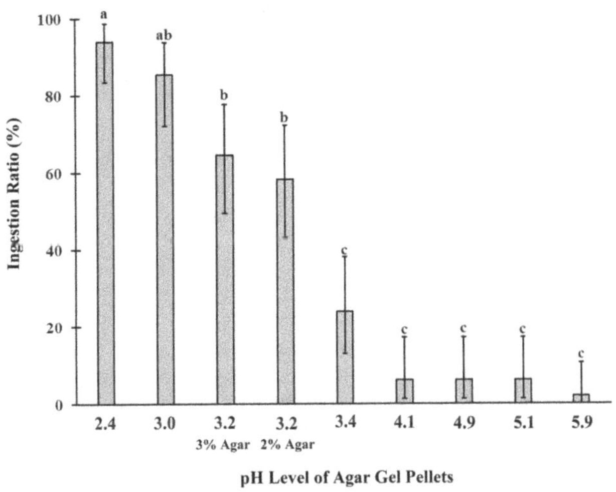

Fig. 1 Ingestion ratio of the agar gel pellets with different pH levels by the *O. marmorata* juveniles. Vertical bars show the 95% confidence interval of the data

the TS pellets increased (pH level decreased), the amount of Cl^- ion was also increased. Fish intraoral taste system can be sensitive to the Cl^- ion as examined electro-physiologically (Konishi and Niwa 1964). Therefore, the high ingestion ratio of the highly acidic TS pellets in the present study also could be contributed by the Cl^- ion. However, such hypothesis could not have the vigor of the experimental trials because NaCl and CaCl were also reported as the taste-indifferent substance for many fish species (Kasumyan and Døving 2003), including *O. marmorata* (Lim et al. 2017).

Conclusions

In conclusion, the preference of *O. marmorata* juveniles for acidic food was confirmed, and the food's pH preferred by the fish was in the range of 2.4–3.0. Evaluation of the weaning performance of *O. marmorata* juveniles with the HCl-acidified feed is recommended in a future study.

Acknowledgements This study was funded by the Research Acculturation Grant Scheme (RAGS) RAG0064-STWN-2015 from the Ministry of Higher Education of Malaysia. The authors thank Prof. Dr. Saleem Mustafa for reviewing this manuscript.

Author contributions C-FT provided ideas to refine the experimental methods, conducted the experiments, collected and analyzed the data, and wrote the manuscript. L-SL contributed the original idea of this study, designed the experimental methods, supervised the experimental data collection and analysis, and revised the manuscript. GK involved in designing the statistical analysis and proof-read the manuscript.

Conflict of interest No competing interests.

References

Adams MA, Johnsen PB, Zhou H (1988) Chemical enhancement of feeding for the herbivorous fish *Tilapia zillii*. Aquaculture 72:95–107

Asia-Pacific Fishery Commission (2014) Regional Overview of Aquaculture Trends in the Asia Pacific Region 2014. RAP Publication 2014/26. Food and Agriculture Organization of the United Nations (FAO), Bangkok, Thailand

Cheah SH, Senoo S, Lam SY, Ang KJ (1994) Aquaculture of a high-value freshwater fish in Malaysia: the marble or sand goby (*Oxyeleotris marmorata*, Bleeker). Naga ICLARM Q 17:22–25

Erteken A, Nezaki G (2002) Effects of feeding stimulants, and diet pH on the growth of Black Sea turbot, *Psetta maxima*. Turk J Fish Aquat Sci 2:19–22

Hidaka I, Ohsugi T, Kubomatsu T (1978) Taste receptor stimulation and feeding Behaviour in the puffer, *Fugu pardalis* I. Effect of single chemicals. Chem Sense Flavor 3:341–354

Kasumyan AO (1997) Gustatory reception and feeding behavior in fish. J Ichthyol 37:72–86

Kasumyan AO, Døving KB (2003) Taste preferences in fish. Fish Fish 4:289–347

Kasumyan AO, Prokopova OM (2001) Taste preferences and the dynamics of behavioural taste response in the tench, *Tinca tinca* (Cyprinidae). J Ichthyol 41:640–653

Konishi J, Niwa H (1964) Some properties of taste receptors in freshwater fish: responses to weak solutions. Jpn J Physiol 14:328–343

Lai SKJ, Lim LS, Yong ASK, Kawamura G, Shapawi R (in press) A preliminary study to determine the potential of a prototype feeding stimulant in improving the weaning of juvenile marble goby (*Oxyeleotris marmoratus*). Songklanakarin J Sci Technol

Lim LS, Lai JSK, Yong ASK, Shapawi R, Kawamura G (2015) A preliminary study on the taste preference of marble goby (*Oxyeleotris marmoratus*) for amino acids. Songklanakarin J Sci Technol 34(4):397–400

Lim LS, Lai JSK, Yong ASK, Shapawi R, Kawamura G (2016) Evaluation on the potential of betaine, taurine, nucleotide and nucleoside as feeding stimulant for juvenile marble goby *Oxyeleotris marmoratus* through behavioural assays. Int Aquat Res 8:161–167

Lim LS, Lai SKJ, Yong ASK, Shapawi R, Kawamura G (2017) Feeding response of marble goby (*Oxyeleotris marmorata*) to organic acids, amino acids, sugars and some classical taste substances. Appl Anim Behav Sci 196:113–118

Lin CK, Kaewpaitoon K (2000) An overview of freshwater cage culture in Thailand. In: Liao IC, Lin CK (eds) Proceedings of the First International Symposium on Cage Aquaculture in Asia, May 2000. Asian Fisheries Society, Manila, Philippine, pp 253–257

Luong VC, Yi Y, Lin CK (2005) Cove culture of marble goby (*Oxyeleotris marmorata* Bleeker) and carps in Tri An reservoir of Vietnam. Aquaculture 244:97–107

Rojtinnakorn J, Rittiplang S, Tongsiri S, Chaibu P (2012) Tumeric extract inducing growth biomarker in sand goby (*Oxyeleotris marmoratus*). In: Proceedings of the 2nd International Conference on Chemical, Biological and Environment Sciences (ICCEBS 2012), Bali, June–July, 2012. Planetary Scientific Research Centre, pp 41–43

Sathe AY (1999) A first course in food analysis. New Age International (P) Limited Publishers, New Delhi

Sauro J, Lewis JR (2012) Quantifying the user experience: practical statistics for user research. Morgan Kaufmann (MK) Publishers, Waltham

Yong ASK, Ooi SY, Shapawi R, Biswas AK, Kenji T (2015) Effects of dietary lipid increments on growth performance, feed utilization, carcass composition and intraperitoneal fat of marble goby, *Oxyeleotris marmorata*, juveniles. Turk J Fish Aquat Sci 15:653–660

15

A metabolic study in hepatopancreas of *Litopenaeus vannamei* response to white spot syndrome virus

Hao Wu · Jianlong Zhang · Yongxing He · Junfang Zhou · Jizhou Yan · Min Jiang

Abstract White spot syndrome virus (WSSV) cause great harm in shrimp aquaculture. To understand the impact of viral infection on the shrimp metabolism, we monitored the culture farms of *Litopenaeus vannamei* and collected the samples on different stages of WSSV infection. The hepatopancreas of shrimp were separated, and then used gas chromatography mass spectrometry to detect the metabolites. Through the mass spectrometric analysis combined with multivariate data analysis, including PCA and OPLS models, metabolism of the shrimp was significantly changed by WSSV infection. The data indicated that in the early stage of WSSV infection, the glycolysis changed significantly, the contents of glucose and lactate increased distinctly. The metabolites of TCA cycle did not show up obviously regularity. The organism of fatty acids showed the same situation with glycolysis. At the early stage of infection, 14 amino acids metabolism were up-regulated, and glycine still increased at later stage of infection and the concentration was increased twice. The data of this study may provide some information to further research of viral disease mechanism.

Keywords Metabolomics · White spot syndrome virus · GCMS

Introduction

White spot syndrome virus (WSSV) often brought disastrous to shrimp farming industry. The virus was reported to infect most aquatic animals including crabs and shrimp (Lo et al. 1996; Sánchez-Paz 2010). So, it is very important to learn the mechanisms of virus attacks the host and the shrimp response to WSSV

Hao Wu and Jianlong Zhang have contributed equally to this work.

H. Wu
Key Laboratory of Freshwater Aquatic Genetic Resources, Ministry of Agriculture, Shanghai Ocean University, Shanghai 201306, China

H. Wu · J. Zhang · J. Yan · M. Jiang (✉)
College of Fisheries & Life Science, Shanghai Ocean Universities, Shanghai 201306, China
e-mail: mjiang@shou.edu.cn

Y. He
Agricultural Service Center of Pudong New Area, Shanghai, China

J. Zhou
Key and Open Laboratory of Marine and Estuarine Fisheries, Ministry of Agriculture, East China Sea Fisheries Research Institute, Chinese Academy of Fishery Sciences, Shanghai, China

infection. Hepatopancrea was an important organ of shrimp metabolize. Some research had focused on genes, biochemical and physiological changes related to WSSV infection in shrimpand differentially expressed proteins in the hepatopancrea of shrimp (Yoganandhan 2003; Leu et al. 2007, 2013; Li et al. 2013) and upregulation and down-regulation proteins were identified. The above studies show that the host's metabolism will be affected after the WSSV virus invades, and this negative effect will lead to the shrimp's immune capacity decline or lead to death. To understand the reaction mechanism of WSSV infection, metabolism data were important supplement to the research. GCMS is a suitable instrumental to detective metabolites, which provides a rapid, steady, and high-throughput methodology. In this study, GCMS technique and multivariate statistics were applied to profile the hepatopancreas and approached for metabolic marker when *Litopenaeus vannamei* were infected by WSSV.

Materials and methods

Sample collection and treatment

There were several shrimp farms under our track, the aquaculture farm in this paper have 15 shrimp pounds. The shrimps would feed 4 or 5 times a day with shrimp fodder. And we monitor the Shrimp of every two weeks contain white spot virus tested and water quality monitored throughout the growing season. When the white spot virus was detected on the shrimp, the testing organisms with the virus and healthy control shrimp would be collected immediately and synchronous, samples were stored at −80 °C ultra-low temperature freezer. The number of samples collected per time is greater than 25, three whole shrimps for virus detection, and 20 shrimps for nutritional analysis and metabolic analysis with different body parts. We got three groups under same track time which contain healthy shrimp without virus as control (group-1), infected virus within 2 weeks as latent infection (group-2) and infected with the virus that causes floating and imminent death which is dominant infection (group-3). The weight range of all the samples is 1.97 ± 0.66 g. Quantitative detection of WSSV by RT-PCR. The hepatopancreas of shrimp would be homogenized in tube after weighted, then 250 μL water–methanol-chloroform (2:5:2 v/v/v) were mixed for metabolite extraction, and then kept at −18 °C for 2 h. After centrifugation at 10000 rpm for 30 min, 100 μL supernatant was dried completely, 80 μL nitrogen-methoxyamine (15 $mg \cdot mL^{-1}$ in pyridine) (Sigma-Aldrich, USA) and 80μL BSTFA (with 1% TMCS) (SUPELCO, USA)were added to the residue to trimethylsilylated at 60 °C for 60 min.

Samples were analyzed by GCMS (7890A + 5975C, Agilent, USA) in non-target detection with a DB-17MS capillary column (J&W Scientific, USA). 1 μL of sample was injected in splitless mode. The initial temperature was kept at 50 °C for 2 min, then raised with four steps, 100 °C at a rate of 8 °C/min, 170 °C at a rate of 10 °C/min and 200 °C at a rate of 5 °C/min, finally to 300 °C at a rate of 8 °C/min and kept for 4 min. The injection and ion source temperatures were 260 and 230 °C, the energy was 70 eV in electron impact mode. The mass spectrometry data were acquired in full-scan mode with the m/z range of 55–600.

Data analyses

The acquired dataset in MS files from GCMS analysis was exported into NetCDF format by MSD ChemStation (Agilent, USA). CDF files were processed using XCMS-online (XCMS, USA). The soft would finish peaks of denoising, smoothing, time-window splitting, multivariate curve resolution and alignment. The results of three-dimensional dataset comprised sample information, peak retention time and intensity. After principal component analysis (PCA) analysis with SIMCA-14 (Umetrics, Sweden) and Metaboanalyst online analysis software (www.metaboanalyst.ca/), projection to latent structures discriminate analysis (PLS-DA) and orthogonal projection to latent structures discriminate analysis (OPLS-DA) were carried out to further analysis. The software SIMCA14 was used to try to identify marker metabolites from three groups. Metabolite identification was determined according to National Institute of Standards and Technology (NIST) spectral library databases. The software of SPSS20 was application to make significant difference analysis and charts. Quantitative normalization within replicates was transformed by logarithmic base of 2 (Xia and Wishart 2011; Wu et al. 2013). Using SPSS 20.0 software for ANOVA analysis of the metabolites' relative content between three groups.

Results

Multivariate analysis

With the multivariate analysis of GCMS data, the score plot (Fig. 1, $R^2X = 98.6\%$, $Q^2 = 0.941$) of PCA was obtained with the first three PCs which represented 49 and 17% variance, respectively. Then the POLS-DA model was used, the quality of model described by the cross-validation parameters Q^2, and the R^2X values show the total number of the variation in the X matrix, and the score plots showed evident distinction between three groups.

Differences in the detected metabolite between 3 groups

The results showed different levels of metabolism including glycolysis, TCA cycle, fatty acid and amino acid. The results showed increase in hepatopancreas glucose and lactate concentration in group 2 compared with

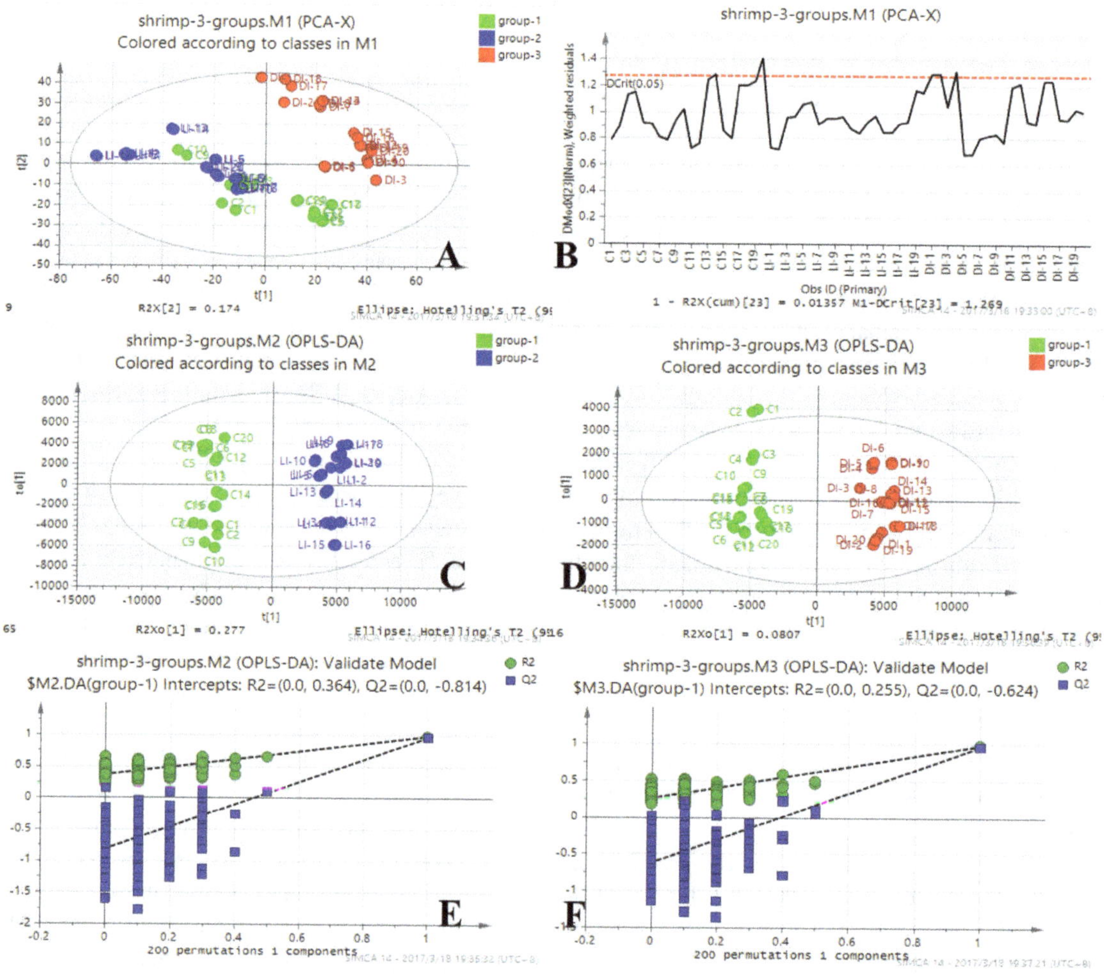

Fig. 1 There were significant differences in metabolomics levels between the three experimental groups by principal component analysis. **a** PCA scores plot on GCMS of shrimp hepatopancreas obtained from 3 groups. *Green dot* control (group-1); *Blue dot* WSSV latent infection stage (group-2); *Red dot* WSSV dominant infection stage (group-3). **b** A moderate outlier of 60 samples. **c** The *scatter plot* of the *scores* indicate that group control and latent infection group (group-2) can be a good distinction under the OPLS-DA model. **d** The *scatter plot* of the *scores* indicate that group control and dominant infection group (group-3) can be a good distinction under the OPLS-DA model. **e** The separation model was tested by permutation test, and the results showed that the model (**c**) between the control and LI groups met the requirements. **f** The separation model was tested by permutation test, and the results showed that the model (**d**) between the control and DI groups met the requirements

control, but subsequently decrease in group 3 which the shrimp were WSSV dominant infection and immediately to death. The concentrations of pyruvate have no significant difference between three groups (Fig. 2). In the metabolic processes of the TCA cycle,the content of oxaloacetate from group 2 showed a significant ($p < 0.05$) increase, contrary succinate and α-ketoglutarate showed a significant decrease. The concentration of fumarate and oxaloacetate changed higher than control, while succinate, malate and α-ketoglutarate were decreased (Fig. 3). Fatty acid metabolism was similar with glycolysis, the concentration of detected fatty acids including palmitic acid, palmitoleic acid, stearic acid, linoleic acid, α-linoleic acid and EPA, changed higher on group 2, and reduced on group 3 except stearic acid which still increased on group 3. Oleic acid did not significant increase on group 2, but changed higher on group 3 significantly (Fig. 4). Amino acid of serine and glutamine in three groups had no evidence difference. The WSSV infected shrimp had significantly higher contents of 14 amino acids on group 2, these metabolites including glycine, alanine, asparagine, valine, threonine, lysine, histidine, tryptophan, leucine, isoleucine, cysteine, arginine, phenylalanine and tyrosine. There were six amino acids changes in the concentration of group 3, the glycine and aspartic acid were upregulation and the contents of alanine; asparagine, proline and glutamate were lower than control (Table 1).

Discussion

To interact with their host cells, viruses have established a dynamic and intricate biological mechanism. They defuse the host defenses and induce extensive changes to cell metabolism (Bruggeman 2007; Heaton and Randall 2011). The results were similar when white spot syndrome virus infected the *L. vannamei* in this study; the results indicated that the WSSV-infected shrimps have a changed hepatopancreas metabolite profile. Compared with healthy controls, glucose of infected shrimp was increased but decreased at the time when the infected shrimp approaching to death. These results of this paper were like some previous research which described a significant increase in plasma glucose concentration of WSSV infected shrimp; they suggested that shrimp infected with virus may need to use glucose as an energy source to maintain of immune response to resist this rapidly replicating virus (Yoganandhan 2003; Galvan-Alvarez et al. 2012). Another research indicates that glucose and lactate concentration increased in WSSV-infected shrimp at the stage of viral genome replication, and the glucose-6-phosphate dehydrogenase showed increased activity which is crucial enzyme of pentose phosphate pathway and lactate was shuttled into gluconeogenesis under normal biochemical situation. The decrease trend of lactate indicated that capacity of hydrogen had been changed and it contributed to the gluconeogenic supply of glucose for energy needing the infected shrimps (Solanky et al. 2003; Chen et al. 2011; Liu et al. 2015). The time of group 2 shrimp which infected by WSSV were less than

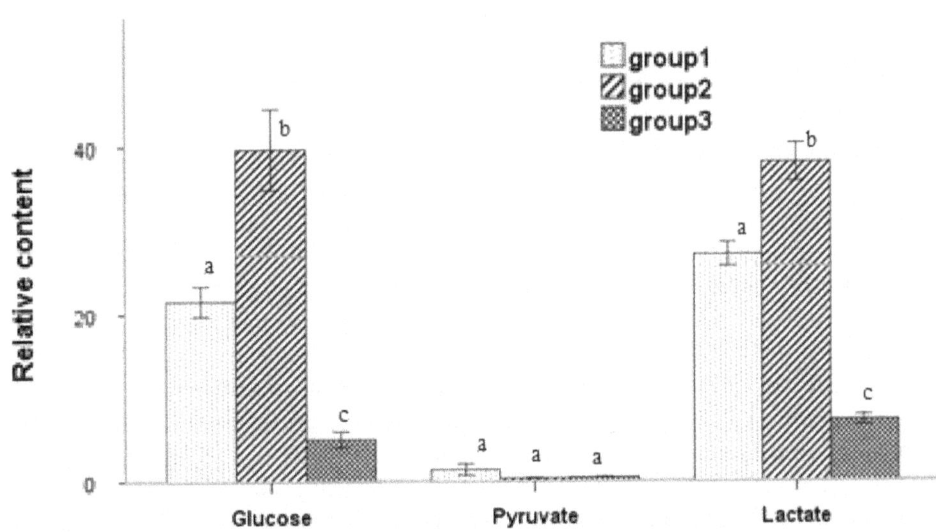

Fig. 2 Relative concentration of glycolysis metabolites in shrimp hepatopancreas after WSSV infected. "*a*", "*b*", "*c*" indicate significant ($p < 0.05$) difference

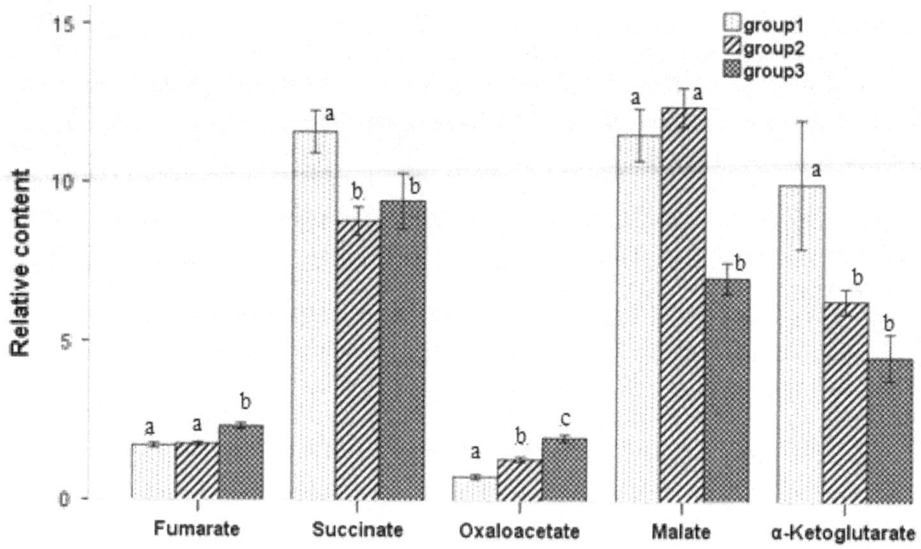

Fig. 3 Relative concentration of TCA metabolites in shrimp hepatopancreas after WSSV infected

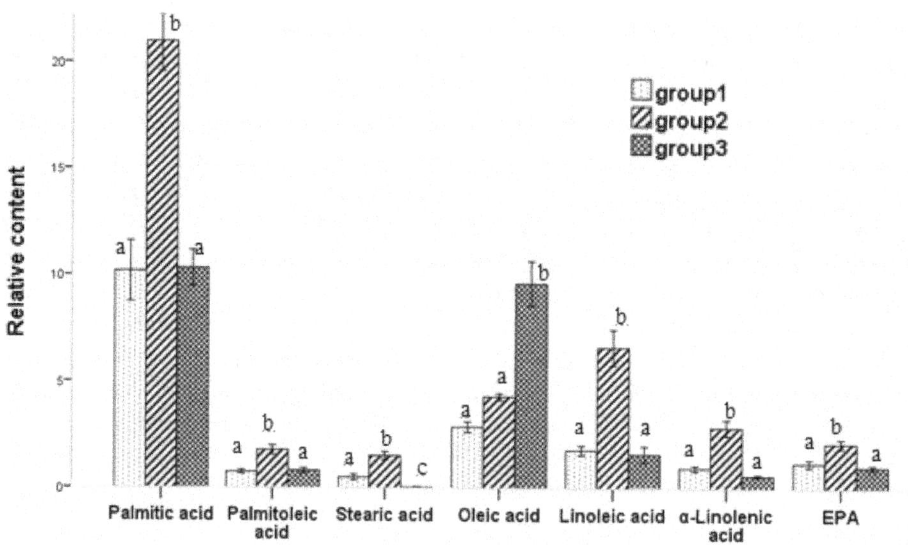

Fig. 4 Relative concentration of fatty acid metabolites in shrimp hepatopancreas after WSSV infected

2 weeks, so the virus reproduction results that quote above can explain the enhanced glycolysis phenomenon in this study. When at the stage of late viral infections, glucose was large consumption and decreased. To maintain the body energy consumption, the gluconeogenesis had intensified in hepatopancreas of shrimp, and the more lactate was consumed.

The previous study (Yoganandhan 2003) had described that the organic acid increased in hepatopancreas was a usual phenomenon, and in this study the fatty acid manifested the same action. α-ketoglutarate was in a constant state of decline, and the downstream product succinate was equal to decrease, and also the content of malate decrease significantly on later infected stage. But the oxaloacetate of shrimp was contrarily, it increases throughout the process of WSSV infected. This may concern TCA cycle enzymes, previous research had reported that proteins were significantly upregulated or downregulated (Chai et al. 2010), and in this study the citrate synthase of shrimp may inhibited, so metabolism of citric acid cycle got disordered. The disorder of amino acid metabolism caused the significant alterations on WSSV infected shrimp that 14 amino acids were upregulation, but it's interesting when at the stage of before death, the concentration of 11 amino acids form above 14 amino acids decrease to normal level of control. An obviously, phenomenon was the content of

Table 1 Relative concentration of amino acid metabolites in shrimp hepatopancreas of group1, group2 and group3 after WSSV infected

Amino acid	Group 1 (control)	Group 2 (infective stage)	Group 3 (death stage)	Regulation direction	
				Group 2/group 1	Group 3/Group 1
Gly	51.34	81.95	106.57	↑*	↑*
Ala	25.65	34.94	14.89	↑*	↓*
Asn	0.82	1.03	0.55	↑*	↓*
Asp	3.30	4.11	4.82	–	↑*
Pro	29.99	29.60	22.22	–	↓*
Glu	5.33	5.18	3.91	–	↓*
Val	7.28	10.08	7.09	↑*	–
The	2.32	2.88	1.97	↑*	–
Lys	9.14	14.70	8.72	↑*	–
His	1.48	2.72	1.39	↑*	–
Try	1.69	2.81	1.50	↑*	–
Leu	12.41	19.66	13.80	↑*	–
Ile	5.01	6.96	5.25	↑*	–
Cys	1.90	2.78	2.42	↑*	–
Arg	6.59	12.99	7.54	↑*	–
Phe	3.01	4.72	4.08	↑*	–
Tyr	6.54	10.77	8.85	↑*	–
Ser	7.88	8.43	7.24	–	–
Gln	3.18	4.31	3.81	–	–

The relative concentration of each metabolite is an average of data from 20 biological replicates using GCMS. The star (*Asterisk*) indicate significant ($p < 0.05$) difference. The *short dash line* indicate no significant difference. *Up arrow* and *Down arrow* represents the upregulation and downregulation respectively

inhibitory amino acid glycine sustained raise, so it's possibility the glycine was the major product of virus cell metabolism. Yoganandhan (2003) support our observation on amino acid metabolism of infected stage, but the trend of amino acid and fatty acid recovers at later stage may concerned with viral reproduction and it need more subsequently study to explain the mechanism.

The results of this study indicated that glucose metabolism and fatty acid metabolism would be abnormal, so that metrics can be used to monitor the shrimp of WSSV infected and determine the extent of the disease.

Acknowledgements This study was supported by the Shanghai Shrimp Industry Technology System Program and Laboratory Technician Team building program of china shanghai education commission (B1-5407-13-0000-5).

References

Bruggeman LA (2007) Viral subversion mechanisms in chronic kidney disease pathogenesis. Clin J Am Soc Nephrol 2(suppl 1):S13–S19

Chai YM, Yu SS, Zhao XF, Zhu Q, Wang JX (2010) Comparative proteomic profiles of the hepatopancreas in Fenneropenaeus chinensis response to white spot syndrome virus. Fish Shellfish Immunol 29:480–486

Chen IT et al (2011) White spot syndrome virus induces metabolic changes resembling the warburg effect in shrimp hemocytes in the early stage of infection. J Virol 85:12919–12928

Galvan-Alvarez D, Mendoza-Cano F, Hernandez-Lopez J, Sanchez-Paz A (2012) Experimental evidence of metabolic disturbance in the white shrimp Penaeus vannamei induced by the Infectious Hypodermal and Hematopoietic Necrosis Virus (IHHNV). J Invertebr Pathol 111:60–67

Heaton NS, Randall G (2011) Multifaceted roles for lipids in viral infection. Trends Microbiol 19:368–375

Leu JH et al (2007) Comparative analysis of differentially expressed genes in normal and white spot syndrome virus infected Penaeus monodon. BMC Genomics 8:120

Leu JH, Lin SJ, Huang JY, Chen TC, Lo CF (2013) A model for apoptotic interaction between white spot syndrome virus and shrimp. Fish Shellfish Immunol 34:1011–1017

Li S, Zhang X, Sun Z, Li F, Xiang J (2013) Transcriptome analysis on Chinese shrimp Fenneropenaeus chinensis during WSSV acute infection. PLoS ONE 8:e58627

Liu PF, Liu QH, Wu Y, Jie H (2015) A pilot metabolic profiling study in hepatopancreas of *Litopenaeus vannamei* with white spot syndrome virus based on (1) H NMR spectroscopy. J Invertebr Pathol 124:51–56

Lo CF et al (1996) White spot syndrome baculovirus (WSBV) detected in cultured and captured shrimp, crabs and other arthropods. Dis Aquat Org 27:215–225

Sánchez-Paz A (2010) White spot syndrome virus: an overview on an emergent concern. Vet Res 41:43

Solanky KS et al (2003) Application of biofluid 1H nuclear magnetic resonance-based metabonomic techniques for the analysis of the biochemical effects of dietary isoflavones on human plasma profile. Anal Biochem 323:197–204

Wu D et al (2013) Tissue metabolic responses to salt stress in wild and cultivated barley. PLoS ONE 8:e55431

Xia J, Wishart DS (2011) Web-based inference of biological patterns, functions and pathways from metabolomic data using MetaboAnalyst. Nat Protoc 6:743–760

Yoganandhan K, Thirupathi S, Hameed ASS (2003) Biochemical, physiological and hematological changes in white spot syndrome virus-infected shrimp, penaeus indicus. Aquaculture 221:1–11

16

Assessment of bacterial communities in skin ulceration in cultured sea cucumber *Apostichopus japonicus* (Selenka)

Yi Yang · Yuchun Li · Zhenlin Liang

Abstract To determine changes in bacterial composition in skin ulcerations in farmed sea cucumber *Apostichopus japonicus*, full-length *16S rRNA* gene sequencing of different skin samples of *A. japonicus* was performed on a Pacific Biosciences (PacBio) RS platform. The fact that one of them varies in bacterial diversity than the other two may suggest different stage of skin ulceration, but the answer is inconclusive due to inefficiency in size of sampling pool. After comparing them with three skin samples from three healthy sea cucumbers, respectively, the majority of the results were distinguishable at the species level. The bacterial composition, which increased in diversity during the possible early stage of skin ulceration, exhibited a decrease in diversity at possible advanced stages, except for the predominant *Saccharicrinis carchari* (846.85 ± 5.75‰). This work may have the first clue implicating *Ornithinibacillus contaminans* as a potential causative pathogen of skin ulcerations in *A. japonicus*.

Keywords Skin ulceration · Third-generation sequencing · 16S rRNA

Introduction

Apostichopus japonicus is an important species for aquaculture in China. In northern China, *A. japonicus* is cultivated on a commercial scale by sea ranching or in shallow ponds, with the latter being more commonly employed in northern China (Choo 2008). Breeding methods for this species have improved in the past decade, and its transcriptome has been sequenced, although extensive studies on the thermal resistance, disease resistance, and growth rates of *A. japonicas* have been conducted, no effective and practical method to prevent the development of skin ulcerations during its cultivation has been established (Du et al. 2012; Gianasi et al. 2017; Zhao et al. 2017). The skin ulceration usually takes place when water temperature gets low from December to March, and once occurs, it spreads to a whole pond within days and brings a lethal death rate higher than 90%. Besides such potential devastating blow to *A. japonicus* cultivation, skin ulceration can be frequently found on couples of day basis. *A. japonicus* only possesses innate immunity, thereby making it relatively vulnerable to harsh cultivation environments. One preventative measure against skin ulcerations is to administer probiotics or prebiotics (Zhang et al. 2010; Ma et al. 2013; Sun et al. 2012; Yang et al. 2014, 2015). However, these lesions continue to occur despite efforts to elucidate the molecular mechanisms

Y. Yang · Y. Li (✉) · Z. Liang (✉)
Marine College, Shandong University, Weihai 264200, China
e-mail: li_yuchun@sdu.edu.cn

Z. Liang
e-mail: liang@sdu.edu.cn

underlying its immune system (Xue et al. 2015). Therefore, an efficient approach in identifying the pathogens that cause skin ulcerations during the cultivation of *A. japonicus* is warranted.

Several studies on the causative pathogens of skin ulcerations have been conducted; however, none have utilized next-generation or third-generation sequencing (Li et al. 2010; Tiruvayipati et al. 2013). The 16S ribosomal RNA (*16S rRNA*) gene has been employed in profiling bacteria and Archaea since the late 1970s (Woese and Fox 1977). Multiple *16S rRNA* databases have accumulated massive amounts of sequence data, and high-throughput sequencing technologies, predominantly the Roche 454 and Illumina sequencing platforms, have facilitated deep microbial community profiling. However, investigations on bacterial composition do not go beyond the family level (Vishnivetskaya et al. 2011; Besemer et al. 2012; Mosher et al. 2012). Recent studies using the Pacific Biosciences (PacBio) RS platform have revealed genus-level discrimination of *16S rRNA* amplicons from microbial communities (Mosher et al. 2013). Because fewer resources are necessary due to improvements in phylogenetic reconstruction, we performed a comparative assessment of *A. japonicus* from a single local breeding spot using the PacBio RS platform to identify candidates responsible for, or linked to, skin ulcerations. To keep skin microbial consortia samples intact, we did not make any effort to exclude host tissues at the sampling stage. Such limitations may also assess the specificity of the polymerase chain reactions (PCRs) and sequencing depth of the PacBio RS platform, as indicated by the generation of good analyzable mega data without contamination.

Materials and methods

Sampling of *A. japonicus*

We sampled three healthy normal (NM) *A. japonica* and three individuals with skin ulceration deterioration (UD) in a same cultivating pond (122°15′N, 37°01′E) in Weifang, Shandong Province, China, on May 5th, 2016. Those with UD were identified by appearances of positive skin ulceration spots. The healthy group (HG) consists of three NM samples with an average weight of 69.0 ± 21.2 g and the three UD samples had an average weight of 41.8 ± 13.0 g. All samples were transported on ice and dissected later for a tiny piece (2 g) of skin tissue (for UD samples, ulceration dots were collected) within 4 h. We did not exclude host tissues or only tried to brush off some of the microbial consortia on the skin during sampling and instead utilized whole-skin tissue samples in PacBio RS II sequencing to precisely reflect the actual skin microbial composition.

PCR amplification of 16S rRNA genes

Genomic DNA was extracted following the Earth Microbiome Project (EMP) protocol (http://www.earthmicrobiome.org/protocols-and-standards/dna-extraction-protocol/). The purity and concentration of DNA in the final 40-μL eluent that was, respectively, extracted from the samples was tested on both NanoDrop NC2000 (Thermo Scientific, Wilmington, DE, USA) and TBS380 Fluorometer (Turner BioSystems, Sunnyvale, CA, USA). The primers employed to amplify the V1–V9 region of the *16S rRNA* were 27F (3′-AGAGTTTGATCMTGGCTCAG) and 1492R (5′-ACCTTGTTACGACTT). The reaction mixture consisted of 5 μL of a 5× reaction buffer, 5 μL of a 5× GC buffer, 2 μL of each dNTP (2.5 mM), 1 μL of the forward primer (10 μM), 1 μL of the reverse primer (10 μM), 2 μL of the DNA template (4–25 ng μL^{-1}), 8.75 μL of ddH_2O, and 0.25 μL of 5U/μL Q5 DNA polymerase. All the samples were subjected to two-step PCR: (1) 25 cycles of the following: initial denaturation at 98 °C for 3 min, denaturation at 98 °C for 30 s, annealing at 52 °C for 30 s, extension at 72 °C for 180 s, and a final extension at 72 °C for 5 min; (2) 16 cycles of the following: initial denaturation at 98 °C for 3 min, denaturation at 98 °C for 30 s, annealing at 62 °C for 30 s, extension at 72 °C for 180 s, and a final extension at 72 °C for 5 min.

PacBio RS single-molecule real-time (SMRT) sequencing

The PCR products (0.2–0.7 μg) were prepared using Pacific Biosciences Template Prep Kit 1.0 for DNA damage repair and end repair. A total of 50 μL of the damage repair mix was, respectively, incubated at 37 °C for 60 min and cooled at 4 °C. By adding 2 μL of the end-repair eluent, the 52-μL mixture was incubated at 25 °C for 5 min. The reaction mix was cleaned using 0.6× Agencourt AMPure XP beads (Beckman Coulter,

Beverly, MA, USA). Then, the mixtures were subjected to a blunt-ligation reaction preparation step at − 25 °C for 1 h, 65 °C for 10 min, and 4 °C overnight. Ligation failure products were removed by adding exonucleases and incubating at 37 °C for 60 min. The resulting 41-μL mixture containing the library was purified using 0.6× Ampure PB beads twice. Once the primers were annealed and DNA polymerase had bound, the SMRT bell-polymerase complexes were sequenced.

Results

PacBio RS II SMRT chip generated a total of 18,195 reads from the V1–V9 amplicons of all six samples. The circular consensus sequencing data were processed through the SMRT pipeline (https://github.com/PacificBiosciences/smrtflow) to make sure that the error rate was < 10% and full passes were > 3. The length distribution of all reads was as expected of typical V1–V9 amplicons of the *16S rRNA* gene (Fig. 1).

From a total of 16,276 optimized full-length *16S rRNA* reads from all six samples, 97% were clustered to the bacterium *16S rRNA* sequence of the Greengenes database (Release 13.8, http://greengenes.secondgenome.com/) (DeSantis et al. 2006) using QIIME (Quantitative Insights Into Microbial Ecology, v1.8.0, http://qiime.org/) (Caporaso et al. 2010) up to species level (Table 1).

The rarefaction curve depicted well-presentable sequencing depth for each sample (Fig. 2). All samples reached saturation as indicated by the plateau.

The composition of each sample at the species level shows a three-stage trend in Fig. 3.

NM1, 2, and 3 comprised HG *A. japonicus*, UD1 showed the early stages of skin ulceration symptom, and UD2 and 3 exhibited advanced stages of skin ulceration.

To identify the bacterium that potentially causes skin ulceration, we compared the bacterial communities between HG and UD1 and that between UD1 and UD2 and 3. Then, we filtered the data matrix to identify strains that showed at least a twofold increase from the HG group to UD1 and still existed at average densities in UD 2 and 3 (Fig. 4). Our results showed that all the filtered strains decreased in number as *A. japonicus* reached the advanced stages of skin ulceration.

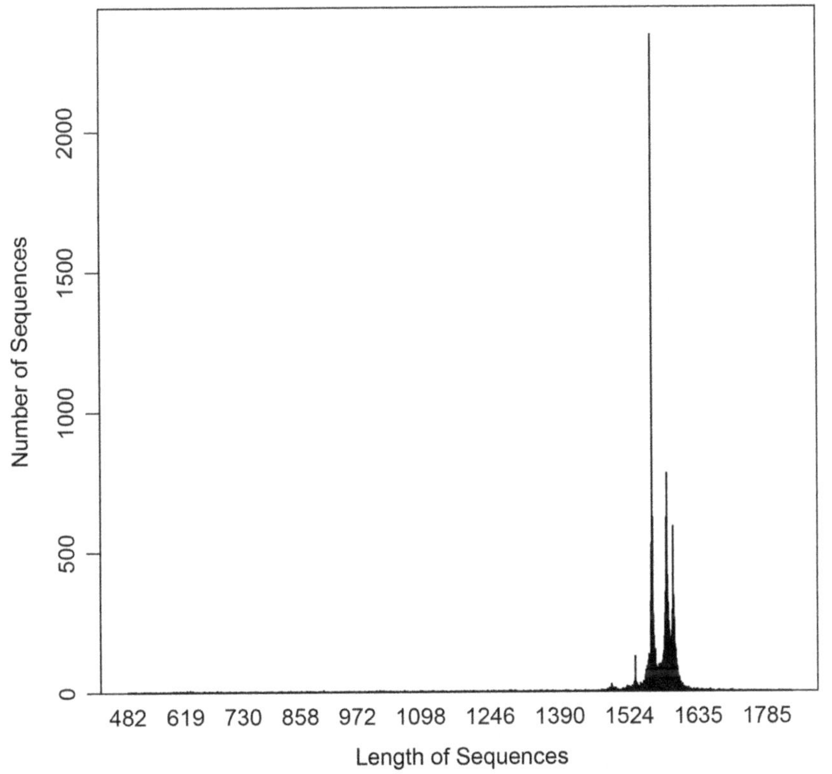

Fig. 1 Length distribution of reads from NM and UD groups using PacBio RS II SMRT chip

Table 1 OTU clustering and classification of outcomes

Sample	Phylum	Class	Order	Family	Genus	Species	Classified vs. unclassified OTU
UD1	8	14	34	64	146	260	824:5
UD2	6	9	18	34	61	94	364:3
UD3	3	5	6	14	27	40	198:2
NM1	8	11	23	42	94	191	646:5
NM2	5	10	17	32	55	123	517:3
NM3	4	8	21	37	59	130	444:6

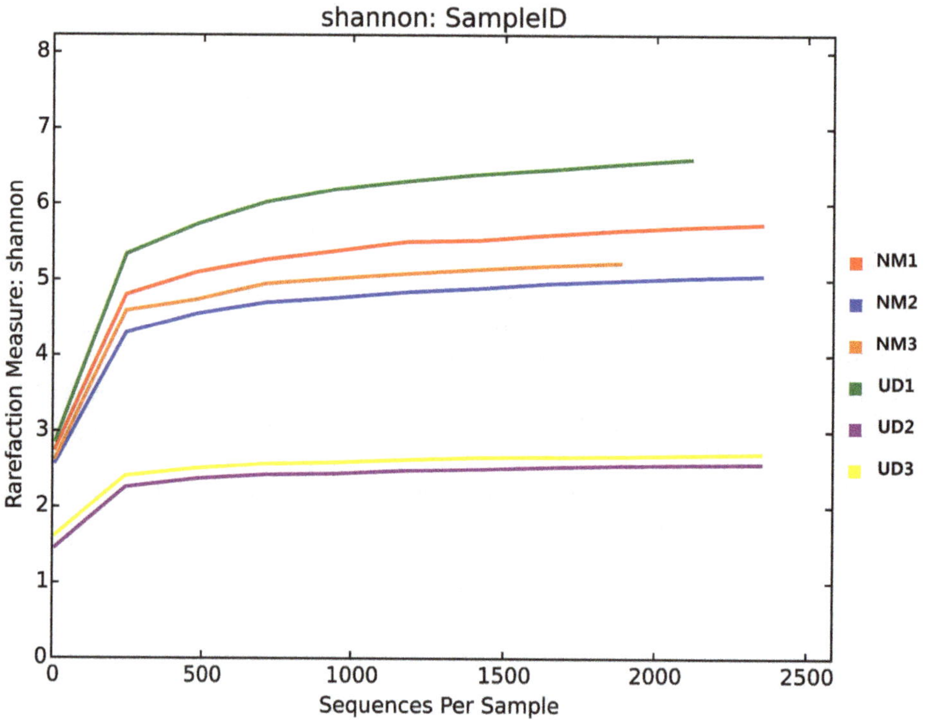

Fig. 2 Rarefaction curves of OTU numbers of the samples

Figure 4A shows a shift in the bacterial community profile from HG to UD1, and that in B depicts similarities in profiles between UD1 and UD 2 and 3. Eight bacterial species exhibited de novo emergence (DE) as reflected in the microbial composition during the early stage of skin ulceration. The specific number of the DEs in UD1 was as follows: *Desulforhopalus singaporensis* (0.44‰), *Lactobacillus fabifermentans* (0.44‰), *Ornithinibacillus contaminans* (0.44‰), *Polaribacter dokdonensis* (0.44‰), *Bizionia argentinensis* (0.87‰), *Rubritalea tangerina* (0.87‰), *Luteolibacter algae* (1.31‰), and *Lutibacter agarilyticus* (4.36‰).

Discussion

D. singaporensis, L. fabifermentans, P. dokdonensis, B. argentinensis, L. agarilyticus, and *L. algae* from the DE group are associated with various metabolic pathways related to skin ulcerations but have not been reported as actual pathogens (Lie et al. 1999; Yoon et al. 2006; Bercovich et al. 2008; De Bruyne et al. 2009; Lanzarotti et al. 2011; Park et al. 2013; Aran et al. 2014; Campanaro et al. 2014; Treu et al. 2014; Yoon et al. 2017).

Approximately seven species have been validated in the genus *Ornithinibacillus* (Mayr et al. 2006). Most are free-living bacteria isolated from natural environments, such as marine sediment and saline soil. *O. scapharcae* TW25T is considered to be the only *Ornithinibacillus* that has pathogenic potential. *O.*

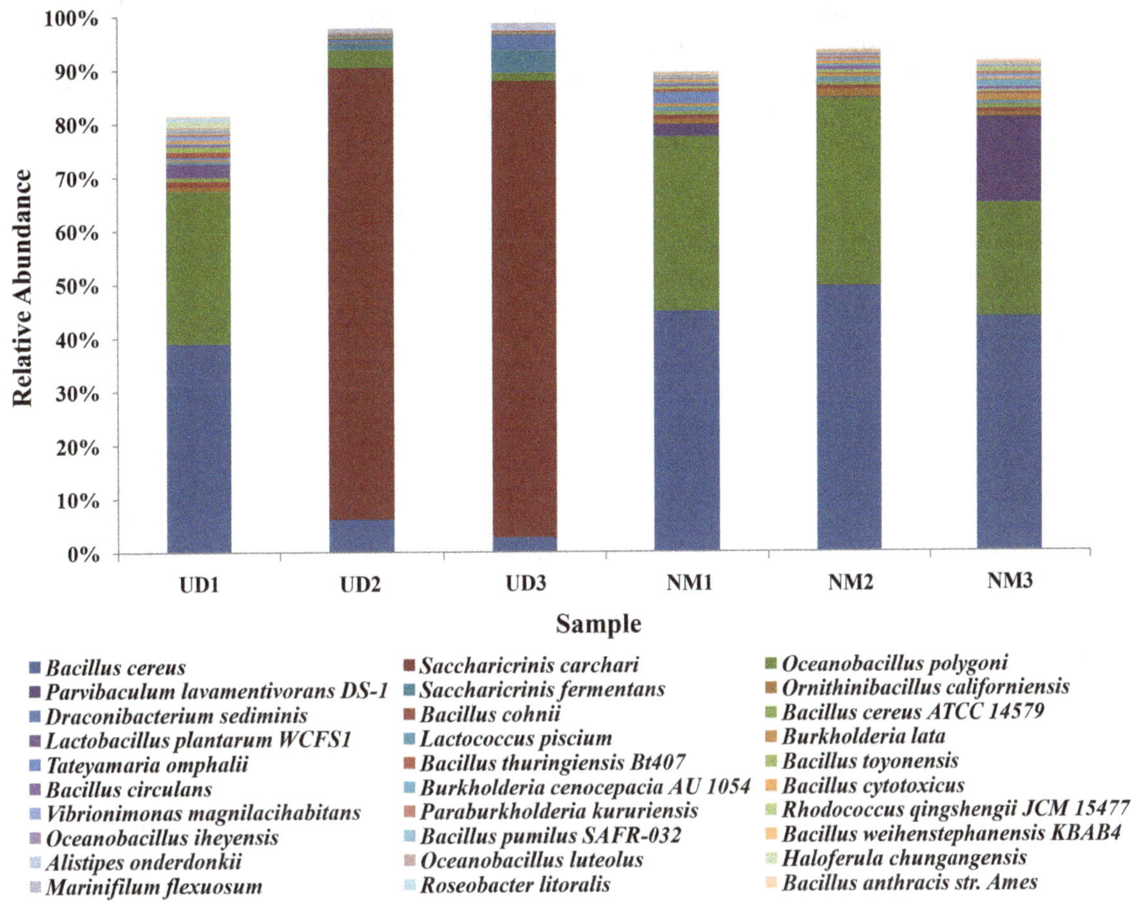

Fig. 3 Main species composition of the samples

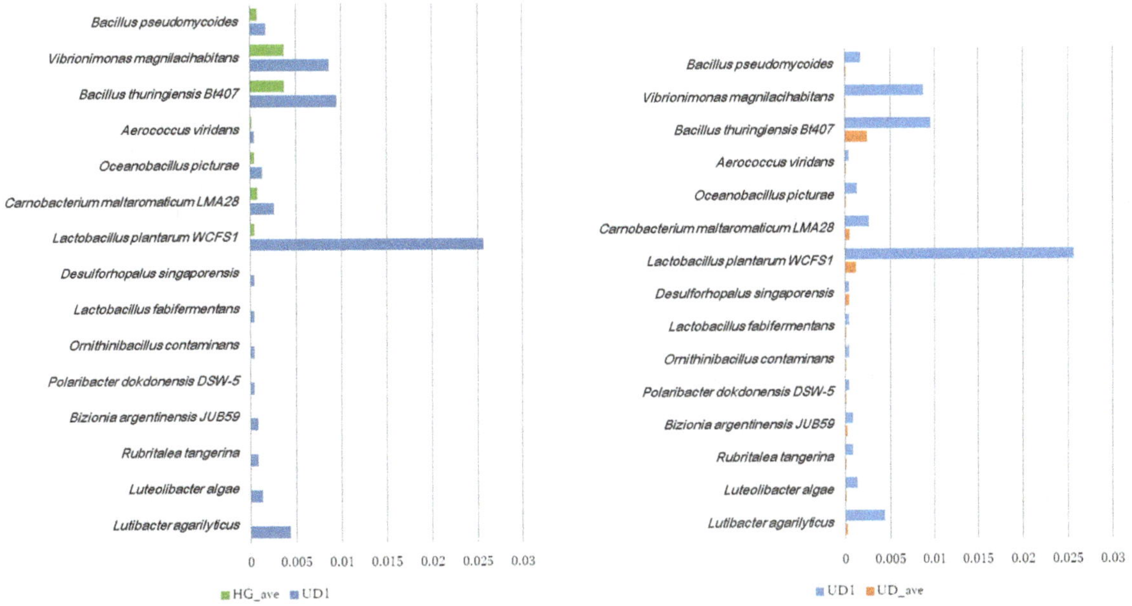

Fig. 4 Strains showing at least a twofold increase in number and a positive average in the UD 2 and 3 groups

contaminans DSM22953T, which emerged in this study, was first isolated from a blood sample of a 75-year-old woman. It is Gram-positive, endospore-forming, and rod-shaped. *O. contaminans* DSM22953T, similar to *O. scapharcae* TW25T, contains a number of virulence-relevant genes, which indicates that it may also be pathogenic. However, no case reports of disease due to *O. contaminans* DSM22953T have been reported to date. The present study thus may suggest a case study of its pathogenicity (Kämpfer et al., 2010; Jiang et al., 2016). Further research is due for conclusions.

R. tangerina is Gram-negative, non-motile, facultatively anaerobic, and coccoid or rod-shaped. *R. tangerina* was isolated from the visceral specimen of an unidentified sea hare collected from Himetsu, Sado, Niigata, Japan, in 2006. Information on this species is limited (Yoon et al. 2007).

Other strains that increased in density during possible different stages of skin ulceration in cultured *A. japonicas* included the following:

1. *Carnobacterium maltaromaticum* LMA28 can be found in the wild and in food products, and can grow anaerobically. This species is pathogenic to fish, whereas cases involving humans have not been reported (Leisner et al. 2007).
2. *Aerococcus viridans* is a member of the bacterial genus *Aerococcus*. It is a causative agent of gaffkaemia, a disease in lobsters (Greenwood et al. 2005).
3. *Bacillus thuringiensis* is a Gram-positive, soil-dwelling bacterium that is commonly used as a biological pesticide. It was first discovered in 1901 by Japanese biologist Shigetane Ishiwata (Roh et al. 2007) and then rediscovered in Germany by Ernst Berliner, who reported that it causes a disease called Schlaffsucht in flour moth caterpillars. *B. thuringiensis* is closely related to *B. cereus*, a soil bacterium, and *B. anthracis*, which causes anthrax; the three organisms differ mainly in their plasmids (Økstad and Kolstø 2011). Besides *B. thuringiensis*, there are a total of 14 members of the genus *Bacillus* that showed an increase in density in UD 1. Among these, *B. thuringiensis* and *B. pseudomycoides* are the only two *Bacillus* species that were observed on UD 2 and 3 at average levels.
4. Seven members of the *Burkholderia* appeared during the early stage of skin ulceration, except for *Burkholderia vietnamiensis* G4, which also emerged in HG. However, none of these species were identified during the advanced stage of skin ulceration. Several members of this genus are considered to be biosecurity threats (Woods and Sokol 2006; Sawana et al. 2014).

The two predominant bacteria of the skin health consortia included *B. cereus* and *Oceanobacillus polygoni*. *B. cereus* averaged 459.93 ± 30.38‰ and *O. polygoni* 295.24 ± 73.53‰ in the HG group. The population of both species dramatically dropped to 43.80 ± 24.20‰ and 24.70 ± 12.17‰, respectively, during the advanced stage of skin ulceration. *B. cereus* is a Gram-positive, rod-shaped, aerobic, motile, β-hemolytic bacterium that is commonly found in soil and food. Some strains of *B. cereus* are harmful, whereas others can be beneficial as probiotics for animals (Charalampopoulos and Rastall 2009). *O. polygoni* is a Gram-positive, peritrichously flagellated, straight, facultatively alkaliphile, rod-shaped bacterium that was first isolated from indigo fermentation fluid in 2013 (Hirota et al. 2013). A strain of *O. polygoni*, MBF-HG6, is salt-tolerant and alkaliphilic and produces bioflocculants (Li et al. 2017). The role of *O. polygoni* in skin ulceration remains unclear.

Saccharicrinis carchari was the predominant bacterium (846.85 ± 5.75‰) during the advanced stage of skin ulceration. However, this bacterium was not observed in all three HG samples and only showed a 1.54-fold increase in percentage from HG to UD1 (0.44‰). *S. carchari* was the only strain that illustrated a shift in the trend wherein it continuously increased. It is a Gram-negative, facultatively anaerobic, gliding, non-endospore-forming, yellow-pigmented, straight or slightly curved, rod bacterium that was initially isolated from shark gill homogenate. This strain is catalase-positive and oxidase-negative (Liu et al. 2014). Its role in skin ulceration remains unclear.

Conclusions

The present study examined changes in the skin microbial consortia of *A. japonicus* with skin ulcerations. PacBio sequencing revealed differences in bacterial communities at the species level. *O. contaminans* and *S. carchari* were the predominant bacterial species in the skin ulcerations. Our approach of keeping skin samples intact and eliminating unrelated sequences by selective PCR for *16S rRNA* sequencing on a PacBio RS platform proved to be sufficient for performing a full-scale, species-discriminating inspection of the skin consortia community. Our technique may thus be potentially employed in future investigations on skin microbial consortia in aquatic invertebrates.

Acknowledgements The authors thank the Technology Department of Shanghai Personal Biotechnology Co., Ltd., for processing and preparing backups of our raw data. We thank LetPub (www.letpub.com) for its linguistic assistance during the preparation of this manuscript.

Author contributions YY designed all the experiments, prepared all the samples, and wrote the manuscript. YY, YL, and ZL discussed the results and commented on the manuscript. All authors reviewed the manuscript.

Funding This research is supported by Natural Science Foundation of Shandong Province, China.

Conflict of interest The authors declare that the research was conducted in the absence of any commercial or financial relationships that could be construed as a potential conflict of interest.

References

Aran M, Smal C, Pellizza L, Gallo M, Otero LH, Klinke S, Goldbaum FA, Ithurralde ER, Bercovich A, Cormack WPM, Turjanski AG, Cicero DO (2014) Solution and crystal structure of BA42, a protein from the Antarctic bacterium *Bizionia argentinensis* comprised of a stand-alone TPM domain. Proteins 82:3062–3078

Bercovich A, Vazquez SC, Yankilevich P, Coria SH, Foti M, Hernandez E, Vidal A, Ruberto L, Melo C, Marenssi S, Criscuolo M, Memoli M, Arguelles M, Cormack WPM (2008) *Bizionia argentinensis* sp. nov., isolated from surface marine water in Antarctica. Int J Syst Evol Microbiol 58:236–2367

Besemer K, Peter H, Logue JB, Langenheder S, Lindström ES, Tranvik LJ, Battin TJ (2012) Unraveling assembly of stream biofilm communities. Microb Popul Community Ecol 6:1459–1468

Campanaro S, Treu L, Vendramin V, Bovo B, Giacomini A, Corich V (2014) Metagenomic analysis of the microbial community in fermented grape marc reveals that *Lactobacillus fabifermentans* is one of the dominant species: insights into its genome structure. Appl Microbiol Biotechnol 98:6015–6037

Caporaso JG, Kuczynski J, Stombaugh J, Bittinger K, Bushman FD, Costello EK, Fierer N, Peña AG, Goodrich JK, Gordon JI, Huttley GA, Kelley ST, Knights D, Koenig JE, Ley RE, Lozupone CA, McDonald D, Muegge BD, Pirrung M, Reeder J, Sevinsky JR, Turnbaugh PJ, Walters WA, Widmann J, Yatsunenko T, Zaneveld J, Knight R (2010) QIIME allows analysis of high-throughput community sequencing data. Nat Methods 7:335–336

Charalampopoulos D, Rastall RA (2009) Prebiotics and probiotics science and technology. Springer, New York

Choo PS (2008) Population status, fisheries and trade of sea cucumbers in Asia. In: Toral-Granda V, Lovatelli A, Vasconcellos M (eds) Sea cucumbers. A global review of fisheries and trade. FAO Fisheries and Aquaculture Technical Paper, Rome, pp 81–118

De Bruyne K, Camu N, De Vuyst L, Vandamme P (2009) Lactobacillus *fabifermentans sp.* nov. and *Lactobacillus cacaonum* sp. nov., isolated from Ghanaian cocoa fermentations. Int J Syst Evol Microbiol 59:7–12

DeSantis TZ, Hugenholtz P, Larsen N, Rojas M, Brodie EL, Keller K, Huber T, Dalevi D, Hu P, Andersen GL (2006) Greengenes, a chimera-checked 16S rRNA gene database and workbench compatible with ARB. Appl Environ Microbiol 72:5069–5072

Du H, Bao Z, Hou R, Wang S, Su H, Yan J, Tian M, Li Y, Wei W, Lu W, Hu XL, Wang S, Hu JJ (2012) Transcriptome sequencing and characterization for the sea cucumber *Apostichopus japonicus* (Selenka, 1867). PLoS One 7:e33311

Gianasi BL, Parrish CC, Hamel JF, Mercier A (2017) Influence of diet on growth, reproduction and lipid and fatty acid composition in the sea cucumber *Cucumaria frondosa*. Aquac Res 48:3413–3432

Greenwood SJ, Keith IR, Després BM, Cawthorn RJ (2005) Genetic characterization of the lobster pathogen *Aerococcus viridans* var. *homari* by 16S rRNA gene sequence and RAPD. Dis Aquat Org 63:237–246

Hirota K, Hanaoka Y, Nodasaka Y, Yumoto I (2013) *Oceanobacillus polygoni* sp. nov., a facultatively alkaliphile isolated from indigo fermentation fluid. Int J Syst Evol Microbiol 63:3307–3312

Jiang XW, Cheng H, Zheng BW, Li A, Lv LX, Ling ZX (2016) Comparative genomic study of three species within the genus *Ornithinibacillus*, reflecting the adaption to different habitats. Gene 578:25–31

Kämpfer P, Falsen E, Lodders N, Langer S, Busse HJ, Schumann P (2010) *Ornithinibacillus contaminans* sp. nov., an endospore-forming species. Int J Syst Evol Microbiol 60:2930–2934

Lanzarotti E, Pellizza L, Bercovich A, Foti M, Coria SH, Vazquez SC, Ruberto L, Hernandez EA, Dias RL, Cormack WPM, Cicero DO, Smal C, Nicolas MF, Vasconcelos ATR, Marti MA, Turjanski AG (2011) Draft genome sequence of *Bizionia argentinensis*, isolated from Antarctic surface water. J Bacteriol 193:6797–6798

Leisner JJ, Laursen BG, Prévost H, Drider D, Dalgaard P (2007) Carnobacterium: positive and negative effects in the environment and in foods. FEMS Microbiol Rev 31:592–613

Li H, Qiao G, Li Q, Zhou W, Won KM, Xu DH, Park SI (2010) Biological characteristics and pathogenicity of a highly pathogenic *Shewanella marisflavi* infecting sea cucumber, *Apostichopus japonicus*. J Fish Dis 33:865–877

Li J, Yun Y, Yun L Xing, Song L (2017) Novel bioflocculant produced by salt-tolerant, alkaliphilic strain *Oceanobacillus polygoni* HG6 and its application in tannery wastewater treatment. Biosci Biotechnol Biochem 81:1018–1025

Lie TJ, Clawson ML, Godchaux W, Leadbetter ER (1999) Sulfidogenesis from 2-aminoethanesulfonate (taurine) fermentation by a morphologically unusual sulfate-reducing bacterium, *Desulforhopalus singaporensis* sp. nov. Appl Environ Microbiol 65:3328–3334

Liu QQ, Wang Y, Li J, Du ZJ, Chen GJ (2014) *Saccharicrinis carchari* sp. nov., isolated from a shark, and emended descriptions of the genus *Saccharicrinis* and *Saccharicrinis fermentans*. Int J Syst Evol Microbiol 64:2204–2209

Ma Y, Liu Z, Yang Z, Li M, Liu J, Song J (2013) Effects of dietary live yeast *Hanseniaspora opuntiae* C21 on the immune and disease resistance against *Vibrio splendidus* infection in juvenile sea cucumber *Apostichopus japonicus*. Fish Shellfish Immunol 34:66–73

Mayr R, Busse HJ, Worliczek HL, Ehling-Schulz M, Scherer S (2006) *Ornithinibacillus gen.* nov., with the species *Ornithinibacillus bavariensis* sp. nov. and *Ornithinibacillus californiensis* sp. nov. Int J Syst Evol Microbiol 56:1383–1389

Mosher JJ, Vishnivetskaya TA, Elias DA, Podar M, Brooks SC, Brown SD, Brandt CC, Palumbo AV (2012) Characterization of the Deltaproteobacteria in contaminated and uncontaminated stream sediments and identification of potential mercury methylators. Aquat Microb Ecol 66:271–282

Mosher JJ, Bernberg EL, Shevchenko O, Kan J, Kaplan LA (2013) Efficacy of a 3rd generation high-throughput sequencing platform for analyses of 16S rRNA genes from environmental samples. J Microbiol Methods 95:175–181

Økstad OA, Kolstø AB (2011) Genomics of bacillus species. In: Wiedmann M, Zhang W (eds) Genomics of foodborne bacterial pathogens. Food microbiology and food safety. Springer, New York, pp 29–53

Park SC, Choe HN, Hwang YM, Baik KS, Seong CN (2013) *Lutibacter agarilyticus* sp. nov., a marine bacterium isolated from shallow coastal seawater. Int J Syst Evol Microbiol 63:2678–2683

Roh JY, Choi JY, Li MS, Jin BR, Je YH (2007) *Bacillus thuringiensis* as a specific, safe, and effective tool for insect pest control. J Microbiol Biotechnol 17:547–559

Sawana A, Adeolu M, Gupta RS (2014) Molecular signatures and phylogenomic analysis of the genus Burkholderia: proposal for division of this genus into the emended genus *Burkholderia* containing pathogenic organisms and a new genus *Paraburkholderia gen.* nov. harboring environmental species. Front Genet 5:429

Sun Y, Wen Z, Li X, Meng N, Mi R, Li Y, Li S (2012) Dietary supplement of fructooligosaccharides and *Bacillus subtilis* enhances the growth rate and disease resistance of the sea cucumber *Apostichopus japonicus* (Selenka). Aquac Res 43:1328–1334

Tiruvayipati S, Bhassu S, Kumar N, Baddam R, Shaik S, Gurindapalli AK, Thong KL, Ahmed N (2013) Genome anatomy of the gastrointestinal pathogen, *Vibrio parahaemolyticus* of crustacean origin. Gut Pathog 5:37

Treu L, Vendramin V, Bovo B, Giacomini A, Corich V, Campanaro S (2014) Genome sequence of *Lactobacillus fabifermentans* strain T30PCM01, isolated from fermenting grape marc. Genome Announc 2:e00060-14

Vishnivetskaya TA, Mosher JJ, Palumbo AV, Yang ZK, Podar M, Brown SD, Brooks SC, Gu B, Southworth GR, Drake MM, Brandt CC, Elias DA (2011) Mercury and other heavy metals influence bacterial community structure in contaminated Tennessee streams. Appl Environ Microbiol 77:302–311

Woese CR, Fox GE (1977) Phylogenetic structure of the prokaryotic domain: the primary kingdoms. Proc Natl Acad Sci 74:5088–5090

Woods DE, Sokol PA (2006) The genus Burkholderia. In: Dworkin M, Falkow S, Rosenberg E, Schleifer KH, Stackebrandt E (eds) The prokaryotes. Springer, New York, pp 848–860

Xue Z, Li H, Wang X, Li X, Liu Y, Sun J, Liu C (2015) A review of the immune molecules in the sea cucumber. Fish Shellfish Immunol 44:1–11

Yang ZP, Sun JM, Xu Z, Zhang CC, Zhou Q (2014) Beneficial effects of *Metschnikowia sp.* C14 on growth and intestinal digestive enzymes of juvenile sea cucumber *Apostichopus japonicus*. Anim Feed Sci Technol 197:142–147

Yang Z, Sun JM, Xu Z (2015) Beneficial effects of *Rhodotorula sp.* C11 on growth and disease resistance of juvenile Japanese spiky sea cucumber *Apostichopus japonicus*. J Aquat Anim Health 27:71–76

Yoon JH, Kang SJ, Oh TK (2006) *Polaribacter dokdonensis* sp. nov., isolated from seawater. Int J Syst Evol Microbiol 56:1251–1255

Yoon J, Matsuo Y, Matsuda S, Adachi K, Kasai H, Yokota A (2007) *Rubritalea spongiae* sp. nov. and *Rubritalea tangerina* sp. nov., two carotenoid-and squalene-producing marine bacteria of the family *Verrucomicrobiaceae* within the phylum '*Verrucomicrobia*', isolated from marine animals. Int J Syst Evol Microbiol 57:2337–2343

Yoon K, Song JY, Kwak MJ, Kwon SK, Kim JF (2017) Genome characteristics of the proteorhodopsin-containing marine flavobacterium *Polaribacter dokdonensis* DSW-5. J Microbiol 55:561–567

Zhang Q, Ma H, Mai K, Zhang W, Liufu Z, Xu W (2010) Interaction of dietary *Bacillus subtilis* and fructooligosaccharide on the growth performance, non-specific immunity of sea cucumber, *Apostichopus japonicus*. Fish Shellfish Immunol 29:204–211

Zhao Y, Zhang Q, Yuan L, Liu X (2017) Effects of dietary taurine on the growth, digestive enzymes, and antioxidant capacity in juvenile sea cucumber, *Apostichopus japonicus*. J World Aquacult Soc 48:478–487

17

Determining the growth and mortality parameters of longtail tuna (*Thunnus tonggol* Bleeker, 1851) using length frequency data in coastal waters of the northern Persian Gulf and Oman Sea, Iran

Mehran Yasemi · Alireza Nazari Bajgan · Mehran Parsa

Abstract Longtail tuna (*Thunnus tonggol*) is one of the most economically important tuna species in Iranian waters of Persian Gulf and Oman Sea. Some population dynamics parameters of this species were investigated in coastal waters of the northern Persian Gulf and Oman Sea from January to December 2014. The fork length (FL) of all specimens ranged from 27 to 107 cm, and total weight (TW) was between 200 and 12,000 g. Among the length frequencies examined, specimens with fork length in the range of 62–65 cm were the dominant group. The equation of length–weight relationship was estimated as TW = 0.00003 × $FL^{2.82}$ ($R^2 = 0.84$) and isometric growth model was determined for longtail tuna in study area. The parameters of Von Bertalanffy growth function were obtained as: $L_\infty = 111.23$ cm, $k = 0.3$ year^{-1} and $t_0 = -0.38$ years and $\phi' = 3.6$. The annual instantaneous rate of total mortality (*Z*), natural mortality (*M*) and fishing mortality (*F*) was estimated to be 1.15, 0.43 and 0.72 year^{-1}, respectively. The exploitation rate (*E*) was calculated as 0.63 indicating that longtail tuna was fully exploited in the study region.

Keywords Growth · Mortality · Longtail tuna · Persian Gulf · Oman Sea

Introduction

Tuna fishes, Scombridae family, are one of the most economically important marine resources for industrial fishing activities in the Persian Gulf and Oman Sea (Kamrani et al. 2010). The Scombridae is a family of 15 genera and about 50 species of epipelagic marine fishes (Block and Stevens 2001). The main tuna species caught in the Iranian waters of Persian Gulf and Oman Sea are: longtail tuna (*Thunnus tonggol* Bleeker, 1851), yellowfin tuna (*Thunnus albacares* Bonnaterre, 1788), skipjack tuna (*Katsuwonus pelamis* Linnaeus, 1758), kawakawa (*Euthynnus affinis* Cantor, 1849), frigate tuna (*Auxis thazard* Lacepede, 1800), Narrow-barred Spanish mackerel (*Scomberomorus commerson* Lacepede, 1800) and Indo-Pacific king mackerel (*Scomberomorus guttatus* Bloch & Schneider, 1801). Tuna fishes constitute around 50% of total landings of

M. Yasemi (✉)
Institute of Applied-Scientific Higher Education of Jahad-e-Agriculture, Agricultural Research, Education and Extension Organization, Tehran, Iran
e-mail: yasemi_m@yahoo.com

A. N. Bajgan
Higher Education Center of Jahad-e-Agriculture Hormozgan, Bandar Abbas, Iran

M. Parsa
Young Researchers and Elite Club, Bandar Abbas Branch, Islamic Azad University, Bandar Abbas, Iran

the Hormozgan province and total annual commercial landings of tuna fishes in Iranian waters of Persian Gulf and Oman Sea were around 175,000 tonnes. Longtail tuna represented 46% of total catch (Iran Fishery Organization 2013).

Longtail tuna is an epipelagic species inhabiting tropical to temperate regions of the Indo-Pacific, found almost exclusively in the neritic waters close to the shore, avoiding estuaries, turbid wasters and open Ocean. This species primarily caught by gillnet fleets operating in coastal waters with the highest reported catches from Iran, Indonesia, Pakistan, Malaysia and, to Oman, Yemen, India and Thailand (Pierre et al. 2014).

Knowledge of growth and mortality parameters is essential for understanding the population dynamics of different species. Several approaches were often used to determine the age and growth of the aquatic species, such as length–frequency analyses, tagging and recapture experiments and observations of the mark on the hard parts (scales, otoliths, spines and vertebrates) (Stequert et al. 1996). Particular morphometric measurements, such as length–weight relationship (LWR), are useful in fishery management and stock assessment models (Barría et al. 2014), and has been widely used in the environmental monitoring programs such as calculation of fish weight at a certain length and the conversion of an equation of growth in weight and vice versa, as well as the calculation of a condition index to allow for morphological comparisons of populations among different regions (Yoon et al. 2013; Oliveira et al. 2014). The general biology and population dynamics of the several tuna species have been researched in Iranian waters of Persian Gulf and Oman Sea (Kaymaram et al. 2010; Taghavi Motlagh et al. 2010; Hosseini and Ehsani 2014). Nevertheless, scanty information is available on growth and mortality characteristics of longtail tuna in this region http://www.jifro.ir/files/site1/user_files_eb12be/admin-A-10-1-11-532319e.pdf. Therefore, the specific objectives of present study were to determine: (1) the length and weight frequency structure (length–weight relationship), (2) the growth parameters (L_∞, K, T_0) using the length-based MULTIFAN method (3) the mortality parameters based on the length-converted catch curve and Pauly's empirical equation, and (4) the exploitation rate of the longtail tuna in coastal waters of the northern Persian Gulf and Oman Sea, Iran.

Materials and methods

Study area and sampling

Using commercial gill nets, specimens of longtail tuna were collected monthly and randomly in coastal waters of the northern Persian Gulf and Oman Sea (Hormozgan province, Iran) from January to December 2014 (Fig. 1). Sampling was carried out from five different landing sites including Bandar lengeh (54°22′E and 26°39′N), Gheshm Island (55°41′E and 26°41′N), Bandar Abbas (56°15′E 27°11′N), Sirik (57°05′E and 26°20′N) and Jask (57°47′E and 25°39′N). A total of 2163 individuals were examined during the study period.

Length–weight relationship

The length frequency distribution of longtail tuna caught in different months was determined. For each specimen, the fork length (FL) was measured with a measuring board to the nearest 1 cm, and total body weight (TW) was measured on a digital scale to 1 g. The relationship between length and weight was calculated using the following equation (Ricker 1975):

$$\mathrm{TW} = a\mathrm{FL}^{b}$$

where TW is the total weight (g), FL is the fork length (cm), a is the intercept of the regression and b is the regression coefficient indicating isometric growth when equal to 3. The best regression parameters were obtained by minimizing the residuals errors by the least square residuals method (Haddon 2011).

Student's t test was utilized to confirm whether the declivity of regression (constant "b") presented a significant difference with 3, indicating the type of growth: isometric ($b = 3$), positive allometric ($b > 3$) or negative allometric ($b < 3$) (Sokal and Rohlf 1987).

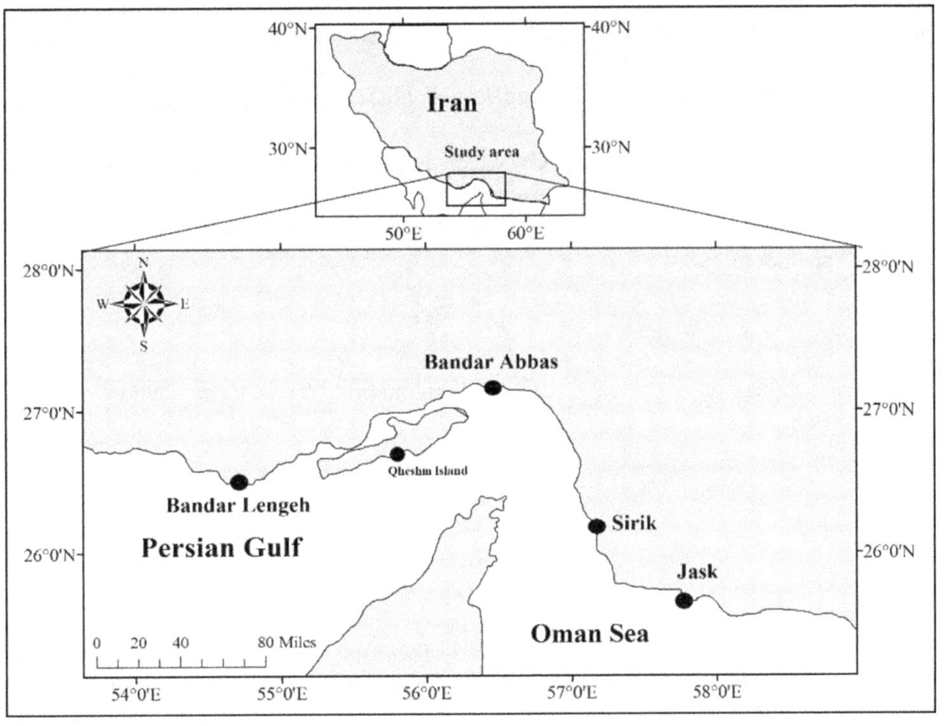

Fig. 1 Map of study area and location of sampling ports for longtail tuna (northern Persian Gulf and Oman Sea, Iran)

$$t_s = \frac{b-3}{sb}$$

where t_s is the t test value, b is the slope and sb is the standard error of the slope (b).

Growth parameters

Length frequency data were used to calculate the von Bertalanffy growth rate (k) and the asymptotic length (L_∞) by model progression analysis using the FiSAT program (Gayanilo et al. 1996) and ELEFAN I method (Pauly 1984). The best value of growth parameter (k) for the given value of L_∞ was identified by Shepherd's method and scan of k values option from Assess menu. A classical von Bertalanffy growth function (VBGF) was fitted to the data using the following formula (Sparre and Venema 1998):

$$L_t = L_\infty(1 - \exp(-k(t - t_0)))$$

In this equation, t_0 represents the theoretical age at length zero, L_∞ is the asymptotic length and L_t is the length at age. Estimation of the t_0 parameter was determined by empirical equation proposed by Pauly (1983):

$$\text{Log10}(-t_0) = -0.392 - 0.275\text{Log10}L_\infty - 1.038k$$

Mortality parameters

The length-converted catch curve was applied for the calculation of the instantaneous annual mortality rate (Z) (Pauly 1983). The growth performance index (ϕ') was calculated using the following formulae (Pauly and Munro 1984):

$$\phi' = \log K + 2\log L_\infty.$$

The natural mortality (M) was calculated by Pauly's empirical equation:

$$\mathrm{Log}M = 0.1228 - 0.192\mathrm{Log}L_\infty + 0.7485\mathrm{Log}K + 0.2391\log T,$$

where T is the mean annual temperature (in °C), which is assumed to reflect the locally sea surface temperature in the survey area (Pauly 1984) (in the present study, $T = 27$ °C).

For the calculation of the fishery mortality (F), the M value was subtracted from the Z value to get the fishing mortality (Appeldoom 1984).

$$F = Z - M$$

With the estimated values of F and Z, the rate of exploitation (U) was calculated using the following formulae (Gulland 1985):

$$E = \frac{F}{Z}$$

Results

The annual length frequency of longtail tuna is shown in Fig. 2. A total of 2163 specimens sampled in five landing sites. The fork length of longtail tuna was in the range of 27–107 cm. Along the length range analyzed, the highest length frequency obtained was in the range of 62–65 cm. The average length obtained was between 59 and 63 cm with no difference between months.

Figure 3 represents the length–weight relationships of longtail tuna during a 1 year period and in different seasons. The length–weight relationship during a 1 year period was estimated to be $\mathrm{TW} = 0.00003 \times \mathrm{FL}^{2.82}$ ($R^2 = 0.84$). The b value of relationship between length and weight was ranged from 2.70 (summer) to 2.89 (spring). According to the obtained b value and using the Student's t test, isometric growth model was determined for longtail tuna in the study area ($P > 0.05$).

The von Bertalanffy growth curves of longtail tuna are shown in Fig. 4. The growth parameters estimated by ELEFAN I were as: $L_\infty = 111.23$ cm fork length, $t_0 = -0.38$ year, $\phi' = 3.6$. The k value was 0.3 year^{-1} (Fig. 5). The estimated age in different fork length classes is represented in Table 1.

The length-converted catch curve of longtail tuna is shown in Fig. 6. The estimated rates of total mortality (Z) and natural mortality rate (M) were 1.15 and 0.43 year^{-1}, respectively. The fishing mortality (F) was 0.72 year^{-1}. The exploitation rate (E) for population of longtail tuna was calculated as $E = 0.63$.

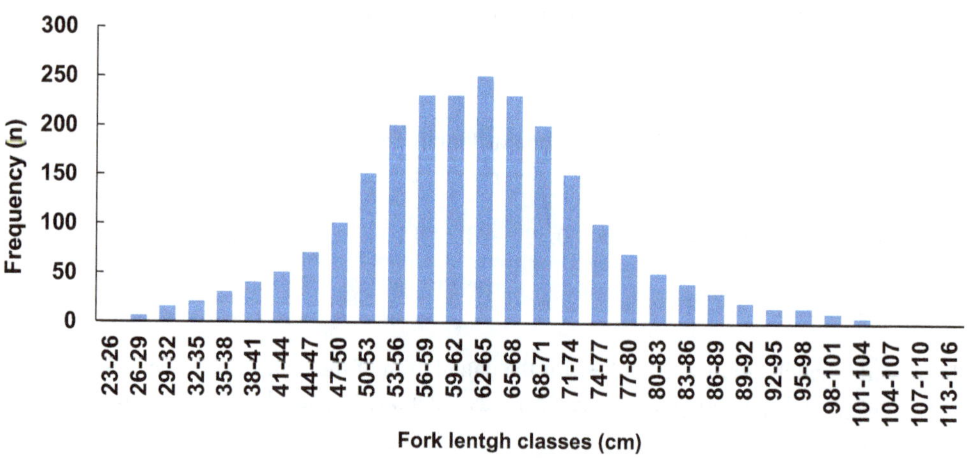

Fig. 2 Annual fork length frequency distribution of longtail tuna in northern Persian Gulf and Oman Sea, Iran

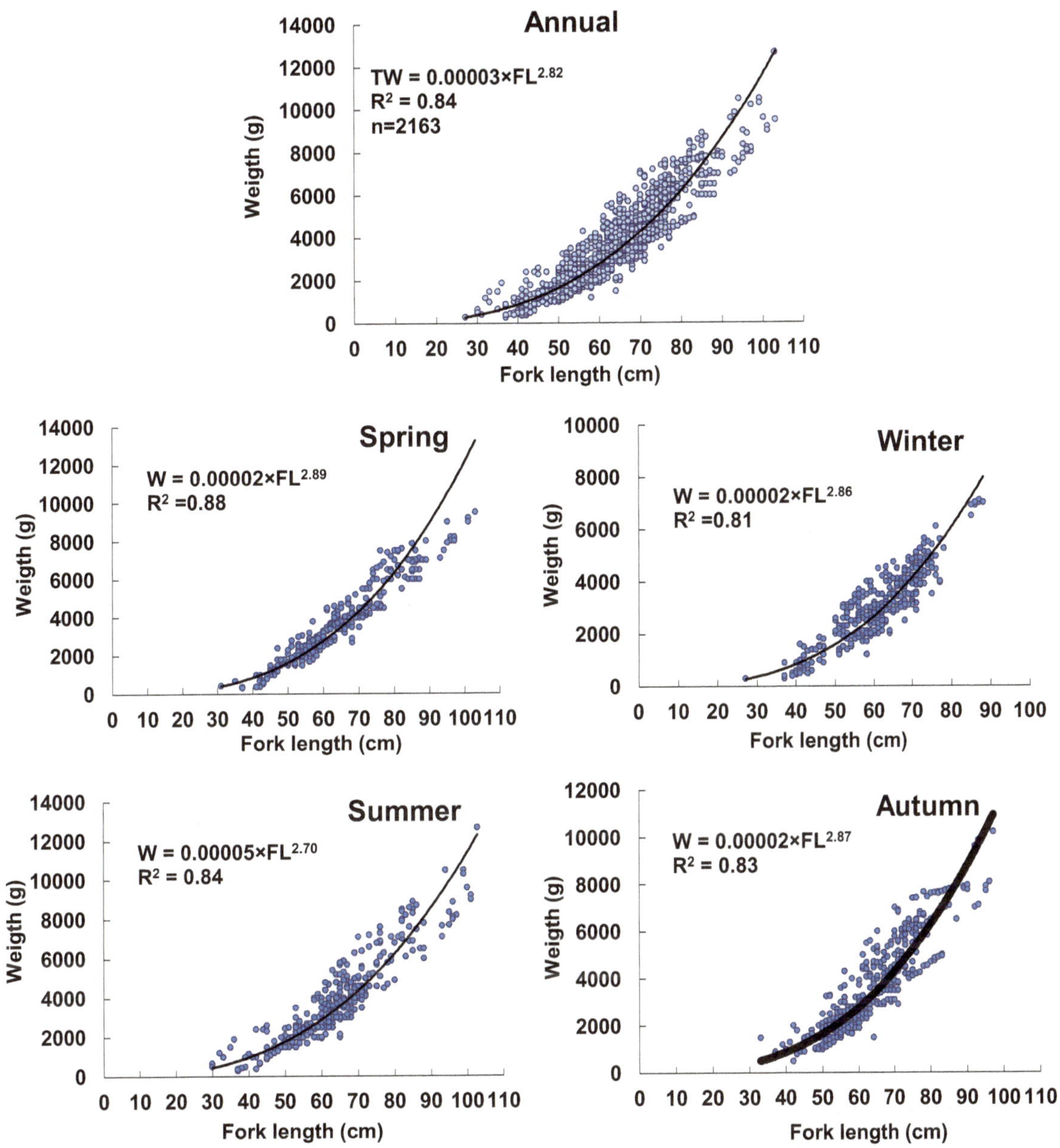

Fig. 3 Length–weight relationships of longtail tuna during a 1 year period and different seasons in northern Persian Gulf and Oman Sea, Iran

Discussion

The length–frequency-based analysis is used widespread at the present time, especially in the tropical and subtropical regions (Sparre and Venema 1998). This method becomes important when other aging techniques are either not possible or very expensive (Wang and Ellis 2005). According to Zhu et al. (2011), the study of growth using length–frequency analysis has long been the most frequently used method.

Gulland and Rosenberg (1992) noted that sampling from large number of smaller samples than sampling from a limited number of larger samples will always give more information, while providing estimates with smaller variance. In this regard, two samples of 150 fish will always give more information than one of 300 fish, but will take more time. Furthermore, a sample of 50–150 fish/month and an annual sample of 1000–1500 fish provide the most reliable image on the biological characteristics of a specific species. To have most

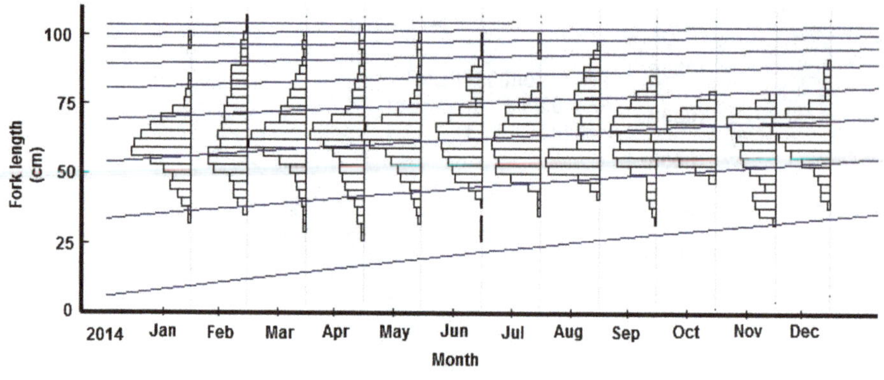

Fig. 4 The von Bertalanffy growth curves of longtail tuna in northern Persian Gulf and Oman Sea, Iran

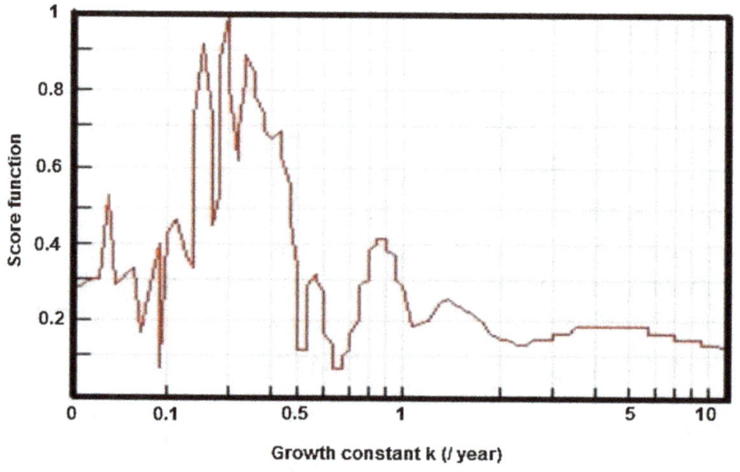

Fig. 5 Shepherd's method with maximum score (S_{max}) appropriate to the best value of growth coefficient of longtail tuna in northern Persian Gulf and Oman Sea, Iran

reliable understanding on the biology of longtail tuna distributed in coastal waters of the northern Persian Gulf and Oman Sea, sampling was done in five different landing sites and covered 12 months (180 fish/month).

According to IGFA (2001), the maximum fork length of longtail tuna was found to be 147 cm. The length frequency of long tail tuna captured from the traditional fishing was 26–130 cm (Darvishi et al. 2003). The author found significant differences in the length frequency of long tail tuna between the traditional and commercial fishing which might probably be governed by the sampling methods and fishing tools. Fish with mean length of 30 cm have been reported in the gillnet catches of long tail tuna in Oman Sea waters in October while the average size was found to be 36 and 46 cm for February and May, respectively.

Abdussamad et al. (2012) have reported that longtail tuna along the Indian Coast is supported by fishery having a length range of 23–111 cm and a mean length of 60.5 cm in gillnets fishery. Khorshidian and Carrara (1993) indicated the differences in size composition of longtail tuna between the Iranian and Omani fisheries on the northern and southern shores of the Oman Sea. Size-related migration or differences in the mesh size of the gill nets used were possibly contributed to this difference. No differences in the monthly length frequency of longtail tuna were observed. Moreover, the average length was lower than that at first sexual maturity, indicating overfishing or the use of small mesh size nets.

Among the biometric relations in fishes, the length–weight relationships are greatly presented by scientists and researchers as useful tools in fish biology (King 2007; Ferreira et al. 2008). Calculation of the length–weight relationship parameter is useful for the prediction of weight from length values, condition of fish, stock assessment, and estimation of biomass (Valset et al. 2007).

Table 1 Detailed data on relationship between fork length classes and estimated age for longtail tuna in northern Persian Gulf and Oman Sea, Iran

Length group (cm)	Age (year)									Numbers
	1	2	3	4	5	6	7	8	9	
26–29	1	–	–	–	–	–	–	–	–	1
29–32	5	–	–	–	–	–	–	–	–	5
32–35	4	–	–	–	–	–	–	–	–	4
35–38	21	–	–	–	–	–	–	–	–	21
38–41	29	–	–	–	–	–	–	–	–	29
41–44	46	–	–	–	–	–	–	–	–	46
44–47	69	–	–	–	–	–	–	–	–	69
47–50	68	–	–	–	–	–	–	–	–	68
50–53	143	–	–	–	–	–	–	–	–	143
53–56	226	–	–	–	–	–	–	–	–	226
56–59	66	116	–	–	–	–	–	–	–	182
59–62	–	268	–	–	–	–	–	–	–	268
62–65	–	227	–	–	–	–	–	–	–	227
65–68	–	217	–	–	–	–	–	–	–	217
68–71	–	227	–	–	–	–	–	–	–	227
71–74	–	–	128	–	–	–	–	–	–	128
74–77	–	–	96	–	–	–	–	–	–	96
77–80	–	–	43	–	–	–	–	–	–	43
80–83	–	–	28	8	–	–	–	–	–	36
83–86	–	–	–	53	–	–	–	–	–	53
86–89	–	–	–	26	–	–	–	–	–	26
89–92	–	–	–	–	5	–	–	–	–	5
92–95	–	–	–	–	16	–	–	–	–	16
95–98	–	–	–	–	–	13	–	–	–	13
98–101	–	–	–	–	–	2	6	–	–	8
101–104	–	–	–	–	–	–	2	3	–	5
104–107	–	–	–	–	–	–	–	–	1	1
Total	678	1055	295	87	21	15	8	3	1	2163

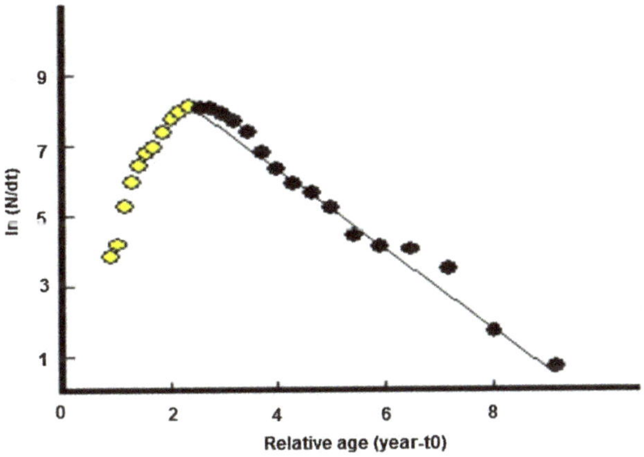

Fig. 6 Length-converted catch curve for longtail tuna in northern Persian Gulf and Oman Sea, Iran

In our study, the length–weight relationship was estimated to be $W = 0.00003 \times FL^{2.82}$ ($R^2 = 0.84$). To verify if b was significantly different from the isometric value ($b = 3$), the Student's t test (H_0: $b = 3$) was utilized. Results showed no difference between obtained b value (2.82) and 3 and isometric growth model was determined for longtail tuna in study area ($P > 0.05$). Biswass (1993) and Froese (2006) presented the acceptable range of b value between 2.5 and 3.5 that is in agreement with our estimation. Length–weight relationship for longtail tuna was estimated as $W = 0.0148 \times L^{3.0}$ in waters of India (Abdussamad et al. 2012), indicating a perfect isometric pattern in growth that is in agreement with the results of our study. Griffiths et al. (2010) estimated the length–weight relationship of longtail tuna as $W = 0.00005 \times L^{2.82}$ in tropical and temperate waters of the central Indo-Pacific that is completely in agreement with our findings.

The growth parameters of longtail tuna estimated by ELEFANI were as: $L_\infty = 111.23$ cm fork length, $t_0 = -0.38$ year, $\phi' = 3.6$. The k value was 0.3 year^{-1}. Studies in different regions have indicated that the L_∞ of longtail tuna was differed from 55 cm in waters of Japan (Itoh et al. 1999) to 135.4 cm in Australia (Griffiths et al. 2010) and the range for k was 0.228 in Oman (Prabhakar and Dudley 1989) to 1.7 in waters of Japan (Itoh et al. 1999). The L_∞ estimated in our study for the longtail tuna was higher than the estimates given by some previous studies (Supongpan and Saikliang 1987; Yesaki 1989; Itoh et al. 1999) but lower than that by Griffiths et al. (2010) and Abdussamad et al. 2012.

Difference in the growth parameters estimated by other authors may be due to the fact that the data used for different analytical methods by them were obtained by different gears such as drift gillnets, hooks and lines, troll and trawls (Pillai et al. 2002). Taghavi Motlagh et al. (2010) declared that different values of L_∞ and k might be associated with sampling error, variation in fishing intensity or environmental conditions. Moreover, slight differences in growth patterns can be the results of differences in genetic structure, differences in temperature, density of food and diseases (Wootton 1998). Age at zero length ($-t_0$) was calculated to be -0.38 year and, according to King (2007), it means juveniles of longtail tuna grow more quickly than adults.

The total mortality for longtail tuna was 1.15 year^{-1}. Abdussamad et al. 2012 presented a total mortality rate as 3.85 year^{-1} for longtail tuna in Indian waters that was considerably higher than our results. The instantaneous natural mortality coefficient (M) was 0.43 year^{-1} in the present study. Our result was lower than the instantaneous natural mortality coefficient estimated for the same species in the Indian waters, 0.77 year^{-1} (Abdussamad et al. 2012) and 0.8 year^{-1} (James et al. 1993). SU et al. (2003) declared that the estimation of natural mortality poses some difficulty because it may be affected by the selection of the estimation method and the study area.

The exploitation rate for population of longtail tuna ($E = 0.63$) in the present study was high. Gulland (1985) stated that in an optimally exploited stock, fishing mortality should be equal to natural mortality, resulting in an exploitation rate of 0.5. Patterson (1992) declared that the fishing rate satisfying optimal E level of 0.5 tended to reduce pelagic fish stock abundance and, hence, the former author suggested that E should be maintained at 0.4 for optimal exploitation of those stocks. The present estimate of E suggests that the population of longtail tuna in the waters of northern Persian Gulf and Oman Sea is being heavily exploited and overfished at a higher level than the optimum and a better management policy is necessary in this area. The decision making on fisheries management of longtail tuna cannot use only biological information; thus, the socioeconomic condition of small-scale fishermen in waters of the Persian Gulf and Oman Sea should be considered to achieve sustainable management of longtail tuna resources. Finally, results of this study are particularly helpful for drift gillnet fisheries management in the Persian Gulf and Sea of Oman.

References

Abdussamad EM, Said Koya KP, Ghosh S, Rohit P, Joshi KK, Manojkumar B, Prakasan D, Kemparaju S, Elayath MNK, Dhokia HK, Sebastine M, Bineesh KK (2012) Fishery, biology and population characteristics of longtail tuna, *Thunnus tonggol* (Bleeker, 1851) caught along the Indian coast. Indian J Fish 59(2):7–16

Appeldoom RS (1984) The effect of size on mortality of small juvenile conchs (*Strombus gigas* L. and *S. costatus* G.). J Shellfish Res 4:37–43

Barría C, Navarro J, Coll M, Fernandez-Arcaya U, Sáez-Llante RS (2014) Morphological parameters of abundant and threatened chondrichthyans of the northwestern Mediterranean Sea. J Appl Ichthyol 31(1):114–119

Biswas SP (1993) Manual of methods in fish biology. South Asian, New Delhi

Block BA, Stevens ED (2001) Tuna (physiology, ecology and evolution). Academic Press, Massachusetts

Darvishi M, Kaymaram F, Talebzadeh SA, Behzadi S (2003) Population dynamics of five Scombrid fish in Hormuzgan Province (Iran). Persian Gulf and Oman Sea. Ecological Research Institute, Iran **(In Persian)**

Ferreira S, Sousa R, Delgado J, Carvalho D, Chada T (2008) Weight–length relationships for demersal fish species caught off the Madeira archipelago (eastern-central Atlantic). J Appl Ichthyol 24(1):93–95

Froese R (2006) Cube law, condition factor and weight–length relationships: history, meta-analysis and recommendations. J Appl Ichthyol 22(4):241–253

Gayanilo FC, Parre PS, Pauly D (1996) The FAO ICLARM stock assessment tools (FISAT). User's guide. FAO Computerized Information Series Fisheries, Rome

Griffiths SP, Fry GC, Manson FJ, Lou DC (2010) Age and growth of longtail tuna (*Thunnus tonggol*) in tropical and temperate waters of the central Indo-Pacific. ICES J Mar Sci 67:125–134

Gulland JA (1985) Fish stock assessment: a manual of basic methods. Wiley, New York

Gulland JA, Rosenberg AA (1992) A review of length-based approaches to assessing fish stock. FAO Fisheries Technical Paper, Rome

Haddon M (2011) Modelling and quantitative methods in fisheries. Chapman and Hall Press, Florida

Hosseini SA, Ehsani E (2014) An investigation of reactive behavior of yellowfin tuna schools to the purse seining process. Iran J Fish Sci 13:330–340

IGFA (2001) Database of IGFA angling records until 2001. IGFA, Fort Laudedale

Iran Fishery Organization (2013) Fisheries data statistics of 2012. Iran Fishery Organization (IFO) Press, Tehran

Itoh T, Yuki Y, Tsuji S (1999) Spawning possibility and growth of longtail tuna, *Thunnus tonggol*, in the water around Japan. Bull Natl Res Inst Far Seas Fish 36:47–53

James PSBR, Pillai PP, Pillai NGK, Jayaprakash AA, Gopakumar G, Kasim HM, Sivadas M, Koya KPS (1993) Fishery, biology and stock assessment of small tunas. In: Sudarsan D, John ME (eds) Tuna research in India. Fishery survey of India, Bombay, pp 123–148

Kamrani E, Sabili AN, Yahyavi M (2010) Stock assessment and reproductive biology of the blue swimming crab, *Portunus pelagicus* in Bandar Abbas Coastal Waters, Northern Persian Gulf. Iran J Persian Gulf 11:11–22

Kaymaram F, Hossainy SA, Darvishi M, Talebzadeh SA, Sadeghi MS (2010) Reproduction and spawning patterns of the *Scomberomorus commerson* in the Iranian coastal waters of the Persian Gulf and Oman Sea. Iran J Fish Sci 9:233–244

Khorshidian K, Carrara G (1993) An analysis of length frequencies of *Thunnus tonggol* in Hormozgan waters. In: Ardill JD (ed) Proceedings of the expert consultation on Indian Ocean Tunas, 5th Session, Mahe, Seychelles

King M (2007) Fisheries biology, assessment and management. Wiley-Blackwell Publishing, New Jersey

Oliveira VA, Loverde-Oliveir SM, Mateus LA, Teixeira-de Mello F (2014) Length–weight relationships of 26 fish species from the streams of the upper section of the Paraguay River basin (Mato Grosso, Brazil). J Appl Ichthyol 31:225–227

Patterson K (1992) Fisheries for small pelagic species: an empirical approach to management targets. Rev Fish Biol Fish 2(4):321–338

Pauly D (1983) Some simple methods for the assessment of tropical fish stocks. Fisheries Technical Paper, Rome

Pauly D (1984) Fish population dynamics in tropical waters: a manual for use with programmable calculators. International Center for Living Aquatic Resources Management, Philippines

Pauly D, Munro JL (1984) Once more on the comparisons of growth in fish and invertebrates. ICLARM Fishbyte 2(1):21

Pierre L, Geehan J, Herrera M (2014) Review of the statistical data available for bycatch species. IOTC, pp 17–19

Pillai NGK, Menon NG, Pillai PP, Ganga U (2002) Management of Scombrid fisheries. Central Marine Fisheries Research Institute, Kochi, pp 56–61

Prabhakar A, Dudley RG (1989) Age, growth and mortality rates of longtail tuna *Thunnus tonggol* (Bleeker) in Omani waters based on length data. Indo-Pac Tuna Dev Manag Progr 16:90–96

Ricker WE (1975) Computation and interpretation of biological statistics of fish populations. Bulletin of the Fisheries Research Board of Canada. The Blackburn Press, New Jersey

Sokal RR, Rohlf FJ (1987) Introduction to biostatistics, 2nd edn. Freeman Publication, New York

Sparre P, Venema SC (1998) Introduction to tropical fish stock assessment. FAO Fisheries Technical Paper, Rome

Stequert B, Panfili J, Dean JM (1996) Age and growth of yellowfin tuna, *Thunnus albacares*, from the western Indian Ocean, based on otolith micro-structure. Fish B-NOAA 94:124–134

Su NJ, Sun CL, Ye SR (2003) Estimation of growth parameters and age composition for yellowfin tuna, *Thunnus albacares*, in the Western Pacific using the length-based MULTIFAN method. J Fish Soc Taiwan 30:171–184

Supongpan S, Saikliang P (1987) Fisheries status of tuna purse seiners (using sonar) in the Gulf of Thailand. Department of Fisheries, Bangkok

Taghavi Motlagh SA, Hashemi SA, Kochanian P (2010) Population biology and assessment of Kawakawa (*Euthynnus affinis*) in coastal waters of the Persian Gulf and Sea of Oman (Hormozgan Province). Iran J Fish Sci 9:315–326

Valset A, Bouchon-Navaro Y, Louis M, Bouchon C (2007) Weight–length relationships for 20 fish species collected in the mangroves of Guadeloupe (Lesser Antilles). J Appl Ichthyol 24(1):99–100

Wang Y, Ellis N (2005) Maximum likelihood estimation of mortality and growth with individual variability from multiple length–frequency data. Fish B-NOAA 103:380–391

Wootton RJ (1998) Ecology of teleost fishes. Kluwer Academic Publishers, Dordrecht

Yesaki M (1989) Estimates of age and growth of kawakawa (*Euthynnus affinis*), longtail tuna (*Thunnus tonggol*) and frigate tuna (*Auxis thazard*) from the Gulf of Thailand based on length data. Indo-Pac Tuna Dev Manag Progr 17:94–108

Yoon HS, An YK, Hwang JH, Lim HS, Lee WK, Han KH, Lee SH, Choi SD (2013) Length–weight relationships for 14 fish species of the Suer River estuary in southern Korea. J Appl Ichthyol 29:468–469

Zhu G, Xu L, Dai X, Liu W (2011) Growth and mortality rates of yellowfin tuna, *Thunnus albacares* (Perciformes: Scombridae), in the eastern and central Pacific Ocean. J Zool 28(2):199–206

18

Induction of DNA damage in the peripheral blood of zebrafish (*Danio rerio*) by an agricultural organophosphate pesticide, monocrotophos

Avelyno H. D'Costa · S. K. Shyama · M. K. Praveen Kumar · Trisha M. Fernandes

Abstract The use of organophosphate pesticides (OPs) in agricultural practices in India has gained popularity due to their broad spectrum, low cost and high potency. Monocrotophos (an OP) which is a systemic and contact broad-spectrum cholinesterase inhibitor is being widely used in India. However, its use entails the damage to other non-target organisms such as fish. In the present study, zebrafish (*Danio rerio*) were exposed to various doses of monocrotophos (0.125, 0.625 and 1.25 μL/L) and genotoxicity studies were carried out employing the comet assay and micronucleus test in the peripheral blood of these fish after 24, 48 and 72 h of exposure. Significant DNA damage in the form of % tail DNA and micronuclei was observed in all the treated fish compared to the control. Both % tail DNA and micronuclei frequency were found to increase significantly as the concentration increased. DNA damage was also observed at all the time intervals for all treatments except in the 0.125 μL/L-treated group. A positive correlation was also observed between the comet assay and micronucleus test. The study thus suggests the use of zebrafish as an experimental model to study the genotoxic effects of agricultural pesticides using the comet assay and micronucleus test.

Keywords Comet assay · Micronucleus test · Genotoxicity

Introduction

Organophosphate pesticides (OPs) are used extensively in agricultural practices around the world due to their high potency and rapid mode of action on a broad spectrum of pests. They are preferred over organochlorines due to their rapid degradation and less environmental persistence. Monocrotophos (MCP), an OP, is used in many parts of the world, particularly in developing countries because of its low cost and is a broad-spectrum pesticide. It is a systemic and contact poison that inhibits cholinesterase and acts as a neurotoxin. Its use, however, entails several undesirable effects particularly in non-target organisms. It is an organic pollutant which is soluble in water and can, therefore, be easily taken up by aquatic fauna. It is labeled as a marine pollutant (FAO/UNEP 1997) and is included under category 1 (highly toxic) by USEPA (1985). It can easily enter the body through oral, dermal or inhalation routes. It is highly toxic to insects, birds and mammals and is moderately toxic to fishes. MCP, as with other pesticides finds its way to the aquatic environment by

---

A. H. D'Costa · S. K. Shyama (✉) · M. K. Praveen Kumar · T. M. Fernandes
Genetic Toxicology Laboratory, Department of Zoology, Goa University, Goa 403206, India
e-mail: skshyama@gmail.com

A. H. D'Costa
e-mail: avelynodc@gmail.com

wastewater discharge and monsoon run-offs from agricultural land and thereafter inevitably enters the aquatic food chain.

A number of studies are available on the inhibitory effect of MCP on acetycholine esterase (AChE) activity in different fish species (de Croux et al. 2002; Rao 2004; Kavitha and Rao 2007).

The uptake of xenobiotics also has the capability to induce damage at the level of DNA. They may either affect the DNA directly by causing DNA strand breaks or indirectly by producing reactive oxygen species (ROS) which then damages the DNA by forming adducts or causing lesions (O'Brien 1988; Bagchi et al. 1995; Marnett 2000). DNA repair systems which try to heal the damaged DNA can do so only up to a certain threshold, beyond which DNA damage can persist, in the form of lesions or misrepaired DNA, leading to mutagenesis and carcinogenesis (Moustacchi 2000). Two reliable genotoxicity endpoints which are used for screening the genotoxicity of such xenobiotics are the micronucleus (MN) test and the single-cell gel electrophoresis (comet assay) (Anbumani and Mohankumar 2012; Praveen Kumar et al. 2014, 2015). The MN test is basically used to check the clastogenicity of a particular compound, whereas the comet assay is used to check the effect of the compound on the integrity of DNA (Hartmann et al. 2001; Bolognesi and Cirillo 2014). A number of studies are available on the genotoxicity of other organophosphate pesticides using these tests. Ali and Kumar (2008) studied the genotoxic effects in *Channa punctatus* exposed to acute treatments of chlorpyrifos and observed DNA damage in the form of micronuclei and percentage tail DNA from the peripheral blood. Similarly, Kumar et al. (2010) also observed the induction of micronuclei and tail DNA in different tissues of *C. punctatus* exposed to malathion. The genotoxicity of MCP has been reported in estuarine bivalves (Revankar and Shyama 2009). However, few studies are available on the genotoxicity of this pesticide in teleost fishes. *Tilapia mossambica* treated with different concentrations of MCP were reported to have significant DNA damage in their erythrocytes as observed by the comet assay (Banu et al. 2001). In another study, Ali and Kumar (2008) reported the DNA damage induced by MCP in the gills, kidneys and lymphocytes of *C. punctatus* using the comet assay. Besides fish, the genotoxicity of MCP has been reported in mice (Mahboob et al. 2002), in rats (Yaduvanshi et al. 2010) and in human peripheral blood lymphocytes (Das et al. 2007).

Although a number of fish models are used for toxicity screening of aquatic environment, the teleost cyprinid, zebrafish (*Danio rerio*) is the most popular choice because of its prominence as a toxicological model (Hill et al. 2005; Sipes et al. 2011). A few studies have been carried out on the different effects of MCP on zebrafish such as the estrogenic and feminizing effects of MCP (Zhang et al. 2013) and cortisol-mediated stress responses (Zhang et al. 2015). In a recent study, different concentrations of MCP were used to evaluate the developmental response of zebrafish embryos (Pamanji et al. 2015). Mortality was observed in all the test concentrations along with a large number of morphological abnormalities. However, there is no information on the DNA damaging effects of MCP in adult zebrafish, and therefore, the present study was carried out to assess the genotoxicity of different concentrations of this OP.

Materials and methods

Quality assurance and quality control

The appropriate quality assurance methods of sample preparation, handling and preservation were carried out in accordance with US EPA procedures. All chemicals used were of analytical grade from Himedia (Himedia, India).

Experimental fish

Zebrafish were handled and maintained as per the European Union directive on the protection of animals used for scientific purposes (EU 2010). Male zebrafish (*D. rerio*) (12-month old and size 4.0 ± 0.3 cm) were procured from a fish rearing and breeding farm from Pune, India and were transported in well-oxygenated polythene bags to the laboratory. They were maintained in filtered tap water in aquaria (25 L) with aerators and temperature regulators at 28 °C, 12 h light/dark cycle, pH 7.0 ± 0.15 and DO 8.5 ± 1.0. They were fed with commercial feed pellets. Foam filters were used to collect the accumulated fecal matter and were cleaned every alternate day. Water was changed once a week. Thus, these fish were acclimatized in the laboratory conditions for 30 days.

Monocrotophos

Monocrotophos [$C_7H_{14}NO_5P$, dimethyl (E)-1-methyl-2-(methylcarbamoyl) vinyl phosphate, MCP], of commercial grade (37% water soluble) was procured from Sds Ramcides Cropscience Pvt Ltd, Chennai, India (Batch No. R/P/879).

Treatment schedule

The fish were distributed into groups; each containing ten fishes in the same aquaria mentioned above and was used for dose–response studies as well as time–response studies. Based on the LC50 value of MCP on *Tilapia mossambicus* as per Mustafa et al. (1982), three sub-lethal doses, i.e., 0.125, 0.625, 1.25 µL/L were selected and exposed in triplicates. Acute toxicity studies were also carried out at 24, 48 and 72 h of exposure without renewal of MCP.

Validation of the comet assay and micronucleus test

A separate group of fish (ten individuals) treated with cyclophosphamide (1, 3 and 5 mg/L in water), a well-known genotoxin was used as a positive control at different time intervals (24, 48 and 72 h) and the comet assay and micronucleus test were validated.

Cell viability

Prior to the comet assay and micronucleus test, the cell count and cell viability of the peripheral blood were checked to ensure that there were enough living cells to perform the assay employing trypan blue dye exclusion test. The samples showing more than 90% viability and a cell count of a minimum of 10^6 cells/mL were used for the tests.

Comet assay (alkaline single-cell gel electrophoresis)

The comet assay was performed according to the protocol of Ferraro et al. (2004) with slight modifications. All the steps were carried out in dim light and at 4 °C to prevent photo-oxidation of DNA. Peripheral blood was collected from the fish by caudal puncture and suspended in phosphate buffered saline (Ca + Mg + free, pH 7.2). 20 µL of this blood suspension was mixed with low-melting agarose (LMA) and overlaid on a base layer of normal-melting agarose on frosted microscopic slides. These slides were placed in a cold lysing solution (2.5 M NaCl, 100 mM Na_2EDTA, 10 Mm Tris, 10% DMSO And 1% Triton X pH 10, 4 °C) for 1 h. The slides were then transferred to an electrophoretic unit containing unwinding buffer (pH > 10) for 20 min to allow the DNA to unwind. Electrophoresis was then performed under alkaline conditions (pH 10) for 20 min at 300 mA, 25 V. After electrophoresis, the slides were placed in neutralization buffer (400 mM Tris base, pH 7.5) for 5 min. About 25 µL of ethidium bromide stain was applied in four-to-five small, equally sized droplets over the gel, covered with a coverslip and examined using a fluorescence microscope (Olympus BX53) with a green filter at 200× magnification. Two slides per fish were prepared and 100 "comets" were screened per slide. Images of the comets of non-overlapping cells were captured using an attached camera and analyzed with the help of computer software, CASP (Konca et al. 2003). The % tail DNA was recorded which is used as a reliable measure of DNA damage (Kumaravel and Jha 2006; Praveen Kumar et al. 2014).

Micronucleus test

The MN test was performed as per Baršiene et al. (2006). Briefly, a small drop of the blood collected by caudal puncture was immediately smeared on clean glass slides, allowed to air-dry, fixed in absolute methanol for 10 min and stained with 5% Giemsa for 30 min. Three slides per fish were prepared. The frequency of micronuclei (MNi) in erythrocytes was evaluated by scoring 5000 intact cells per fish at 1000× magnification. MNi were identified as structures with the following morphological features: (1) spherical or ovoid-shaped extra-nuclear bodies in the cytoplasm (2) a diameter of 1/3–1/20 of the main nucleus (3) non-refractory bodies (4) color texture resembling that of the nucleus, and (5) the bodies completely separated from the main nucleus (Fenech et al. 2003).

Statistical analysis

Statistical analyses of the data were carried out using IBM SPSS 23 statistical software package. Since the data are expressed as percentage values, arc sine transformation of the variables were computed and were tested to check the normality and homogeneity of variance using Shapiro–Wilk test and Welch's test, respectively, before being analyzed. A two-way ANOVA was applied to test the effect of dose as well as time on the DNA damage parameters and also the interactions between them followed by a post hoc Dunnet's test to compare the different groups with the control within the same treatment group or time interval group. Pearson's correlation with a scatter plot was also used to test the relationship between the frequency of MNi and % tail DNA. The data were considered to be statistically significant at $p < 0.05$.

Results

The DNA damage (% tail DNA) induced by different concentrations of CP (1, 3, and 5 mg/L) in the blood cells of zebrafish at various time intervals (24, 48 and 72 h) is shown in Fig. 1a ,b. All the doses of CP induced a significant increase in % tail DNA and % MNi compared to the control (Figs. 1, 2).

A significant increase of dose-dependent damage ($p < 0.05$) was observed in the form of % tail DNA in the erythrocytes of *D. rerio* compared to the control (Fig. 3). Tail DNA (%) was also found to increase significantly with advancing time.

A significant increase in the frequency of micronuclei (MNi) was also observed in the erythrocytes for the 0.62 and 1.25 μL/L treatment groups (Fig. 4). However, in the time-dependent studies, there was no significant increase of MNi formation after 24 and 48 h in the 0.125 μL/L treatment group although there was a significant increase after 72 h.

The results of the two-way ANOVA (Table 1) indicate the effect of treatments and time as well as their interactions (treatment × time) on the % tail DNA and MNi. The treatment variable contributed the most to the % tail DNA ($F = 1739.89$, $p < 0.001$) and to a lesser extent, MNi ($F = 697.98$, $p < 0.001$). On the other

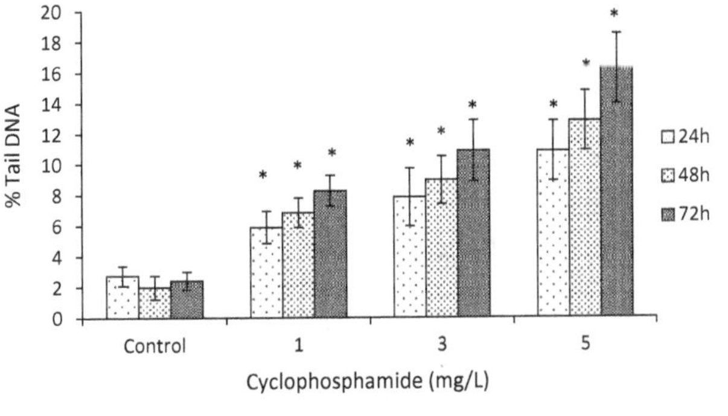

Fig. 1 Tail DNA (%) in erythrocytes of *D. rerio* exposed to different treatments of cyclophosphamide at different intervals of time. Statistical differences (*$p < 0.05$) are in relation to the control of the respective time interval

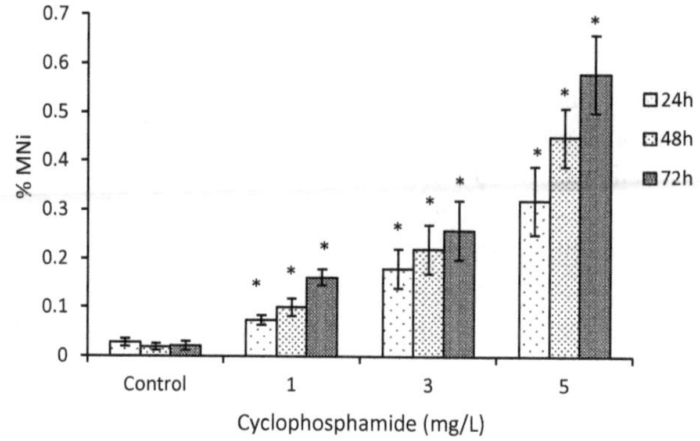

Fig. 2 MNi (%) in erythrocytes of *D. rerio* exposed to different treatments of cyclophosphamide at different intervals of time. Statistical differences (*$p < 0.05$) are in relation to the control of the respective time interval

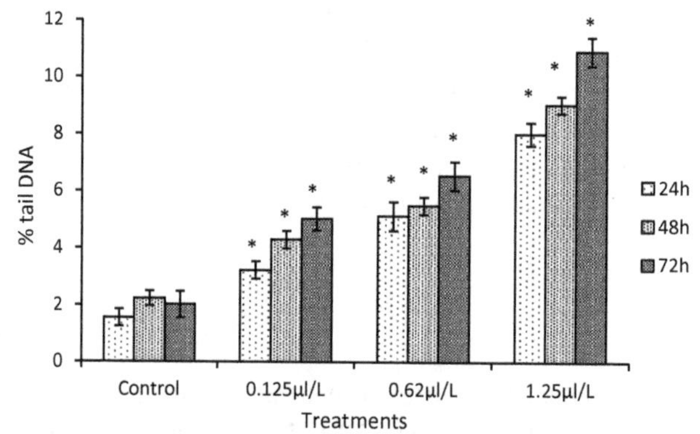

Fig. 3 Tail DNA (%) in erythrocytes of *D. rerio* exposed to different treatments of monocrotophos at different intervals of time. Statistical differences (*$p < 0.05$) are in relation to the control of the respective time interval

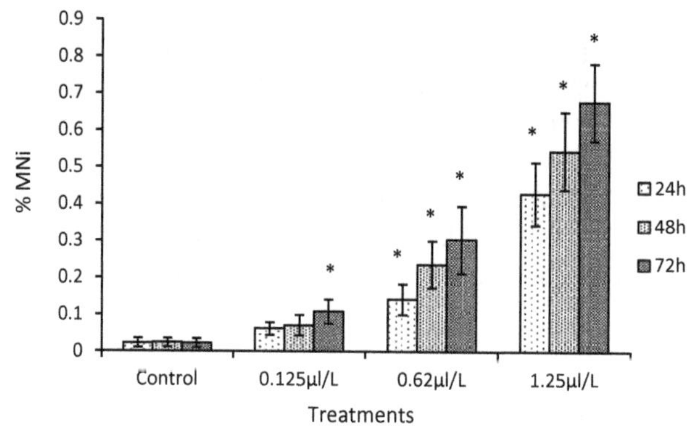

Fig. 4 MNi (%) in erythrocytes of *D. rerio* exposed to different treatments of monocrotophos at different intervals of time. Statistical differences (*$p < 0.05$) are in relation to the control of the respective time interval

hand, the time variable had the most effect on MNi ($F = 25.8$, $p < 0.001$) followed by the % tail DNA ($F = 10.96$, $p < 0.001$). The interaction of the two independent variables (treatment × time), however, influenced only the formation of MNi ($F = 4.56$, $p < 0.01$).

The data of the arc sine-transformed % tail DNA and % MNi were found to be highly correlated by Pearson's correlation analysis ($R = 0.92$) and the relationship is plotted in the form of a scatter plot in Fig. 5. A significant positive correlation is, therefore, found to exist between these two parameters of DNA damage ($R^2 = 0.86$).

Table 1 Two way ANOVA testing the influence of treatments and time as well as the interaction between them (treatment × time) on the DNA damage parameters in erythrocytes of *D. rerio* exposed to monocrotophos

Dependent variable	Independent variables					
	Factors				Interaction	
	Treatment		Time		Treatment × time	
	F value	*p*	*F* value	*p*	*F* value	*p*
% tail DNA	1739.89	< 0.001	10.96	< 0.001	1.36	ns
MNi	697.98	< 0.001	25.80	< 0.001	4.56	< 0.01

ns non-significant

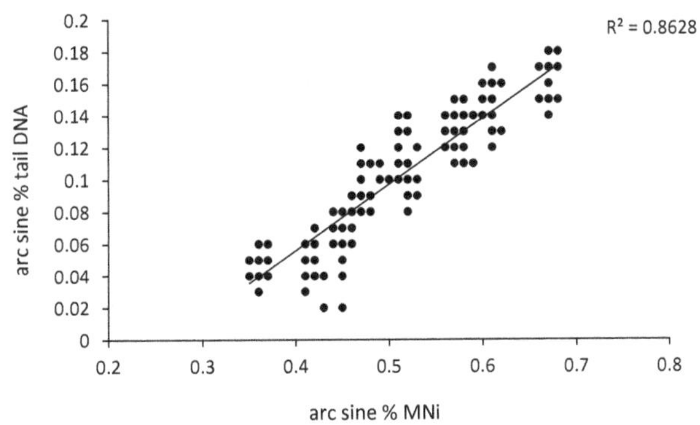

Fig. 5 Scatter plot showing the relationship between arc sine-transformed data of % tail DNA and MNi

Discussion

The present study revealed the genotoxicity of MCP in the peripheral blood cells of zebrafish as indicated by the comet assay and the micronucleus test. Further, genetic damage increased with increasing doses and time intervals. Our results are in agreement with that of Anbumani and Mohankumar (2015) in which they observed that freshwater fish (*Catla catla*) exposed to various doses of MCP exhibited a dose and time-dependent increase of DNA damage in the form of micronuclei. Our results also find similarity with the studies of Ali and Kumar (2008) wherein significantly high % tail DNA was induced in the erythrocytes of *C. punctatus* exposed to MCP. They reported that the DNA damage was highest on the fourth day, i.e., 96 h after exposure and the same was also evident in the present study in which DNA damage was highest on the third day, i.e., 72 h after exposure for all the doses. However, Banu et al. (2001) observed maximum DNA damage in the erythrocytes of *Tilapia mossambica* exposed to different doses of MCP at 24 h after exposure and a subsequent decrease of DNA damage at later time intervals. These differences in DNA recovery may be attributed to the rate of metabolite removal/detoxification or DNA repair capacity of different species. In the time-dependent study, MCP did not induce significant DNA damage at 0.125 μL/L at 24 and 48 h of exposure. This may probably be due to the initial low concentration of the toxicant which could not induce any immediate noticeable effects but did affect DNA at the 72 h time interval. This subsequent increase in DNA damage may have occurred due to gradual bioaccumulation of MCP over the span of 72 h. Further, the results of the two-way ANOVA suggest that both time and treatment play an important role in the formation of DNA strand breaks and micronuclei.

Our results also find similarities with studies on other related organophosphate pesticides by several workers. Ali et al. (2009) reported the induction of micronuclei and tail DNA by various treatments of chlorpyrifos, in the erythrocytes and lymphocytes of *C. punctatus*. They also reported that DNA damage was highest on the third day after exposure at all the concentrations of chlorpyriphos. Similar observations were reported by Mohanty et al. (2011) in which the erythrocytes of *Labeo rohita* fingerlings exposed to another OP, phorate exhibited significant DNA damage which declined after 72 h of exposure. Sub-lethal doses of profenofos, an OP pesticide were found to induce DNA strand breaks in the gill cells of *C. punctatus* and was found to increase in a dose-dependent manner (Pandey et al. 2011). DNA damage was also observed in the gill cells of the European eel (*Anguilla anguilla*) exposed to the OP glyphosate (Guilherme et al. 2012). Ali et al. (2014) also reported the time-dependent and dose-dependent increase of genotoxicity of an OP insecticide, dimethoate in *C. punctatus,* which were found to be highly correlated with oxidative stress. Our results along with those of Pamanji et al. (2015) mentioned earlier, provide information on the toxic effects of MCP in both the adults and embryos of zebrafish suggesting that MCP can affect zebrafish at any time during its life cycle.

Organophosphate pesticides such as MCP and those cited above cause the alkylation of DNA bases either directly or indirectly by protein alkylation leading to DNA disintegration (Bhinder and Chaudhry 2013). The presence of a methyl ester group is more likely to cause alkylation than ethyl esters or other higher esters and thus MCP is more reactive than other organophosphate pesticides (Yaduvanshi et al. 2010). Alternatively, the phosphorous group in OPs may be a favorable substrate for nucleophilic attack which in turn may cause DNA phosphorylation and consequently, DNA damage (Rahman et al. 2002; Das et al. 2007). Another reason for damaged DNA could be oxidative stress due to the formation of reactive oxygen species (ROS). Oxidative stress may be caused due to the production of ROS above the level of detoxification and repair. The free radicals thus formed can affect DNA indirectly by lipid peroxidation which in turn causes single-stranded breaks (SSB) (Yaduvanshi et al. 2010; Nwani et al. 2011). Lipid peroxidation products such as 4-hydroxynonenal and malondialdehyde (MDA) which are the most genotoxic and mutagenic, respectively, form adducts with DNA thereby causing DNA damage (Luczaj and Skrzydlewska 2003). Thus, DNA damage prevailing in the erythrocytes of zebrafish may be due to the combined direct or indirect effects.

We also observed that the induction of tail DNA had a strong positive correlation with induction of micronuclei in zebrafish erythrocytes. The reason for correlating these two genotoxicity tests is that they reflect different forms of environmental stress. MCP, as mentioned earlier can induce SSBs which may be easily repaired and these transient DNA breaks can be easily detected by the comet assay. However, DNA damage which is difficult to repair tends to form extra-nuclear bodies or micronuclei which can be detected by the micronucleus test (Hartmann et al. 2001). These micronuclei are formed when a whole chromosome or a fragment of a chromosome do not get incorporated into either of the two daughter cells during cell division due to aneugenic agents that affect the spindle apparatus or clastogenic agents that damage and break the chromosome (Udroiu 2006). The comet assay and micronucleus test can, therefore, be applied on zebrafish erythrocytes and be used as an integrated part of a monitoring program of water bodies that might be contaminated with organophosphate pesticides.

Conclusions

The present study thus revealed the ability of MCP to induce DNA damage in zebrafish erythrocytes. The formation of MNi and tail DNA is highly correlated with each other and may occur in a dose-dependent and/or time-dependent manner. The MN test and comet assay can, therefore, be reliably used to assess the genotoxicity of MCP in the aquatic environment using zebrafish as a toxicological model. Despite several reports of induced toxicity mentioned earlier, MCP still continues to be used in agricultural practices and may affect aquatic fauna at the DNA level. Therefore, it is suggested to limit its use in agricultural practices to avoid the potential contamination and of nearby water bodies. Although there is a lack of data concerning the health implications of consuming fish contaminated with organophosphate pesticides, care and caution should be taken due to probable persistence of these residues.

Conflict of interest The authors report no conflicts of interest. The authors alone are responsible for the content and writing of the paper.

References

Ali D, Kumar S (2008) Long-term genotoxic effect of monocrotophos in different tissues of freshwater fish *Channa punctatus* (Bloch) using alkaline single cell gel electrophoresis. Sci Total Environ 405:345–350

Ali D, Nagpure NS, Kumar S, Kumar R, Kushwaha B (2009) Genotoxicity assessment of acute exposure of chlorpyrifos to freshwater fish *Channa punctatus* (Bloch) using micronucleus assay and alkaline single-cell gel electrophoresis. Chemosphere 71:1823–1831

Ali D, Kumar PG, Kumar S, Ahmed M (2014) Evaluation of genotoxic and oxidative stress response to dimethoate in freshwater fish *Channa punctatus* (Bloch). Chem Speciat Bioavailab 26:111–118

Anbumani S, Mohankumar MN (2012) Gamma radiation induced micronuclei and erythrocyte cellular abnormalities in the fish *Catla catla.* Aquat Toxicol 122–123:125–132

Anbumani S, Mohankumar MN (2015) Cytogenotoxicity assessment of monocrotophos and butachlor at single and combined chronic exposures in the fish *Catla catla* (Hamilton). Environ Sci Pollut Res 7:4964–4976

Bagchi D, Bagchi M, Hassoun EA, Stohs SJ (1995) In vitro and in vivo generation of reactive oxygen species, DNA damage and lactate dehydrogenase leakage by selected pesticides. Toxicology 104:129–140

Banu BS, Danadevi K, Rahman MF, Ahuja YR, Kaiser J (2001) Genotoxic effect of monocrotophos to sentinel species using comet assay. Food Chem Toxicol 39:361–366

Baršiene J, Dedonyte V, Rybakovas A, Andreikenait L, Andersen OK (2006) Investigation of micronuclei and other nuclear abnormalities in peripheral blood and kidney of marine fish treated with crude oil. Aquat Toxicol 78S:S99–S104

Bhinder P, Chaudhry A (2013) Mutagenicity assessment of organophosphates using polymerase chain reaction-restriction fragment length polymorphism assay. Toxicol Int 20:254–260

Bolognesi C, Cirillo S (2014) Genotoxicity biomarkers in aquatic bioindicators. Curr Zool 60:273–284

Das PP, Shaik AP, Jamil K (2007) Genotoxicity induced by pesticide mixtures: in vitro studies on human peripheral blood lymphocytes. Toxicol Ind Health 23:449–458

de Croux MJP, Loteste A, Cazenave J (2002) Inhibition of plasma cholinesterase and acute toxicity of monocrotophos in a neotropical fish *Prochilodus lineatus* (Pisces, Curimatidae). Bull Environ Contam Toxicol 69:356–363

EU (2010) Directive 2010/63/EU of the European parliament and of the council of 22 September 2010 on the protection of animals used for scientific purposes. Off J EU L 276:33–79

FAO/UNEP (1997) Decision Guidance Documents: Methamidophos - Methyl parathion - Monocrotophos - Parathion - Phosphamidon

Fenech M, Chang WP, Kirsch-Volders M, Holland N, Bonassi S, Zeiger E (2003) HUMN project: detailed description of the scoring criteria for the cytokinesis-block micronucleus assay using isolated human lymphocyte cultures. Mutat Res 534:65–75

Ferraro MV, Fenocchio AS, Mantovani MS, Cestari MM, Ribeiro CAO (2004) Mutagenic effects of tributyltin (TBT) and inorganic lead (PbII) on the fish *H. malabaricus* as evaluated using the comet assay, piscine micronucleus and chromosome aberrations tests. Genet Mol Biol 27:103–107

Guilherme S, Gaivão I, Santos MA, Pacheco M (2012) DNA damage in fish (*Anguilla anguilla*) exposed to a glyphosate-based herbicide—elucidation of organ-specificity and the role of oxidative stress. Mutat Res 743:1–9

Hartmann A, Elhajouji A, Kiskinis E, Poetter F, Martus HJ, Fjällman A, Frieauff W, Suter W (2001) Use of the alkaline assay for industrial genotoxicity screening: comparative investigation with the micronucleus test. Food Chem Toxicol 39:843–858

Hill AJ, Teraoka H, Heideman W, Peterson RE (2005) Zebrafish as a model vertebrate for investigating chemical toxicity. Toxicol Sci 86:6–19

Kavitha P, Rao JV (2007) Oxidative stress and locomotor behaviour response as biomarkers for assessing recovery status of mosquito fish, *Gambusia affinis* after lethal effect of an organophosphate pesticide, monocrotophos. Pestic Biochem Physiol 87:182–188

Konca K, Lankoff A, Banasik A, Lisowska H, Kuszewski T, Góźdź S, Koza Z, Wojcik A (2003) A cross-platform public domain PC image-analysis program for the comet assay. Mutat Res 534:15–20

Kumar R, Nagpure NS, Kushwaha B, Srivastava SK, Lakra WS (2010) Investigation of the genotoxicity of malathion to freshwater teleost fish *Channa punctatus* (Bloch) using the micronucleus test and comet assay. Arch Environ Contam Toxicol 58:123–130

Kumaravel TS, Jha AN (2006) Reliable comet assay measurements for detecting DNA damage induced by ionising radiation and chemicals. Mutat Res 605:7–16

Łuczaj W, Skrzydlewska E (2003) DNA damage caused by lipid peroxidation products. Cell Mol Biol Lett 8:391–413

Mahboob M, Rahman MF, Danadevi K, Banu BS, Grover P (2002) Detection of DNA damage in mouse peripheral blood leukocytes by the comet assay after oral administration of monocrotophos. Drug Chem Toxicol 25:65–74

Marnett LJ (2000) Oxyradicals and DNA damage. Carcinogenesis 21:361–370

Mohanty G, Mohanty J, Nayak AK, Mohanty S, Dutta SK (2011) Application of comet assay in the study of DNA damage and recovery in rohu (*Labeo rohita*) fingerlings after an exposure to phorate, an organophosphate pesticide. Ecotoxicology 20:283–292

Moustacchi E (2000) DNA damage and repair: consequences on dose-responses. Mutat Res 464:35–40

Mustafa AM, Anjum F, Qadri SSH (1982) A technique to evaluate acute toxicity of insecticide (technical and formulation) to freshwater fish, *Tilapia mossambica*. Int Pest Control 24:4

Nwani CD, Nagpure NS, Kumar R, Kushwahab B, Kumar P, Lakra WS (2011) Mutagenic and genotoxic assessment of atrazine-based herbicide to freshwater fish *Channa punctatus* (Bloch) using micronucleus test and single cell gel electrophoresis. Environ Toxicol Pharmacol 31:314–322

O'Brien PJ (1988) Radical formation during the peroxidase catalyzed metabolism of carcinogens and xenobiotics: the reactivity of these radicals with GSH, DNA, and unsaturated lipid. Free Radic Biol Med 4:169–183

Pamanji R, Bethu MS, Yashwanth B, Leelavathi S, Rao VJ (2015) Developmental toxic effects of monocrotophos, an organophosphorous pesticide, on zebrafish (*Danio rerio*) embryos. Environ Sci Pollut Res Int 10:7744–7753

Pandey AK, Nagpure NS, Trivedi SP, Kumar R, Kushwaha B (2011) Profenofos induced DNA damage in freshwater fish, *Channa punctatus* (Bloch) using alkaline single cell gel electrophoresis. Mutat Res 726:209–214

Praveen Kumar MK, Shyama SK, Sonaye BS, Naik RU, Kadam SB, Bipin PD, D'costa A, Chaubey RC (2014) Evaluation of γ-radiation induced DNA damage in two species of bivalves and their relative sensitivity using comet assay. Aquat Toxicol 150:1–8

Praveen Kumar MK, Shyama SK, Sonaye BS, D'costa A, Chaubey RC (2015) The effect of gamma radiation on the Common carp (*Cyprinus carpio*): in vivo genotoxicity assessment with the micronucleus and comet assays. Mutat Res 792:19–25

Rahman MF, Mahboob M, Danadevi K, Banu SB, Grover P (2002) Assessment of genotoxic effects of chloropyriphos and acephate by the comet assay in mice leucocytes. Mutat Res 516:139–147

Rao JV (2004) Effects of monocrotophos and its analogs in acetylcholinesterase activity's inhibition and its pattern of recovery on euryhaline fish, *Oreochromis mossambicus*. Ecotoxicol Environ Saf 59:217–222

Revankar PR, Shyama SK (2009) Genotoxic effects of monocrotophos, an organophosphorous pesticide, on an estuarine bivalve, *Meretrix ovum*. Food Chem Toxicol 47:1618–1623

Sipes NS, Padilla S, Knudsen TB (2011) Zebrafish: as an integrative model for twenty-first century toxicity testing. Birth Defects Res C Embryo Today 93:256–267

Udroiu I (2006) The micronucleus test in piscine erythrocytes. Aquat Toxicol 79:201–204

USEPA (1985) Pesticide fact sheet no 72: monocrotophos. USEPA, Washington D.C.

Yaduvanshi SK, Ojha A, Pant SC, Lomash V, Srivastava N (2010) Monocrotophos induced lipid peroxidation and oxidative DNA damage in rat tissues. Pestic Biochem Physiol 97:214–222

Zhang X, Gao L, Yang K, Tian H, Wang W, Ru S (2013) Monocrotophos pesticide modulates the expression of sexual differentiation genes and causes phenotypic feminization in zebrafish (*Danio rerio*). Comp Biochem Physiol C Toxicol Pharmacol 157:33–40

Zhang X, Zhong Y, Tian H, Wang W, Ru S (2015) Impairment of the cortisol stress response mediated by the hypothalamus–pituitary–interrenal (HPI) axis in zebrafish (*Danio rerio*) exposed to monocrotophos pesticide. Comp Biochem Physiol C Toxicol Pharmacol 176–177:10–16

19

Effects of salinity on the growth and mucous cells of the abalone *Haliotis diversicolor* Reeve, 1846

Lota Alcantara Creencia · Tadahide Noro

Abstract This study was conducted to determine the influence of salinity on the growth of abalone *Haliotis diversicolor* Reeve, including the density and size of mucous cells. Abalone individuals were reared in the laboratory at salinities of 20, 25, 31, 35 and 40 ppt. The mucous cells of the lips, gills and digestive gut of *H. diversicolor*, which react to some forms of stress such as suboptimal salinity, were characterized following staining with Alcian Blue–Periodic Acid–Schiff's Reagent (AB–PAS). The specific growth rate in wet weight and shell length of *H. diversicolor* were highest at 31 ppt and lowest at 20 ppt (0.52 vs 0.15% d^{-1}, and 0.058 vs 0.021 mm d^{-1}, respectively). The abalone *H. diversicolor* tolerated salinity fluctuations within the range of 20–40 ppt, but growth was optimum at 25–35 ppt. Mucous cells of the lips and gills showed significant differences (ANOVA, df = 4, P = <0.001) in cell density and cell size, being less dense and larger at 31 ppt than at 40 ppt, which could be an effect of osmotic and ionic regulation. Consistent with reports in literature, salinity ranges of 25–35 ppt are suitable for growth of *H. diversicolor*. Results of this study indicated that areas with such salinity are favorable for stock enhancement and mariculture of the abalone *H. diversicolor*.

Keywords Acclimation · Fishery resource · *Haliotis diversicolor* · Japan · Mariculture · Stock enhancement

Background

The abalone *Haliotis diversicolor* Reeve, 1846, commonly called 'tokobushi' in Japan, inhabit the rocky littoral to sublittoral zone (Geiger and Poppe 2000). In Tanegashima, Kagoshima Prefecture, Japan, *H. diversicolor* is an important fishery commodity with declining yield. Hence, the local government and fisheries cooperative associations are managing this fishery resource. Interventions include regulating the minimum

L. A. Creencia (✉)
College of Fisheries and Aquatic Sciences, Western Philippines University, Sta. Monica, 5300 Puerto Princesa City, Philippines
e-mail: lotacreencia@gmail.com

L. A. Creencia · T. Noro
Faculty of Fisheries, Kagoshima University, 4-50-20 Shimoarata, Kagoshima 890-0056, Japan
e-mail: tadnoro@gmail.com

T. Noro
Kagoshima Prefectural College, 1-42-1 Shimo-ishiki, Kagoshima City 890-0005, Japan

size of harvest and releasing cultured juveniles in designated fishing grounds for stock enhancement (Alcantara and Noro 2005, 2006). In the Philippines, stock enhancement of abalone *H. asinina* resulted in spillovers of abalone outside the release site (Salayo et al. 2016). On the other hand, the mariculture of abalone *H. diversicolor* involves hatchery, nursery and on-growing in inland facilities and grow-out in nearshore ponds and cages (Chen and Lee 1999; Creencia et al. 2016).

Fishers tend to gather larger individuals of *H. diversicolor* near estuaries (pers. obs.) which may be related to salinity. Salinity has some adverse effect on various aspects of the physiology of abalone. *H. diversicolor* at salinities lower or higher than optimum have reduced immunity and resistance to infection (Cheng et al. 2004), survival (Chen and Chen 2000) and growth (Chen et al. 2000). According to Chen et al. (2000), growth rate of *H. diversicolor* is optimum at 30–35 ppt with variable rates for different sizes of juveniles. We hypothesized that the seasonal inflow of freshwater in estuaries and the resulting decrease in marine salinity from 34–38 ppt to 31–33 ppt might have promoted faster growth of *H. diversicolor* (Chen et al. 2000). In general, faster growth of marine animals in intermediate salinities has been correlated with reduced metabolic rate and elevated assimilation efficiency (Boeuf and Payan 2001).

The lips, gills and digestive gut may react to changes in salinity level by adjusting the density and size of mucous cells and other epithelial cells. In varying salinities, epithelial mucous cells may adjust to local osmotic demands (Allen 1961; Gilles 1972; Yiyan et al. 2004; Di et al. 2012). Since mucous cells mainly contain water (Davies and Hawkins 1998), we hypothesized that the size and density of mucous cells in the lips, gills and digestive gut might reflect acclimation and adaptation to the prevalent water salinity. Acclimation to salinity in marine mollusks occurs via cellular mechanisms such as reversible changes of protein and RNA synthesis, alteration of molecular forms of enzymes, and regulation of ionic content and cell volume (Berger and Kharazova 1997). Various mucous cells are partly responsible for ionic regulation as well as lubrication of different organs (Crofts 1929; Davies and Hawkins 1998; Luchtel et al. 1997). However, there is limited information on the effect of acclimation to changes in salinity on mucous cell number and size.

Because of the regulatory activity of mucous cells, alteration of salinity may have effects on the overall physiology of abalone, as reflected in growth rate. Therefore, in this study we investigated the growth (wet weight and shell length) of abalone and their mucous cells (characterized by density and size) when reared at different salinities. The results will be useful for selecting sites for stock enhancement or mariculture and in designing inland culture systems for *H. diversicolor*. Abalone raised from the hatchery are released to fishing grounds to replenish the overfished wild population (Body 1987; Braje et al. 2007) and cultured to increase abalone products in the market (Troell et al. 2006, Wu and Zhang 2012).

Methods

One-year-old individuals of the abalone *H. diversicolor* produced from the hatchery facilities of Kagoshima Mariculture Society in Tarumizu City, Kagoshima, Japan, with total wet weight of 1.67 ± 0.49 g (mean ± SD) and shell length of 23.26 ± 1.90 mm, were used in this study. In the laboratory, the animals were maintained in 10-L plastic tanks with static seawater. Each tank was provided with a $7 \times 7 \times 6$ cm^3 portable filter-aerator and a shelter made from an inverted length of plastic guttering 15×5 cm^2, with a central hole 6 cm in diameter. A shelter was necessary to provide a substrate for attachment and to cover the nocturnal abalone during daytime. Except for the control, salinity in each tank was gradually adjusted to the desired level within 1 week to acclimatize the test individuals. Dechlorinated tap water was added to decrease seawater salinity. Dissolved sea salt (Hagata, Japan sea salt) was added to increase salinity. The experiment began after an acclimation period of 2 weeks. Two replicate tanks for each of five treatments [salinity levels of 20, 25, 31(control), 35 and 40 ppt] were arranged in a completely randomized design. These salinities were chosen because they span the possible salinity ranges in coastal areas where *H. diversicolor* lives. The control was the salinity of the nearby sea where the seawater in the laboratory was pumped.

Ten individuals of *H. diversicolor* (2 tanks × 5 salinities) were maintained in each tank for 2 months. Rehydrated blades of kelp (*Laminaria* sp., from Kagoshima City local supermarket) were provided ad libitum and replaced every 3 or 4 days, simultaneous with change of the tank water. Water temperature, dissolved oxygen (DO) (YSI 85, YSI, USA) and pH (Yokogawa pH82, Japan) of the tank water in each tank were measured weekly. Values of percent DO saturation were computed using the DOTABLES on-line program (available at: water.usgs.gov). There was no management of bicarbonate levels during the experiment.

Growth was measured weekly by recording wet weight (WW) (electronic balance, Shimadzu EL-120 W/AC, Japan) and shell length (SL) (Vernier caliper, Mitotuyo Corp., Japan). The shells were blotted with paper towels before measuring the wet weight. Repeat weighing after re-immersion and blotting indicated an accuracy of ± 0.02 g. Specific growth rate was calculated as percentage increase in wet weight per day relative to original weight. This measure was used rather than absolute growth rate to reduce variability with size and age. On the other hand, to measure growth rate in shell length we used the absolute increase per day, because shell length is not affected as much by size or age.

The formulae used were:

Specific growth rate ($\% \ d^{-1}$) = $(\ln wetW_2 - \ln wetW_1)/(d_2 - d_1) \times 100$

Growth rate in shell length ($mm \ d^{-1}$) = $(SL_2 - SL_1)/(d_2 - d_1)$

where ln is the natural logarithm, $wetW_1$ (g) is the wet weight of *H. diversicolor* at the start of the experiment, $wetW_2$ (g) is the wet weight of *H. diversicolor* at the end of the experiment, SL_1 (mm) is the initial shell length of *H. diversicolor*, SL_2 (mm) is the final shell length of *H. diversicolor*, d1 is the initial day of the experimental culture period, and d2 is the final day of the experimental culture period.

At the end of the tank period, two individuals were randomly chosen from each tank for histological analyses. These individuals were relaxed in 5% ethanol then fixed in 10% seawater–formalin solution for 48–72 h and stored in 70% ethanol. About 1–2 cm of the lips, gills (lamellae) and digestive gut (stomach and intestine) from each individual were taken and dehydrated in a graded ethanol series (70, 80, 85, 90, 95, and 100%) with two changes at 10–12 h each. This was followed by gradual infiltration and embedding with the cold resin Glycol Methacrylate (GMA) (Technovit 7100®, HeraeusKulzer, Germany) at ethanol:GMA ratios of 4:1, 2:1, 1:1, 1:3, 0:1,0:1 for 10–12 h each. Sections of tissues that were 2 μm thin were stained with Alcian Blue–Periodic Acid–Schiff's Reagent (AB–PAS) to test for the presence of mucopolysaccharides and glycoconjugates. The density (number per 0.01 mm^2) and length (in μm) of mucous cells positive to AB–PAS (blue or purple to pink or magenta) in each organ were compared among the treatments. Using photomicrographs of ten different areas per individual sample, the number of teardrop-shaped mucous cells was counted and their corresponding maximum dimension of length was measured. The density and size (average of 20 cells) of mucous cells were measured using ImageJ (Image Processing and Analysis in Java). The ten areas per individual were blocked in the statistical analysis. An average density was calculated from 10 observations × 2 individuals × 5 salinities = 200 observations (40 per salinity treatment). For size, an average was calculated from 20 cells × 2 individuals × 2 tanks × 5 salinities = 400 observations (80 per salinity treatment).

To determine if the differences in the growth rates of *H. diversicolor*, and the density and size of mucous cells positive to AB–PAS among the salinity treatments were significant, single-factor analysis of variance (ANOVA) followed by pairwise multiple comparisons (Tukey Test) was performed using SIGMASTAT (Systat Software Inc., California, USA).

Results

Environmental conditions

The average measured salinity was usually a little higher than the nominal salinity (Table 1). Average DO in tanks ranged from 6.13 to 6.76 mg L^{-1}, the average DO saturation ranged from 84.96 to 86.48%, and the average pH values ranged from 7.65 to 7.79. There were no significant differences (ANOVA, $P > 0.05$) between the two tanks in each treatment for salinity, DO or pH. Indoor water temperature during the experimental period decreased from 24.2 °C (October) to 12.4 °C (December) as the ambient temperature decreased with the progress of winter, which was similar to what would be experienced by abalone in their natural marine habitat. The temperature was the same in all treatments and did not affect the readings of salinity, DO or pH for the duration of the experiment. Survival rate of abalone was 100% in all treatments for the duration of the experiment.

Table 1 Measured salinity, measured DO, DO saturation, pH and temperature (mean ± SD) of the tank water from two tanks of *H. diversicolor* for 2 months at five different nominal salinity levels

Treatment salinity		Measured DO	DO saturation	pH	Temperature		
Nominal (ppt)	Measured (ppt)	(ppm)	(%)		(°C)		
					Oct	Nov	Dec
20	20.8 ± 0.5	6.69 ± 0.48	86.05 ± 6.93	7.79 ± 0.22	24.2 ± 0.58	20.3 ± 1.30	12.4 ± 2.47
25	25.5 ± 0.5	6.76 ± 0.55	85.75 ± 7.79	7.65 ± 0.29			
31 (control)	31.6 ± 0.9	6.58 ± 0.50	86.40 ± 7.77	7.69 ± 0.17			
35	35.4 ± 0.7	6.43 ± 0.54	86.48 ± 8.12	7.78 ± 0.39			
40	40.6 ± 0.6	6.13 ± 0.45	84.96 ± 7.37	7.75 ± 0.29			

No significant differences were found between tanks and among treatments (ANOVA, df = 4, $P > 0.05$)

Growth rate

There were significant differences in specific growth rate of abalone *H. diversicolor* among the salinity treatments (ANOVA, df = 4, $P = <0.001$). The specific growth rate was highest at 31 and lowest at 4 and 20 ppt (Table 2). Moreover, there were significant differences in SL growth rate of abalone among the salinity treatments (ANOVA, df = 4, $P = <0.001$). The highest SL growth rate was obtained at 31 ppt and lowest at 40 and 20 ppt (Table 2).

Density of mucous cells

There were significant differences in the density of AB–PAS-positive mucous cells (Fig. 1) of the lips of abalone among treatments (ANOVA, df = 4, $P = <0.001$). The density of mucous cells from the lips of abalone reared at 40 ppt was significantly higher ($P = 0.03$) than in those raised at 31 ppt (Table 3). In the gills, there were significant differences in the density of AB–PAS-positive mucous cells (Fig. 2) of abalone among treatments (ANOVA, df = 4, $P = <0.001$). The density of mucous cells from gills of abalone reared at 40 and 35 ppt was significantly higher ($P = <0.001$) than cell densities of abalone raised at 20, 25 and 31 ppt (Table 3). In the digestive gut, there were significant differences in density of AB–PAS-positive mucous cells (Fig. 3) of abalone among treatments (ANOVA, df = 4, $P = <0.001$). Density of mucous cells from digestive gut of abalone reared at 25 ppt (19.0 ± 1.9 cells per 0.1 mm^2) were significantly higher ($P = 0.04$) than those reared at 20 ppt (Table 3).

Sizes of mucous cells

There were significant differences in sizes of AB–PAS-positive mucous cells from the lips of abalone among treatments (ANOVA, df = 4, $P = 0.021$). The lengths of the mucous cells of lips of abalone raised at 31 ppt were significantly greater than those from abalone reared at 40 ppt (Table 4). In the gills, there were

Table 2 Specific growth rate in wet weight and growth rate in shell length (mean ± SE) of individuals of abalone *H. diversicolor* reared for 2 months at five different salinity levels

Treatment (ppt)	Specific growth rate in wet weight (% d^{-1})	Growth rate in shell length (mm d^{-1})
20	0.15 ± 0.03^{d}	0.021 ± 0.002^{d}
25	0.36 ± 0.02^{bc}	0.034 ± 0.002^{c}
31	0.52 ± 0.03^{a}	0.058 ± 0.003^{a}
35	0.42 ± 0.04^{ab}	0.044 ± 0.003^{b}
40	0.29 ± 0.02^{c}	0.026 ± 0.002^{d}

Different letters superscript indicate significant differences (ANOVA, df = 4, $P = <0.001$, Tukey test), common letters indicate no significant differences ($P = 0.06$)

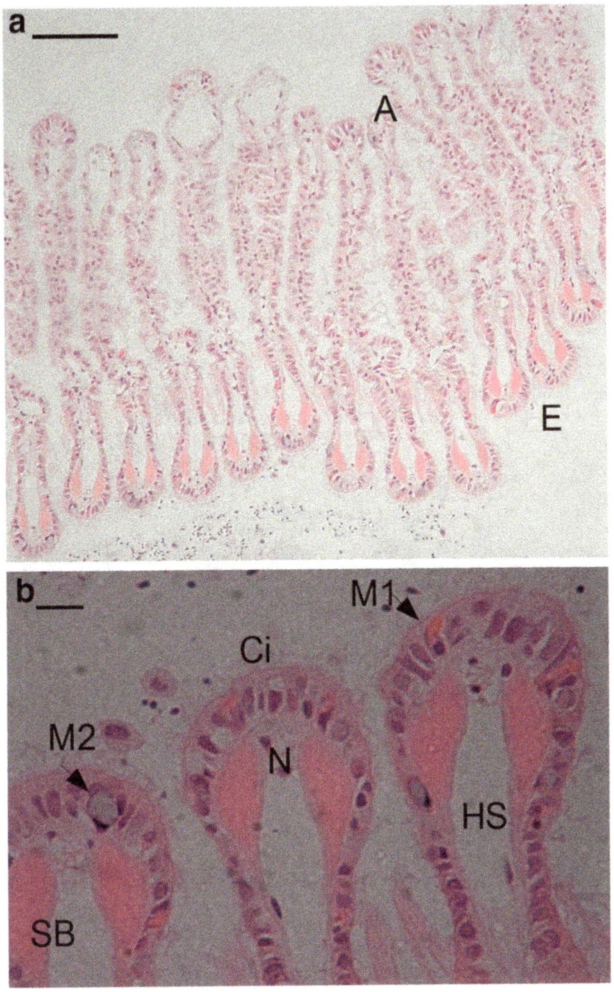

Fig. 1 Photomicrograph of the lip of *H. diversicolor* stained with Hematoxylin and Eosin: **a** longitudinal section of folds of epithelial cells and supporting muscle, scale bar = 100 μm; **b** enlarged boxed fold showing mucous-1 (M1), mucous-2 (M2) and mucous-3 (M3), scale bar = 10 μm

Table 3 Cell density (mean ± SE) of mucous cells from the lips, gills and digestive gut of abalone *H. diversicolor* stained with AB–PAS reared at five different salinity levels

Treatment (ppt)	In lips (mucous cells per 0.1 mm^2)	In gills (mucous cells per 0.1 mm^2)	In digestive gut (mucous cells per 0.1 mm^2)
20	8.8 ± 1.5^{ab}	9.2 ± 1.2^{b}	12.1 ± 1.3^{b}
25	8.6 ± 1.3^{ab}	12.9 ± 1.3^{b}	19.0 ± 1.9^{a}
31	7.2 ± 1.5^{b}	9.8 ± 1.1^{b}	16.6 ± 0.9^{ab}
35	8.7 ± 1.3^{ab}	23.1 ± 2.0^{a}	13.1 ± 1.3^{ab}
40	17.0 ± 1.9^{a}	20.6 ± 0.9^{a}	13.9 ± 1.7^{ab}

Different letters superscript indicate significant differences (ANOVA, df = 4, $P = <0.001$, Tukey test), common letters indicate no significant differences ($P = 0.06$)

significant differences in sizes of AB–PAS-positive mucous cells among treatments (ANOVA, df = 4, $P = <0.001$). The mucous cells of gills of individuals reared at 31 ppt were significantly longer than those reared at 40 ppt (Table 4). In the digestive gut, there were significant differences in sizes of AB–PAS-positive mucous cells among treatments (ANOVA, df = 4, $P = <0.001$). The AB–PAS-positive mucous cells from

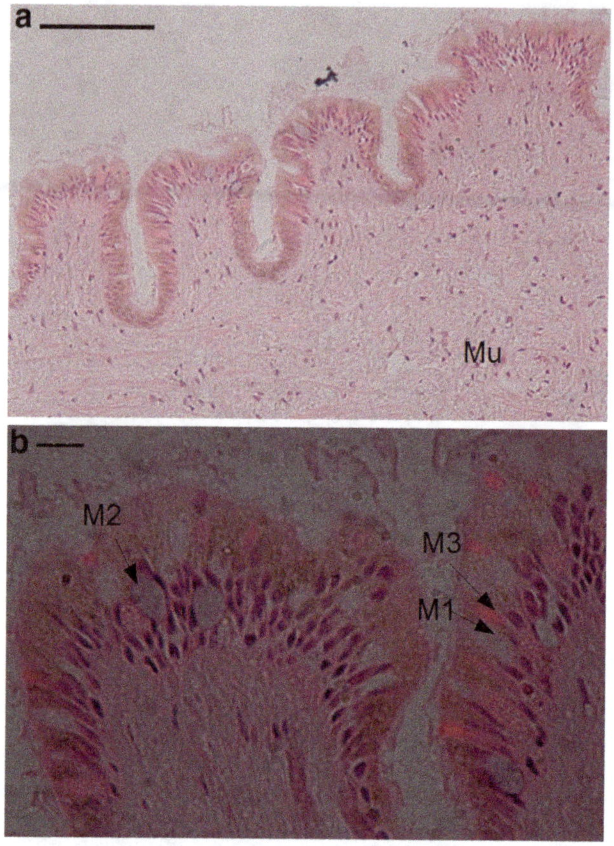

Fig. 2 Photomicrograph of the gill of *H. diversicolor* stained with H–E: **a** longitudinal section of gill filaments showing the afferent end (A) and efferent end (E), scale bar = 100 µm; **b** enlarged efferent end showing mucous-1 (M1), mucous-2 (M2) and cilia (Ci), hemocoelic space (HS) and nerve (N), bar = 10 µm

abalone raised at 20 ppt were significantly longer than those from abalone raised at 35, 31 and 25 ppt (Table 4).

Discussion

Growth

Salinity in coastal areas where the abalone *H. diversicolor* lives may drop sharply during heavy rains and large discharge events from rivers and streams. Alternatively, salinity may rise in tide pools during low tides with high rates of evaporation. The normal range of salinity in *H. diversicolor* habitat is 32 ± 3 ppt. The 100% survival in five different salinity levels tested indicates that the range of 20–40 ppt is tolerated by *H. diversicolor* in culture. This is an important finding and supports the occurrence of *H. diversicolor* in some estuarine areas where our results show a wider range than the 25–35 ppt marine water range suitable for the growth of *H. diversicolor* reported by Chen and Chen (2000). In the earlier farming studies of *H. diversicolor* in Taiwan, the reported salinity range is 30–35 ppt (Chen 1984). There is a wide range of salinity tolerance across the genus *Haliotis* in general. The salinity tolerance of *Haliotis laevigata* Leach *and Haliotis rubra* Leach is from 25 to 40 ppt and mortality occurs at 2 ppt outside this range (Edwards 2003). Individuals of *Haliotis asinina* Linnaeus survive at lower salinity levels, 20.5 ppt without acclimation and down to 12.5 ppt with gradual acclimation, but die at 10 ppt (Singhagraiwan et al. 1992).

During the conduct of this experiment, water temperature was decreasing due to the onset of winter (October–December) in the study site. Temperature was not considered to affect the growth of the abalone *H. diversicolor* reared at five different salinities because they would experience such water temperatures in their

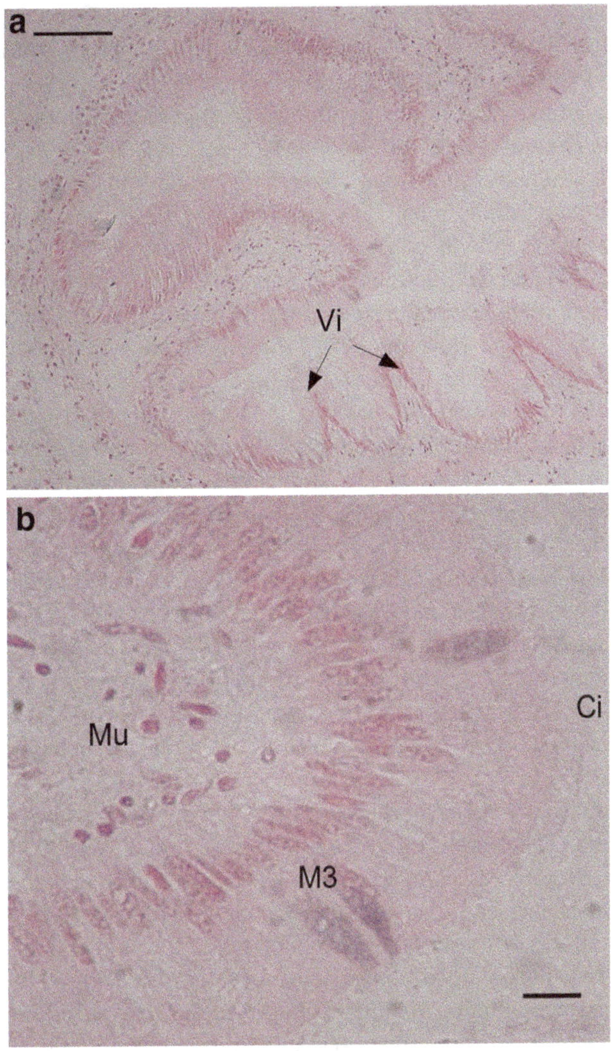

Fig. 3 Photomicrograph of the digestive gut (intestine) of *H. diversicolor* stained with H–E: **a** transverse section of intestine showing villi, scale bar = 100 μm; **b** enlarged villi showing mucous-3 (M3) and muscle (Mu), scale bar = 10 μm

Table 4 Cell length (mean ± SE) of mucous cells from the lips, gills and digestive gut of abalone *H. diversicolor* stained with AB–PAS reared at five different salinity levels

Treatment (ppt)	In lips (μm)	In gills (μm)	In digestive gut (μm)
20	15.2 ± 0.7^{ab}	12.7 ± 1.0^{ab}	72.0 ± 6.0^{a}
25	16.3 ± 0.1^{ab}	12.7 ± 0.9^{ab}	39.2 ± 5.0^{c}
31	16.3 ± 0.8^{a}	16.6 ± 1.3^{a}	40.0 ± 6.9^{bc}
35	13.8 ± 1.2^{ab}	12.2 ± 1.1^{ab}	56.0 ± 7.7^{bc}
40	12.8 ± 0.6^{b}	9.1 ± 0.6^{b}	64.6 ± 7.3^{ab}

Different letters superscript indicate significant differences (ANOVA, df = 4, P = <0.001, Tukey test), common letters indicate no significant differences (P = 0.06)

natural habitat. Generally, the effect of temperature on growth of various abalone species would depend on what temperature range they are exposed to in their natural environment. For instance, in California the abalone *Haliotis rufescens* Swainson that inhabits deep cool water has been reported to decrease growth at increasing temperature, whereas the abalone *Haliotis fulgens* Philippi that inhabits warm water has increased growth at higher temperature (Vilchis et al. 2005). On the other hand, in Chile the abalones *Haliotis discus hannai* Ino and *H. rufescens* cultured in tank systems have low growth during austral winter and high growth during the austral summer Mardones et al. (2013).

As indicated by both the growth rates in wet weight and shell length, 31 ppt is the most favorable salinity for *H. diversicolor* reared in culture, closely followed by 35 and 25 ppt. This implies that a salinity fluctuation of ± 4–6 ppt from 31 ppt is still favorable for growth of *H. diversicolor* in the wild or in a culture facility. Our results indicate a wider optimum salinity range than that of Chen (1984) and Chen et al. (2000) who reported that the optimum salinity for the growth of *H. diversicolor* is from 30 to 35 ppt. In our study, we used *Laminaria* as food, whereas Chen et al. (2000) gave formulated diet. Moreover, we have related the effect of salinity to mucous cells which might have influenced the growth performance of abalone. Since mucous cells principally contain water (Davies and Hawkins 1998), these cells are responsible for volume regulation of mucous in the body to better adapt to salinity stress (Drew et al. 2001); hence, the changes in density and size of mucous cells directly affect the growth of abalone. In another study, individuals of *H. diversicolor* of a similar age raised in similar tanks, densities, and temperatures but at slightly higher salinity (32–34 ppt) have a specific growth rate of 0.44% d^{-1}, (Alcantara and Noro 2006), similar to that of abalone in this study raised at 35 ppt but lower than those raised at 31 ppt. The optimal range of salinity across marine mollusks having similar natural habitats generally appears to be similar. For example, the scallop *Argopecten purpuratus* Lamarck, the pearl-oyster *Pinctada maxima* Jameson and the clam *Laternula marilina* Reeve show the highest growth between 27 and 30 ppt (Navarro and Gonzalez 1998; Taylor et al. 2004; Zhuang 2005). However, there are often differences among local habitats of other mollusks in general and among abalone in particular. The higher growth rate exhibited at 40 ppt compared to 20 ppt (for weight) and at 35 ppt over 25 ppt (for shell length) may suggest that individuals of *H. diversicolor* can better adapt to higher salinity. On the contrary, individuals of *H. laevigata* and *H. rubra* can adapt better at lower rather than higher salinity within the range of 25–40 ppt (Edwards 2003).

Although abalone are exclusively marine species, results of previous studies (i.e. Singhagraiwan et al. 1992; Edwards 2003) and this study reveal that their growth performance is enhanced at intermediate salinity. The control treatment seawater (31 ppt) was sampled near (~ 800 m) a river mouth and thus was slightly diluted by river influx. Several studies have speculated on the physiological advantages of lower salinity on growth. Greater growth of some mollusks at lower salinities is due to the lower energetic cost of ionic and osmotic regulation, high food intake, efficient food conversion ratio, high retention of nutrients, low excretion of metabolites and low oxygen consumption (Ghiretti 1966). On the other hand, at higher salinities abalone spends more energy in counteracting stress rather than investing in growth, thus resulting in reduced growth rate (Morash and Alter 2016). It has been implicated that during seasonal salinity fluctuations the chemical balance between the environment and the body fluids is disrupted and efforts to regain homeostatic balance is an energetically costly process for abalone (Martello et al. 1998; Morash and Alter 2016). Further, suboptimal low salinity causes added stress to the abalone *Haliotis varia* Linnaeus when exposed to toxic substances (Lasut 1999). As in other marine mollusks, cell volume is regulated during variation in external salinity using intracellular free amino acids as osmotic solutes (Baginski and Pierce 1975; Amende and Pierce 1980) causing changes in water content and weight of abalone (Morash and Alter 2016). In our study, the lowest salinity (20 ppt) is the least favorable to the growth of *H. diversicolor*. This can be explained by their adaptive capacity at this particular salinity (Javanshir 2013) which affects their feeding behavior, metabolism and growth (Riisgard et al. 2012). Favorable feeding behavior, metabolism and growth of *H. diversicolor aqualitis*, a subspecies of *H. diversicolor*, all benefit at 25–37 ppt (Yan et al. 2009).

Mucous cells

The lips, gills and digestive gut of abalone are some of the organs that may show pronounced reactions to different salinity levels. The lip is directly exposed to the environment and contains many mucous and sensory cells. Changes in the cells of the lip may affect the abalone's ability to graze and ingest. Salinity directly influences feeding of mollusks (Broom 1985; Riisgard et al. 2012). According to Chen et al. (2000), *H. diversicolor* feeds on artificial diets at salinity range of 20–38 ppt which means that they feed on a wide salinity range. In another study, the gastropod *Lithopoma tectum* has increased ingestion of algae after exposure to reduced-salinity water (Irlandi et al. 1997). In the case of abalone, highest growth of those reared at 31 ppt could be due to high intake of algal food compared to those at 40 and 20 ppt.

The bipectinate gills are positioned under the shell openings and are thus directly exposed to the salinity of the environment because the water passes through them. This organ is very sensitive because each filament is innervated and contains several mucous and ciliated epithelial cells (Wanichanon et al. 2004). In the gills, mucous cells are important in detecting salinity stress. According to Drew et al. (2001), the abalone *Haliotis rubra* Leach are able to regulate the swelling of tissues due to exposure to hyposaline stress. Thus, salinity stress are able to affect the growth of abalone by diverting energy usage to addressing the stressful condition. In this study, salinity stress caused by prolonged exposure to hypersaline (40 ppt) culture environment results in higher density and smaller sizes of mucous cells in externally located lips and gills but not in internally located digestive gut. The digestive gut is located internally, hence its function is not influenced by salinity. However, mucous cells in the digestive system may produce hormones such as hydrolases that are active in osmoregulation, control of food intake and growth regulation (Gilles 1972; Harris et al. 1998; Di et al. 2012).

Implications on stock enhancement and mariculture

Results of this study has application in choosing areas with average salinity of 31 ± 4 ppt for stock enhancement and open sea culture system of abalone *H. diversicolor*. To ensure favorable survival and growth of abalone released or reared in the site, one of the important considerations is the potential seasonal fluctuation of salinity due to rainy or dry season and other factors that may cause drastic change in water salinity. For land-based mariculture system of abalone, the findings will be helpful in designing and developing mariculture facilities and protocols that will minimize salinity stress and other forms of stress to optimize production and produce high-quality abalone product.

Conclusions

Generally, the growth (wet weight and shell length) of *H. diversicolor* was greatest at 31 ppt and lowest at 40 and 20 ppt with those at 35 and 25 ppt having average growth performance. Mucous cells positive to AB–PAS of lips and gills had their highest density and smallest cell size at 40 ppt and lowest density and largest cell size at 31 ppt. These findings suggest that differences in density and size of mucous cells could be an effect of osmotic and ionic regulation of *H. diversicolor*. It appears that mucous cells proliferate at the least favorable salinity level (40 ppt) compared with the salinity level suitable for growth (31 ppt). In non-optimal salinities, a high level of mucous cell production may be needed to maintain the ionic concentration of the blood in the anisosmotic extracellular regulation of some mollusks (Florkin 1966), which may also happen in *H. diversicolor*.

Acknowledgements Thanks to Kyushu Electric Company for research funds and Kagoshima Mariculture Society in Tarumizu City, Kagoshima, Japan, for individuals of *Haliotis diversicolor*. We are also grateful for the inputs of peers that improved the manuscript.

References

Alcantara L, Noro T (2005) Effects of macroalgal type and water temperature on macroalgal consumption rates of the abalone *Haliotis diversicolor* Reeve. J Shellfish Res 24:1169–1177

Alcantara L, Noro T (2006) Growth of the abalone *Haliotis diversicolor* (Reeve) fed with macroalgae in floating net cage and plastic tank. Aquac Res 37:708–717

Allen K (1961) The effect of salinity on the amino acid concentration in *Pangia cuneata* (Pelecypoda). Biol Bull 121(3):419–424

Amende LM, Pierce SK (1980) Cellular volume regulation in salinity stressed mollusks: the response of *Noetia ponderosa* (Arcidae) red blood cells to osmotic variation. J Comp Physiol 138:283–289

Baginski RM, Pierce SK (1975) Anaerobiosis: a possible source of osmotic solute for high-salinity acclimation in marine mollusks. J Exp Biol 62:589–598

Berger VJ, Kharazova AD (1997) Mechanisms of salinity adaptations in marine mollusks. In: Naumov AD, Hummel H, Sukhotin AA, Ryland JS (eds) Interactions and adaptation strategies of marine organisms. Hydrobiologia, vol 355. Kluwer Academic Publishers, Belgium, pp 115–126

Body AGC (1987) Abalone culture in Japan. Mar Fish Rev 49(4):75–76

Boeuf G, Payan P (2001) How should salinity influence fish growth? Comp Biochem Physiol C Toxicol Pharmacol 130(4):411–423

Braje TJ, Erlandson JM, Rick TC (2007) An historic Chinese abalone fishery in California's Northern Channel Islands. Hist Archaeol 41(4):117–128

Broom MJ (1985) ICLARM studies and reviews. The biology and culture of marine bivalve mollusks of the Genus *Anadara*. International Center for Living Aquatic Resources Management, Manila

Chen HC (1984) Recent innovations in cultivation of edible mollusks in Taiwan with special reference to the small abalone *Haliotis diversicolor* and the hard clam *Meretrix lusoria*. Aquaculture 39:11–27

Chen JC, Chen WC (2000) Salinity tolerance of *Haliotis diversicolor supertexta* at different salinity and temperature levels. Aquaculture 181:191–203

Chen JC, Lee WC (1999) Growth of Taiwan abalone *Haliotis diversicolor supertexta* fed on *Gracilaria tenuistipitata* and artificial diet in a multiple-tier basket system. J Shellfish Res 18:627–635

Chen C, Zhong Y, Wu Y, Cai H, Guo C, Zheng L (2000) The effect of salinity on food intake, growth and survival of *Haliotis diversicolor supertexta*. J Fish China 24:41–45

Cheng W, Juang F-M, Chen J-C (2004) The immune response of Taiwan abalone *Haliotis diversicolor supertexta* and its susceptibility to *Vibrio parahaemolyticus* at different salinity levels. Fish Shellfish Immunol 16:295–306

Creencia LA, Noro T, Fukumoto M (2016) Composition, size and relative density of diatoms in the stomach of 4–75 day-old juvenile abalone *Haliotis diversicolor* (Reeve). Palawan Sci 8:1–12

Crofts DR (1929) Liverpool marine biology committee. *Haliotis*. The University Press of Liverpool, UK

Davies MS, Hawkins J (1998) Mucus from marine molluscs. Adv Mar Biol 34:1–71

Di G, Ni J, Zhang Z, Yon W, Wang B, Ke C (2012) Types and distribution of mucous cells of the abalone *Haliotis diversicolor*. Afr J Biotechnol 11(37):9127–9140

Drew B, Miller D, Toop T, Hanna P (2001) Identification of expressed HSP's in blacklip abalone (*Haliotis rubra* Leach) during heat and salinity stress. J Shellfish Res 20(2):695–703

Edwards S (2003) Assessment of the physiological effect of altered salinity on greenlip (*Haliotis laevigata*) and blacklip (*Haliotis rubra*) abalone using respirometry. Aquac Res 34:1361–1365

Florkin M (1966) Nitrogen metabolism. In: Wilbur KU, Yonge CM (eds) Physiology of mollusca, vol II. Academic Press, New York, pp 309–351

Geiger D, Poppe G (2000) The family haliotidae, A conchological iconography. Conch Books, Hackenheim, p 135

Ghiretti F (1966) Respiration. In: Wilbur KU, Yonge CM (eds) Physiology of mollusca, vol II. Academic Press, New York, pp 175–208

Gilles R (1972) Osmoregulation in three molluscs: *acanthochitona discrepans* (Brown), *Glycymeris glycymeris* (L.). and *Mytilus edulis* (L.). Biol Bull 142:25–35

Harris JO, Burke CM, Maguire GB (1998) Characterization of the digestive tract of greenlip abalone, *Haliotis laevigata* Donovan. I. Morphology and histology. J Shellfish Res 17(4):979–988

Irlandi E, Macia S, Seraf J (1997) Salinity reduction from freshwater canal discharge: effects on mortality and feeding of an urchin (*Lytechinus variegatus*) and a gastropod (*Lithopoma tectum*). Bull Mar Sci 61(3):869–879

Javanshir A (2013) Low salinity changes in an intertidal zone may affect population dynamics of *Littorina scabra* (Linaeus 1758) in northern coasts of Persian gulf. Turkish J Fish Aquat Sci 13(1):133–138

Lasut MT (1999) Effects of salinity-cyanide interaction on the mortality of abalone *Haliotis varia* (Haliotidae: Gastropoda). In: Proceedings of the 9th workshop of the tropical marine molluscprogramme (TMMP), Phuket Marine Biological Center, Indonesia, 19–29 August 1998. Part 1. Special publication, Vol. 19, No. 1, pp. 165–168

Luchtel DL, Martin AW, Deyrup-Olsen I, Boer HH (1997) Gastropoda: pulmonata. In: Harrison FW, Kohn AJ (eds) Microscopic anatomy of invertebrates mollusca. Wiley-Liss, New York

Mardones A, Augsburger A, Vega R, de Los Rios-Escalante P (2013) Growth rates of *Haliotis rufescens* and *Haliotis discus hannai* in tank culture systems in southern Chile (41.5 °C). Lat Am J Aquat Res 41(5):959–967

Martello LB, Tjeerdemab RS, Smithc WS, Kautenc RJ, Crosbyd DG (1998) Influence of salinity on the actions of pentachlorophenol in *Haliotis* as measured by in vivo 31P NMR spectroscopy. Aquat Toxicol 41(3):229–250

Morash AJ, Alter K (2016) Effects of environmental and farm stress on abalone physiology: perspectives for abalone aquaculture in the face of global climate change. Rev Aquac 8:342–368

Navarro JM, Gonzalez CM (1998) Physiological responses of the Chilean scallop *Argopecten purpuratus* to decreasing salinities. Aquaculture 167:315–327

Riisgard HM, Battiger L, Pleissoner D (2012) Effect of salinity on growth of mussels, *Mytilus edulis*, with special reference to Great Belt (Denmark). Open J Mar Sci 2:167–176

Salayo ND, Castel RJG, Barrido RT, Tormon DHM, Azuma T (2016) Community-based stock enhancement of abalone, *Haliotis asinina* in Sagay marine reserve: achievements, limitations and directions. In: Hajime K, Iwata T, Theparoonrat Y, Manajit N, Sulit VT. Consolidating the strategies for fishery resources enhancement in southeast Asia. Proceedings of the symposium on strategy for fisheries resources enhancement in the southeast Asian Region, Pattaya, Thailand, 27–30 July 2015. pp 131–135

Singhagraiwan T, Doi M, Sasaki M (1992) Salinity tolerance of juvenile Donkey's ear abalone, *Haliotis asinina* Linne. Thailand Mar Fish Res Bull 3:71–77

Taylor JJ, Southgate PC, Rose RA (2004) Effects of salinity on growth and survival of silver-lip pearl oyster, *Pinctada maxima*, spat. J Shellfish Res 23:375–377

Troell M, Robertson-Anderson D, Anderson R, Bolton JJ, Maneveldt G, Halling C, Probyn T (2006) Abalone farming in South Africa: an overview in perspectives in kelp resources, abalone feed, potential for on-farm seaweed production and socio-economic importance. Aquaculture 257:266–281

Vilchis LI, Tegner MJ, Moore JD, Friedman CS, Riser KL, Robbins TT, Dayton PK (2005) Ocean warming effects on growth, reproduction and survivorship of southern California abalone. Ecol Appl 15(2):469–480

Wanichanon C, Laimek P, Linthong V, Sretarugsa P, Kruatrachue M, Upatham ES, Poomtong T, Sobhon P (2004) Histology of hypobranchial gland and gill of *Haliotis asinina* Linnaeus. J Shellfish Res 23:1107–1112

Wu F, Zhang G (2012) Suitability of cage culture for Pacific abalone *Haliotis discus hannai* Ino production in China. Aquac Res 44:485–494

Yan XZ, Wang GZ, Li SJ (2009) Effects of water salinity on energy budget of *Haliotis diversicolor aquatilis*. Chin J Ecol 28(8):1520–1524

Yiyan W, Hushan S, Meiyu Z (2004) Types and distribution of mucous cells in the digestive tract in abalone *Haliotis discus hannai* Ino. Fish Sci 23(5):1–4

Zhuang S (2005) Influence of salinity, diurnal rhythm and daylength on feeding in *Laternula marilina* Reeve. Aquac Res 36:130–136

20

Locus-specific EPIC-PCR primers for four distinct calmodulin genes of the Pacific bluefin tuna *Thunnus orientalis*

Seinen Chow · Nobuaki Suzuki · Yoji Nakamura · Motoshige Yasuike · Kenji Saitoh · Takashi Yanagimoto

Abstract Nucleotide sequences of four distinct calmodulin genes (designated as *CaM*-A to -D) of the Pacific bluefin tuna (*Thunnus orientalis*) were compared. Nucleotide sequence divergence between the coding regions of four genes ranged from 16.0 to 21.6%, and locus-specific eight primer pairs targeting introns were designed. These primer pairs amplified expected size of fragments in five congeneric tuna species, and utility of these primer pairs was investigated using distant fish species.

Keywords *Thunnus orientalis* · Four calmodulin genes · Locus-specific primers · Cross-species amplification · DNA marker

Introduction

The Pacific bluefin tuna (*Thunnus orientalis*) is one of the largest bony fish and most commercially valuable species. Because of the overexploitation and the declining stock status, this species has been listed as vulnerable by the IUCN. To date, polymorphic DNA markers developed and used for the Pacific bluefin tuna are microsatellite and mitochondrial DNA (Takagi et al. 1999; Nomura et al. 2014; Yagishita et al. 2014).

Introns rapidly accumulate mutations, which may be used as selectively neutral nuclear genetic marker complementing microsatellite and mitochondrial DNAs. Exon-primed intron-crossing (EPIC) PCR strategy employs conserved exon sequences to design universal primers (Lessa 1992). The efficiency of isolating a single copy of the target nuclear genome may be dependent on the number of paralogs and pseudogenes. Targeting highly conserved single copy of nuclear gene such as ribosomal protein gene has been successful to design universal primers applicable to wide variety of marine animals (Chow and Hazama 1998; Chow and Yanagimoto 2016). Calmodulin (CaM) is one of the most highly conserved molecules. Two to four distinct genes have been detected in vertebrates (Chien and Dawid 1984; Simmen et al. 1985; Nojima and Sokabe 1987; SenGupta et al. 1987; Nojima 1989; Koller et al. 1990; Matsuo et al. 1992; Rhyner et al. 1994; Friedberg and Rhoads 2002), and three to four distinct CaM-encoding cDNAs were reported in fish (Matsuo et al. 1992; Friedberg and Rhoads 2002). Despite considerable amount of nucleotide substitutions between

S. Chow (✉) · Y. Nakamura · M. Yasuike · K. Saitoh · T. Yanagimoto
National Research Institute of Fisheries Science, 2-12-4 Fukuura, Kanagawa 236-8648, Japan
e-mail: chow@affrc.go.jp

N. Suzuki
National Research Institute of Far Seas Fisheries, 5-7-1 Orido, Shizuoka 424-8633, Japan

distinct *CaM* genes, amino acid sequences of the metazoan CaMs are extremely conserved (Yuasa et al. 1999, 2001).

The draft genome sequence of the Pacific bluefin tuna has been reported (Nakamura et al. 2013), in which we detected four distinct *CaM* genes. In this study, we compared whole structure of these four tuna *CaM* genes and designed locus-specific primers for EPIC-PCR. We also investigated the utility of these primer pairs in other fish species.

Materials and methods

Fish samples used in this study were derived from our laboratory stock collected by scientific sampling. Matsuo et al. (1992) reported cDNAs of four *CaM* genes (*CaM*-A to -D) in medaka *Oryzias latipes* (DDBJ/EMBL/GenBank accession numbers D10363 to D10366). We used these sequences as in silico probes to detect similar sequences in the draft genome of the Pacific bluefin tuna (http://nrifs.fra.affrc.go.jp/ResearchCenter/5_BB/genomes/Tuna_DNAmicroarray/index.html). BLAST searches (Altshul et al. 1990) using these probes were performed, and we detected four scaffolds (Ba00000990, Ba00001092, Ba00001784, and Ba00003491) and confirmed these to contain whole *CaM* genes. *CaM* gene sequence in one scaffold (Ba00000990) was complete, while partial intron sequences were not determined in the other three scaffolds (Ba00001092, Ba00001784 and Ba00003491). We designed oligo nucleotide primers at flanking regions of these undetermined sequences for PCR. PCR mixtures were preheated at 94 °C for 4 min, followed by 35 amplification cycles (94 °C for 30 s, 55 °C for 30 s, and 72 °C for 50 s), with a final extension at 72 °C for 7 min. PCR products were cloned using TOPO TA Cloning Kit (Invitrogen). Transformed colonies were subjected to colony direct PCR using M13 forward and M13 reverse primers, and PCR products were purified via ExoSAP-IT (GE Healthcare Life Sciences). Nucleotide sequencing reaction was performed using the BigDye Terminator Cycle Sequencing Kit Ver3.1 (Applied Biosystems) with the PCR primers and M13 forward and M13 reverse primers, and the nucleotide sequences were determined with an automated DNA sequencer (ABI PRISM 3730XL Genetic Analyzer) (Applied Biosystems).

Exon sequences of medaka and Pacific bluefin tuna determined were subjected to BLAST search in GenBank to find similar sequences of teleostei. Among sequences shown by BLATS search, those from distant fish taxa but common to medaka and Pacific bluefin tuna were selected. Three species and the sequences selected were barramundi *Lates calcarifer* (accession no. XM_018694615, XM_018694749, XM_018695169, XM_018702063), mummichog *Fundulus heteroclitus* (XM_012861547, XM_012862386, XM_012868907, XM_012877456), and mangrove killifish *Kryptolebias marmoratus* (XM_017404200, XM_017409196, XM_017424744, XM_017426177). Barramundi belongs to Perciformes, while mummichog and mangrove killifish belong to Cyprinodontiformes.

Alignment of exon sequences was performed using ClustalW algorithm implemented in MEGA6 (Tamura et al. 2013), and K2 + G + I was selected as the best fit model. Kimura 2-parameter distance (K2P) between the exon sequences was calculated and phylogenetic tree was constructed using MEGA6.

Results

Nucleotide sequences of the four tuna *CaM* genes determined in the present study are available in DDBJ/EMBL/GenBank (LC183900 to LC183903). All these *CaM* genes are consisted of six exons and five introns. Exon sequences of these four genes were identical in length (450 bp), and K2P distances between the exon sequences ranged from 16.0 to 21.6%. All nucleotide substitutions among the sequences were silent. Phylogenetic tree constructed using exon sequence data of medaka, Pacific bluefin tuna, barramundi, mummichog and mangrove killifish is shown in Fig. 1. Four large clades appeared in the tree, and mean K2P ± S.E. within and between these clades ranged from 5.8 ± 0.9 to 13.4 ± 1.7% and from 16.7 ± 2.2 to 23.5 ± 3.0%, respectively. Four tuna *CaM* genes were designated as A to D according to the affinity with medaka *CaM* genes. Structures of four tuna *CaM* genes (A to D) are shown in Fig. 2, and introns of 300 to 1500 bp long were selected for EPIC-PCR (closed areas of horizontal bars). As the 1st exon was three bases long for all genes, the 1st introns were not included for the target. We designed eight locus-specific oligo-primer (20 nt)

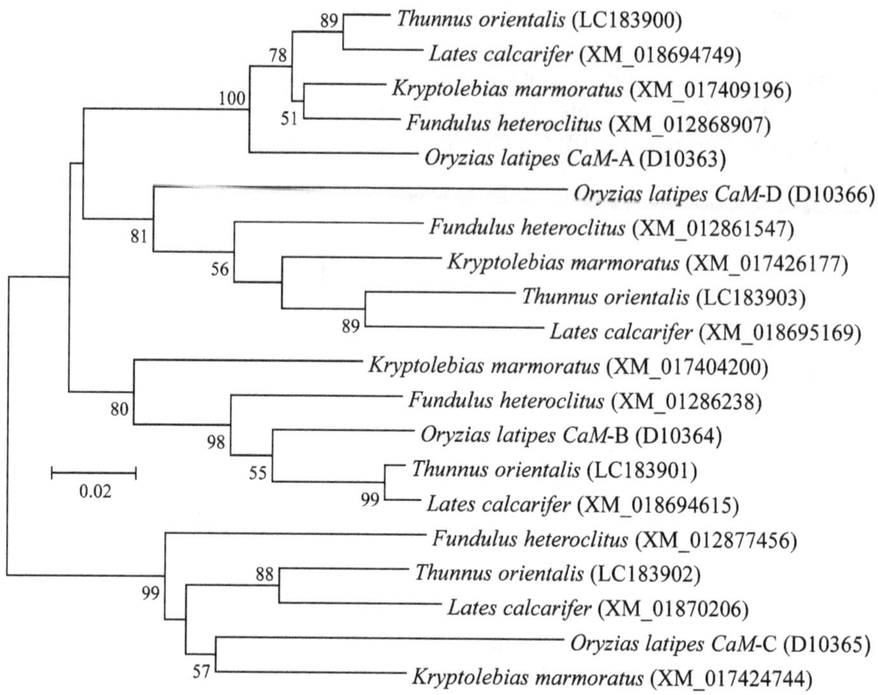

Fig. 1 Unrooted neighbor-joining phylogenetic tree (K2 + G + I) using exon sequences of distinct *CaM* genes of the Pacific bluefin tuna (*Thunnus orientalis*), medaka (*Oryzias latipes*), barramundi (*Lates calcarifer*), mummichog (*Fundulus heteroclitus*), and mangrove killifish (*Kryptolebias marmoratus*). Bootstrap values >50% (out of 1000 replicates) are shown at the nodes. Accession *numbers* are shown in the *parenthesis*. Essentially the same tree topology was obtained in maximum likelihood and maximum parsimony trees

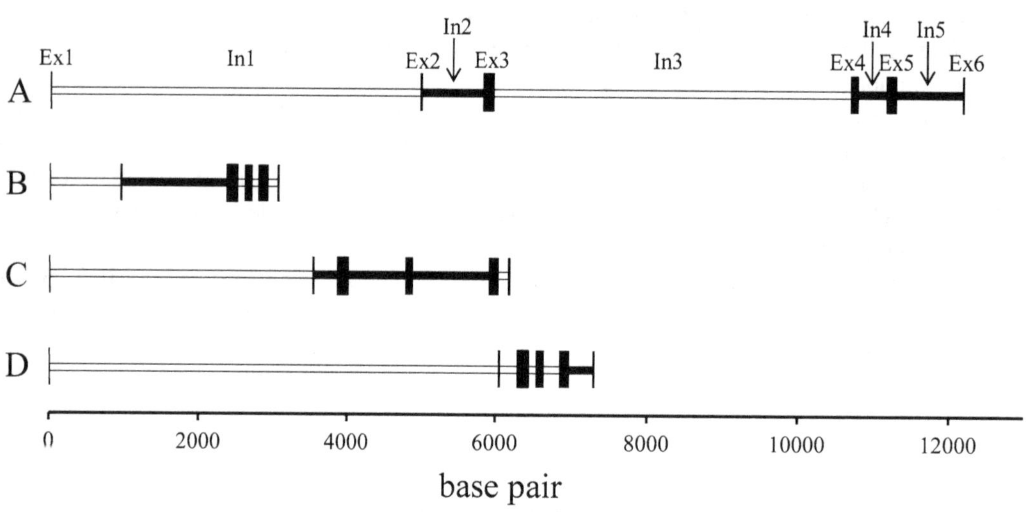

Fig. 2 Genomic structures of four *CaM* genes (A to D) of the Pacific bluefin tuna. *Vertical bars* represent exons (Ex), and *horizontal bars* are introns (In). Introns of 300 to 1500 bp long (*black horizontal bars*) are chosen as targets for exon-primed intron-crossing PCR

pairs based on the nucleotide substitutions between exon sequences of the Pacific bluefin tuna (Fig. 3, underlined), using which fish species shown in Table 1 were subjected to PCR amplification under above-mentioned condition. Since the size of amplified fragments in all *Thunnuns* spp. used was the same, the Pacific bluefin tuna (2nd lane) was shown as a representative for all *Thunnus* spp. In tuna, a single fragment amplification was observed with seven primer pairs (Fig. 4a–c, e–h) and two fragments were observed in one primer pair (Fig. 4d), in which fragment size amplified by seven primer pairs (Fig. 4a–c, e–h) and longer

```
                ex2F                  2nd intron             ex3R
CaM-A   GCTGACCAACTAACAGAGGA[11nt]    ~    [12nt]CTTTCTCCTTATTCGACAAG   882†
CaM-B   GCTGATCAGCTTACAGAAGA[11nt]    ~    [12nt]CATTTTCGCTCTTTGACAAG   1464
CaM-C   GCTGATCAGCTGACTGAAGA[11nt]    ~    [12nt]CATTCTCGCTGTTTGACAAG   369
CaM-D   GCTGACCAACTAACAGAAGA[11nt]    ~    [12nt]CTTTCTCCCTGTTTGATAAG   281
        *****  **  **  **  **  **                *  **  **   *  **  **  ***

                ex3F                  3rd intron             ex4R
CaM-A   AGCTTGGCACCGTCATGAGG[64nt]    ~    [7nt]AACCATTGACTTCCCGGAGT    4835
CaM-B   AGCTGGGCACAGTCATGCGT[64nt]    ~    [7nt]AACGATAGACTTCCCGGAGT    206
CaM-C   AGCTTGGGACTGTGATGCGC[64nt]    ~    [7nt]TACAATTGATTTTCCTGAGT    886
CaM-D   AGCTGGGTACAGTCATGAGG[64nt]    ~    [7nt]CACCATCGACTTCCCTGAGT    206
        ****  **  **  **  ***  *                 **  **  **  **  **  ****

                ex4F                  4th intron             ex5R
CaM-A   GGAGATCCGCGAGGCTTTCC[14nt]    ~    [64nt]AACAGACGAGGAGGTGGACG   490
CaM-B   GGAGATCAGAGAAGCATTCC[14nt]    ~    [64nt]GACTGATGAAGAAGTGGATG   196
CaM-C   GGAGATCAGAGAAGCCTTCA[14nt]    ~    [64nt]CACTGATGAAGAGGTGGACG   1135
CaM-D   AGAGATCAGAGAGGCATTCA[14nt]    ~    [64nt]GACAGACGAGGAAGTAGACG   317
         ******  *  **  **  ***                  **  **  **  **  **  **  *

                ex5F                  5th intron             ex6R
CaM-A   AGAAGCTAACAGACGAGGAG[45nt]    ~    [17nt]TGTACAGATGATGACTGCAA   968
CaM-B   AGAAGCTGACTGATGAAGAA[45nt]    ~    [17nt]CGTACAAATGATGACGGCGA   225
CaM-C   AGAAGCTCACTGATGAAGAG[45nt]    ~    [17nt]CGTCCAGATGATGACTGCCA   223
CaM-D   AGAAGCTGACAGACGAGGAA[45nt]    ~    [17nt]CGTTCAGATGATGACCGCCA   416
        *******  **  **  **  **                  **  **  ********  **  *
```

Fig. 3 Alignments of partial exon sequences of four *CaM* genes (A to D) of the Pacific bluefin tuna, flanking 2nd to 5th introns. Introns are shown by tilde, and number of nucleotides between primer and intron is presented in *bracket*. Identical nucleotides among genes (A to D) are indicated by *asterisk*, and nucleotide sequences adopted for designing primers are *underlined*. †Predicted size (bp) of amplified fragment for the Pacific bluefin tuna

Table 1 Bony fish species used in this study

Species	Common name	Order	Catch locality (n: sample size)
Thunnus orientalis	Pacific bluefin tuna	Perciformes	Northwest Pacific (3)
Thunnus thynnus	Atlantic bluefin tuna	Perciformes	Northwest Atlantic (3)
Thunnus maccoyi	Southern bluefin tuna	Perciformes	Southwest Indian Ocean (2)
Thunnus alalunga	Albacore	Perciformes	Northwest Pacific (2)
Thunnus albacares	Yellowfin tuna	Perciformes	Northwest Pacific (2)
Sardinops melanostictus	Japanese pilchard	Clupeiformes	Tosa Bay, Japan (1)
Takifugu rubripes	Japanese puffer	Tetraodontiformes	Japan Sea (1)
Platichthys stellatus	Starry flounder	Pleuronectiformes	Shibetsu, Japan (1)
Sebastolobus macrochir	Broadbanded thornyhead	Scorpaeniformes	Northwest Pacific (1)
Anguilla japonica	Japanese eel	Anguilliformes	Tone River, Japan (1)

fragment by one primer pair (Fig. 4d) matched well with those predicted from the sequence data (see Fig. 3). We cloned and sequenced the shorter fragment (c.a. 400 bp) amplified by *CaM*-C: ex3F × ex4R (Fig. 4d). This fragment (376 bp, accession no. LC157866) had a same primer sequence (Fig. 3, *CaM*-C: ex3F) at both 5′ and 3′ ends and no *CaM*-like sequence was observed, indicating non-specific amplification.

Amplification efficiency varied among the other fish species. No amplicon was observed in four primer pairs (*CaM*-B: ex2F × ex3R, *CaM*-C: ex2F × ex3R, *CaM*-C: ex3F × ex4R, *CaM*-C: ex4F × ex5R) (Fig. 3) for the Japanese pilchard (3rd lane of Fig. 4b–d, f) and one primer pair (*CaM*-B: ex2F × ex3R) for the Japanese eel (7th lane of Fig. 4b), while amplicons were obtained in all primer pairs in the other species. Single fragment amplification was obtained in 21 out of 40 occasions (5 species by 8 primer pairs).

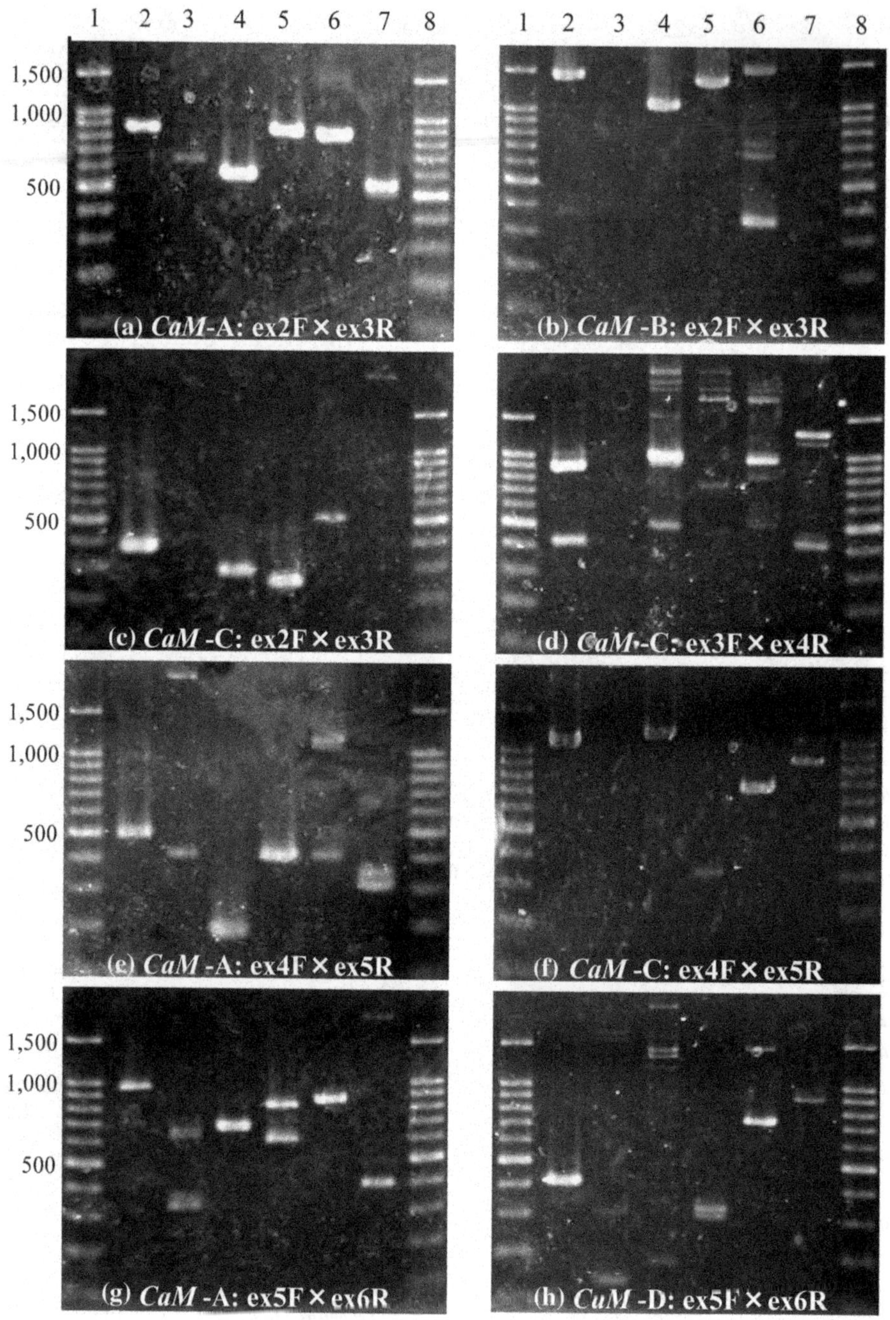

Fig. 4 Agarose gel electrophoresis images of PCR products amplified by eight primer pairs (see Fig. 3). 1st and 8th *lanes* are molecular marker, and the size (bp) is presented at the *left* margin. 2nd to 7th *lanes* are the Pacific bluefin tuna, Japanese pilchard, Japanese puffer, starry flounder, broadbanded thornyhead, and Japanese eel, respectively (see Table 1)

Discussion

One or two CaM-encoding genes have been reported in invertebrates (Hardy et al. 1988; Doyle et al. 1990; Swanson et al. 1990; Côrte Real et al. 1994; Karabinos and Riemer 1997; Yuasa et al. 1999, 2001). Using locus-specific primers, Côrte-Real et al. (1994) successfully isolated one (*CaM-1*) of two *CaM* genes of mussel *Mytilus edulis*, and length polymorphism found in the 3rd intron was used for population genetic analysis. A high restriction fragment length polymorphism (RFLP) was observed in the *CaM* intron of the Atlantic bluefin tuna *Thunnus thynnus* (Nakadate and Chow 2008). Single-nucleotide polymorphism (SNP) found in the 4th intron of swordfish *Xiphias gladius* was used to investigate population structure (Chow and Takeyama 2000; Chow et al. 2007). Partial exon sequence (200 bp) of the Atlantic bluefin tuna was observed to have higher homology (99% homologous) with that of *CaM*-A in *T. orientalis* than with the others (*CaM*-B, -C and -D) (83–86%). Similarity of exon sequence (105 bp) of the swordfish *CaM* to medaka *CaM*-A was suggested previously (Chow and Takeyama 2000). *CaM*-A of *T. orientalis* showed much higher homology with the swordfish exon sequence (94% homologous), while *CaM*-B to D of *T. orientalis* were much less homologous (83–86%) with the swordfish exon sequence. *CaM*s of the Atlantic bluefin tuna and swordfish were 94% homologous. These indicate that the primer pair used in these previous studies exclusively amplified homologous intron of *CaM*-A gene. Locus-specific primer pairs presented in this study are designed based on genome data of the Pacific bluefin tuna. Nevertheless, success rate of amplifying single fragment among distant fish species was relatively high, indicating the possibility to develop a genetic assay to target a specific locus in wide variety of fish taxa.

Acknowledgements This work was supported in part by a Grant-in-Aid for Scientific Research on Priority Areas (C) (No. 25450292) from the Ministry of Education, Science, Sports, and Culture of Japan. We thank Takahiro Ohtsuka, Ayako Kondo and Haruka Itoh for their excellent technical assistance.

References

Altshul SF, Gish W, Miller W, Myers EW, Lipman DJ (1990) Basic local alignment search tool. J Mol Biol 215:403–410

Chien YH, Dawid IB (1984) Isolation and characterization of calmodulin genes from *Xenopus laevis*. Mol Cell Biol 4:507–513

Chow S, Hazama K (1998) Universal PCR primers for S7 ribosomal protein gene introns in fish. Mol Ecol 7:1255–1256

Chow S, Takeyama H (2000) Nuclear and mitochondrial DNA analyses reveal four genetically separated breeding units of the swordfish (*Xiphias gladius*). J Fish Biol 56:1087–1098

Chow S, Yanagimoto T (2016) Universal PCR primers for ribosomal protein gene introns of fish. Int Aquat Res 8:29–36

Chow S, Clarke S, Nakadate M, Okazaki M (2007) Boundary between the north and south Atlantic populations of the swordfish (*Xiphias gladius*) inferred by a single nucleotide polymorphism at calmodulin gene intron. Mar Biol 152:87–93

Côrte-Real HBSM, Dixon DR, Holland PWH (1994) Intron-targeted PCR: a new approach to survey neutral DNA polymorphism in bivalve populations. Mar Biol 120:407–413

Doyle KE, Kovalick GE, Lee F, Beckingham K (1990) *Drosophila melanogaster* contains a single calmodulin gene: further structure and expression studies. J Mol Evol 213:599–605

Friedberg F, Rhoads AR (2002) Multiple calmodulin genes in fish. Mol Biol Rep 29:377–382

Hardy DO, Bender PK, Kretsinger RH (1988) Two calmodulin genes are expressed in *Arbacia punctulata*. An ancient gene duplication is indicated. J Mol Biol 199:223–227

Karabinos A, Riemer D (1997) The single calmodulin gene of the cephalochordate *Branchiostoma*. Gene 195:229–233

Koller M, Schnyder B, Strehler EE (1990) Structural organization of the human *CAMIII* calmodulin gene. BBA-Gene Struct Expr 1087:180–189

Lessa EP (1992) Rapid surveying of DNA sequence variation in natural populations. Mol Biol Evol 9:323–330

Matsuo K, Sato K, Ikeshima H, Shimoda K, Takano T (1992) Four synonymous genes encode calmodulin in the teleost fish, medaka (*Oryzias latipes*): conservation of the multigene one-protein principle. Gene 119:279–281

Nakadate M, Chow S (2008) Isolation and characterization of single copy nuclear DNA markers in the Atlantic bluefin tuna *Thunnus thynnus*. Fish Sci 74:1333–1335

Nakamura Y, Mori K, Saitoh K, Oshima K, Mekuchi M, Sugaya T, Shigenobu Y, Ojima N, Muta S, Fujiwara A, Yasuike M, Oohara I, Hirakawa H, Chowdhury VS, Kobayashi T, Nakajima K, Sano M, Wada T, Tashiro K, Ikeo K, Hattori M, Kuhara S, Gojobori T, Inoue K (2013) Evolutionary changes of multiple visual pigment genes in the complete genome of Pacific Bluefin tuna. Proc Natl Acad Sci USA 110:11061–11066

Nojima H (1989) Structural organization of multiple rat calmodulin genes. J Mol Biol 208:269–282

Nojima H, Sokabe H (1987) Structure of a gene for rat calmodulin. J Mol Biol 193:439–444

Nomura S, Kobayashi T, Agawa Y, Margulies D, Scholey V, Sawada Y, Yagishita N (2014) Genetic population structure of the Pacific bluefin tuna *Thunnus orientalis* and the yellowfin tuna *Thunnus albacares* in the North Pacific Ocean. Fish Sci 80:1193–1204

Rhyner JA, Ottiger M, Wicki R, Greenwood TM, Strehler EE (1994) Structure of the human *CALM1* calmodulin gene and identification of two *CALM1*-related pseudogenes *CALM1P1* and *CALM1P2*. Eur J Biochem 225:71–82

SenGupta B, Friedberg F, Detera-Wadleigh SD (1987) Molecular analysis of human and rat calmodulin complementary DNA clones: evidence for additional active genes in these species. J Biol Chem 262:16663–16670

Simmen RC, Tanaka T, Ts'ui KF, Putkey JA, Scott MJ, Lai EC, Means AR (1985) The structural organization of the chicken calmodulin gene. J Biol Chem 280:907–912

Swanson ME, Sturner SF, Schwartz JH (1990) Structure and expression of the *Aplysia californica* calmodulin gene. J Mol Biol 216:545–553

Takagi M, Okamura T, Chow S, Taniguchi N (1999) PCR primers for microsatellite loci in tuna species of the genus *Thunnus* and its application for population genetic study. Fish Sci 65:571–576

Tamura K, Stecher G, Peterson D, Filipski A, Kumar S (2013) MEGA6: molecular evolutionary genetics analysis version 6.0. Mol Biol Evol 30:2725–2729

Yagishita N, Sawada Y, Agawa Y, Kobayashi T (2014) Isolation and characterization of 25 microsatellite loci from the Pacific bluefin tuna *Thunnus orientalis*. Conserv Genet Resour 6:189–191

Yuasa HJ, Yamamoto H, Takagi T (1999) The structural organization of the ascidian, *Halocynthia roretzi*, calmodulin genes: The vicissitude of introns during the evolution of calmodulin genes. Gene 229:163–169

Yuasa HJ, Yamamoto H, Takagi T (2001) Structural organization of lower marine nonvertebrate calmodulin genes. Gene 279:205–212

21

Endocrine disrupting effects of 4-nonylphenol on plasma vitellogenin, reproductive system and histology in koi carp (*Cyprinus carpio*)

Parisa Amaninejad · Homayoun Hosseinzadeh Sahafi · Mehdi Soltani · Seyed Pezhman Hosseini Shekarabi

Abstract This study aimed to assess the effect of 4-nonylphenol (4-NP), a ubiquitous aquatic micro-chemical pollutant substance on some on some reproductive parameters and histology in koi carp. Fish were intraperitoneally subjected to ascending doses of 4-NP at 10, 50 and 100 μg/g body weight (BW) on days 7, 14 and 21 after the initiation of the experiment. Also, one group was treated with 17β-estradiol (E2) at 2 μg/g BW, positive control group (C2) received the vehicle (50 μL of coconut oil + 50 μL Ethanol) per fish and a negative control group (C1) was considered without injection. The results showed that 4-NP induced significantly an increase in E2, progesterone, and plasma vitellogenin at 50 μg/g in both male and female, while level of testosterone significantly decreased in the treated fish compared to control fish ($P < 0.05$). Conversely, the levels of these parameters at the higher concentration of 4-NP (100 μg/g) were significantly reduced compared to control one ($P < 0.05$). Also, some sever histopathological changes were observed in ovary, testicular and liver of treated fish, reflecting fish sensitivity to 4-NP at higher doses. This study showed that 4-NP has an estrogenic potency in the sexual development, hepatotoxicity and gonado-toxicity of koi carp.

Keywords Nonylphenol · Reproductive system · Vitellogenin · Histopathology · Koi carp

Introduction

Both natural and synthetic endocrine disrupting chemicals (EDCs) are able to cause many disorders and interferences with animal endogenous hormones systems including adverse effects on hormone receptors, normal hormone synthesis and secretion (Maradonna and Carnevali 2007; Zaccaroni et al. 2009). Huge volume of EDCs is released to the global environment influencing serious potential hazardous effects on

P. Amaninejad (✉) · S. P. Hosseini Shekarabi
Department of Fisheries Science, Science and Research Branch, Islamic Azad University, Tehran, Iran
e-mail: parisaamaninejad@yahoo.com

H. Hosseinzadeh Sahafi
Iranian Fisheries Science Research Institute Agriculture Research, Education and Extension Organization, Tehran, Iran

M. Soltani
Department of Aquatic Animal Health, Faculty of Veterinary Medicine, University of Tehran, Tehran, Iran

Present Address:
M. Soltani
Freshwater Fish Group and Fish Health Unit, School of Veterinary and Life Science, Murdoch University, Perth, Australia

animal health condition and reproductive functions via the food chain and contaminated water resources (Vos et al. 2000). Among classes of EDCs, 4-nonylphenol (4-NP) is considered as an estrogen-like and anti-androgenic chemical and captured the attention of many investigators to its effect on the animal reproductive system (Sayed et al. 2012). 4-NP is an ultimate degradation product of nonylphenol ethoxylates (NPEs) and both are the most widely used members of alkylphenol and alkylphenol ethoxylate family of non-ionic detergents used in a wide variety of industrial applications including pesticides, spermicides, paints, wetting agents, textiles, plastics and paper products (John et al. 2000; Uguz et al. 2003; Sone et al. 2004; Rivero et al. 2008).

Reproduction in vertebrates such as fish is mainly regulated by the endocrine system primarily affected along the hypothalamic–pituitary–gonadal (HPG) axis via neural and hormonal feedback (Yaron et al. 2003; Choi et al. 2010). The role of 17 β-estradiol (E2) as a neuroendocrine feedback control on the HPG axis has been shown in fish reproductive process (Yadetie et al. 1999). 4-NP has the ability to compete with E2 receptor binding and exerts an estrogen-like transcriptional activity on expression of estrogen receptor gene in liver, gill, pituitary and brain (Christensen et al. 1999; Yadetie et al. 1999). Several studies have been demonstrated an impairment of natural action of primary steroid hormones responsible for regulating gametogenesis including E2, androgens testosterone (T), and 11-ketotestosterone in fish collected from highly contaminated sites with alkylphenols or in fish after being exposed to nonylphenol ethoxylates (Sone et al. 2004; Rankouhi et al. 2004; Mitchelmore and Rice 2006). An increase in vitellogenin level in liver and blood of male common carp previously exposed to 4-NP has been also reported (Cardinali et al. 2004; Mitchelmore and Rice 2006). Moreover, histological alterations and lesions in reproductive organs as well as hepatic and kidney tissues have been reported in some fish species after being exposed to 4-NP (Weber et al. 2003; Zha et al. 2008; Yang et al. 2008). However, the harmful effects of 4-NP on fish reproductive system are still a controversial issue (Giesy et al. 2000; Matsumura et al. 2005; Yang et al. 2008).

This study aimed to investigate the effects of 4-NP as the biological breakdown product of widely used non-ionic surfactant on some reproductive parameters and histology of reproductive tissues in male and female of koi carp (*Cyprinus carpio*). As the HPG axis of this extensively farmed species is activated in small size (50–100 g), therefore, it is a good biological model for such physio-toxicological assays.

Materials and methods

Fish

One hundred and eighty apparently healthy immature koi carp including 90 females (55 ± 0.5 g) and 90 males (54 ± 0.7 g) were obtained from a local hatchery of an ornamental fish center (Tehran, Iran); and were transported to fisheries laboratory at Tehran Science and Research University. Fish were randomly divided into 18 glass aquaria (100 × 30 × 50 cm) with 10 fish per aquarium. Water quality condition (conductivity 400 L/min, pH 7.5, dissolved oxygen 5 ± 0.2 mg/L, temperature 23–24 °C and photoperiod of 12:12 h light:dark) was maintained constant during the trail with a whole water exchange per day and fish were fed with a commercial carp pellet at 3% body weight/day and waste products were siphoned daily. The experiment was performed according to the standard guidelines for the care and use of experimental animals by Science and Research Branch University Committee (Collegiate Committee on Ethics Matched Biomedical Researchers, Tehran).

Exposure conditions

Branched chain isomers of 4-NP (CAS No. 84852-15-3 with purity > 95%) and E2 (Sigma-Aldrich, Deisenhofen, Germany) were mixed with coconut oil (Sigma-Aldrich, Deisenhofen, Germany) and absolute ethanol (Merck, Darmstadt, Germany) as the chemical vehicle (Pait and Nelson 2003) to prepare treatment doses of 10, 50 and 100 μg/g body weight (BW) (Christensen et al. 1999; Yadetie et al. 1999; Casini et al. 2002; Carrera et al. 2007). E2 at 2 μg/g BW was also used as a separated treatment. Feeding was stopped 24 h prior to injection and ten fish per treatment were intraperitoneally injected with the mentioned dosages after being anesthetized with 2-phenoxyethanol at 0.1% (Merck, Darmstadt, Germany). The positive control group (C2) received the vehicle (50 μL of coconut oil + 50 μL Ethanol per fish) and a negative control group (C1) was considered without injection. Fish were injected on days 7, 14 and 21 after the initiation of the experiment.

Blood sampling

On day 22, the blood samples were collected from the caudal vein after fish being anaesthetized with clove oil (150 mg/L), and sera samples were separated by centrifugation at 3000×*g* for 10 min (SIGMA 3–30 K, Osterode am Harz, Germany) before being frozen at − 80 °C until analysis. Fish were then dissected and separated by sex gonad characters.

Steroid hormones assay

Steroid hormone levels including E2, T, and P were measured using radio immune assay (RIA) kit (Immunotech, Marseille, France) and gamma counter (LKB-Wallac RackGamma, Turku, Finland). Samples and standards were incubated in monoclonal antibody-coated tubes with 125I-labeled tracer T, E2 and P (50 mL sample or standard and 500 mL of tracer). After one h incubation with horizontal shaking at 400 rpm, the liquid contents of the tubes were removed and the radioactivity bound to the antibody was measured by JNG 403 (Multichannel gamma counter for RIA). The samples were analyzed using two kits in two analyses' forms of the calibration curve ($r^2 > 0.98$).

Vitellogenin assay

The plasma vitellogenin (VTG) concentration was determined in the sera samples of five females and 5 males (totally 10 fish) in each treatment. VTG concentration was detected using carp vitellogenin ELISA kit (Biosense Laboratories AS, Bergen, Norway) following manufacturer's instructions by an A3 plate reader (DAS, Rome, Italy) at 492 nm. VTG concentrations were detected based on a standard curve developed from the absorbance values for VTG standards ($r^2 > 0.99$).

Histological study

Five fish of each treatment were randomly removed and tissue samples of ovary, testis and liver were taken prior to be fixed in 10% neutral buffered formalin. The fixed tissues were processed for routine histotechnique, embedded in paraffin blocks before obtaining 5 μm thickness sections by a Leica RM2255 microtome (Leica Microsystems, Germany). The obtained sections were stained with Harris's hematoxylin and eosin stain (H&E), dehydrated through a graded alcohol series and mounted onto the glass slides (Bancroft and Stevens 1982) before being examined under light microscope (Olympus DP71, Tokyo, Japan).

Transmission electron microscope (TEM)

To evaluate TEM, the ovaries were cut into approximately 1 × 1 mm piece, fixed in 3% glutaraldehyde and washed in cacodylate buffer prior to being fixed in 1% osmium tetroxide prepared in the same buffer for one h at 4 °C. Specimens were then dehydrated in ascending grades of acetone with two changes of 15 min each after being cleared by propylene oxide for 30 min. Infiltration was performed gradually in different proportion of propylene oxide with liquid resin. Tissue samples were embedded in the araldite embedding medium using BEEM® capsules (BEEM Inc., Bronx, USA) before being incubated at 50 and 60 °C in an embedding oven for 24 and 48 h, respectively, to complete polymerization. The obtained thin sections (1 μm) were stained with toluidine blue before being used for examination under a light microscope. The ultrathin sections collected on copper grids were also stained with uranyl acetate followed by lead nitrate prior to examination under scanning electron microscopy (Oxford INCA 200, Oxford Instruments Group, High Wycombe, UK) at an accelerating voltage of 80 kV.

Statistical analysis

Data were subjected to SPSS Version 16.0 for windows (SPSS Inc., USA), and checked for normal distribution (Shapiro–Wilk's test) and homogeneity of variances (Levene's test) using P–P plot analysis. One-way analysis of variance (ANOVA) followed by the NP exposure for each endpoint relative Duncan's test was used to examine the effect of the control group and significance was set at $P < 0.05$.

Results

Steroid hormones

Female

The value of E2 in female fish received 10 and 50 μg 4-NP were significantly ($P < 0.05$) higher than control groups (Fig. 1), while fish exposed to 100 μg 4-NP showed significantly ($P < 0.05$) a lower value of E2 than both 10 and 50 μg 4-NP treatments (Fig. 1). Also, the value of T was decreased with an increase in 4-NP dose ($P < 0.05$) with the lowest value obtained in fish treated with 100 μg 4-NP ($P < 0.05$) (Fig. 1). Also, E2 value in E2-treated fish group was significantly lower than fish treated with 4-NP at 50 μg/g BW, while it increased to higher levels than 10 and 50 μg 4-NP groups. Also, no significant difference was seen in the level of P between 10 and 50 μg 4-NP groups, while this was significantly lower in the 100 μg 4-NP group. No significant difference was seen in level of T between E2 treated fish and 4-NP treated with 100 μg/g BW ($P < 0.05$), while T level in E2-treated fish was lower than 10 and 50 μg/g BW groups. Also, E2 value in fish treated with 2 μg E2 was significantly ($P < 0.05$) higher than control groups.

Male

The highest value of E2 was obtained in fish treated with 4-NP at 50 μg compared to other groups and control ones ($P < 0.05$). Also, this value in fish treated with 2 μg E2 was significantly higher than fish treated with 10

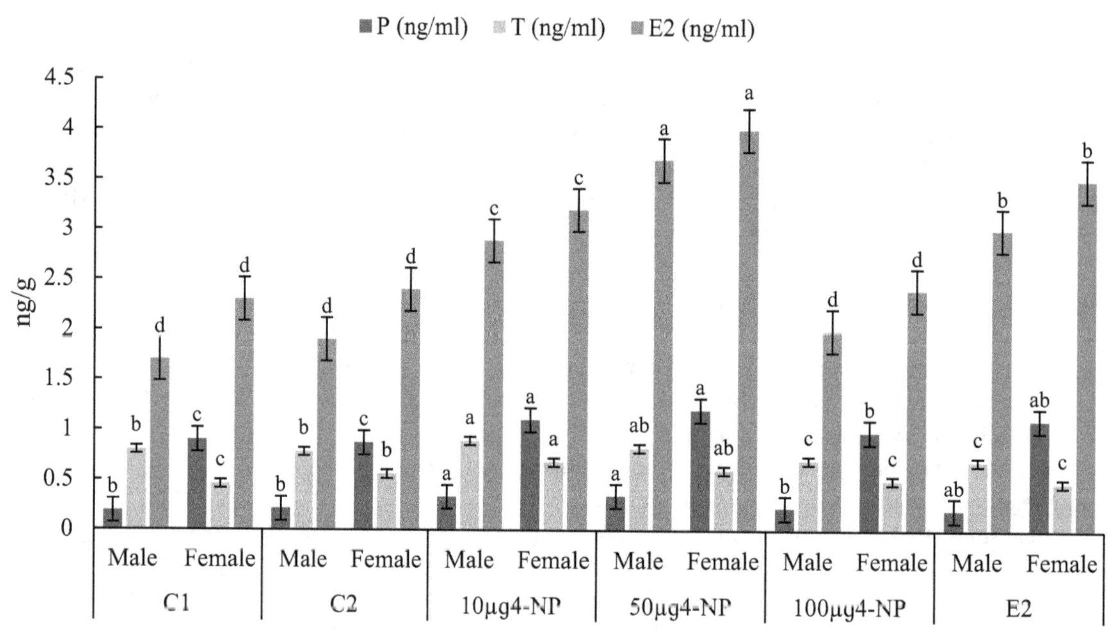

Fig. 1 Steroid hormones' changes in koi carp after 21 days exposure to different concentrations of 4-nonylphenol. E2, 17β-Estradiol; T, testosterone; P, progesterone; C1, control 1; C2, control 2. Different letters indicate significant difference between experimental groups ($n = 3$, $P < 0.05$)

and 100 µg4-NP ($P < 0.05$) (Fig. 1). The concentration of E2 in fish treated with 10 and 50 µg 4-NP and 2 µg E2 showed significantly a higher value than both control groups. In addition, the value of P increased significantly ($P < 0.05$) with an increase in 4-NP concentration up to 50 µg, but its concentration at 100 µg 4-NP gave a significant decrease compared to both 10 and 50 µg treatments (Fig. 1). No significant difference was seen in level of P among fish treated with 100 µg 4-NP, 2 µg E2 and control groups ($P > 0.05$). Value of T in fish treated with 10 and 50 µg 4-NP was significantly higher than fish treated with 100 µg 4-NP and 2 µg E2. Also, this value in both fish treated with 100 µg 4-NP and 2 µg E2 was significantly lower than control groups ($P < 0.05$).

Plasma vitellogenin

Female

The plasma VTG concentration was significantly higher in fish treated with 10 and 50 µg 4-NP/g BW compared to controls groups ($P < 0.05$). Also, fish received 100 µg 4-NP significantly showed a decrease in VTG value compared to both 10 and 50 µg 4-NP trails ($P < 0.05$). Further, fish treated with 2 µg E2 significantly gave an increase in VTG level compared to other groups (Fig. 2).

Male

A similar trend in plasma VTG concentration was observed in male fish (Fig. 2). VTG concentration significantly ($P < 0.05$) increased in fish treated with 10 µg 4-NP compared to control groups. Considerable elevations were also observed in plasma concentration of male fish exposed to 50 µg 4-NP/g BW compared to other treatments ($P < 0.05$), but a significant decrease was observed in 100 µg 4-NP trail compared to both 10 and 50 µg 4-NP trails (Fig. 2).

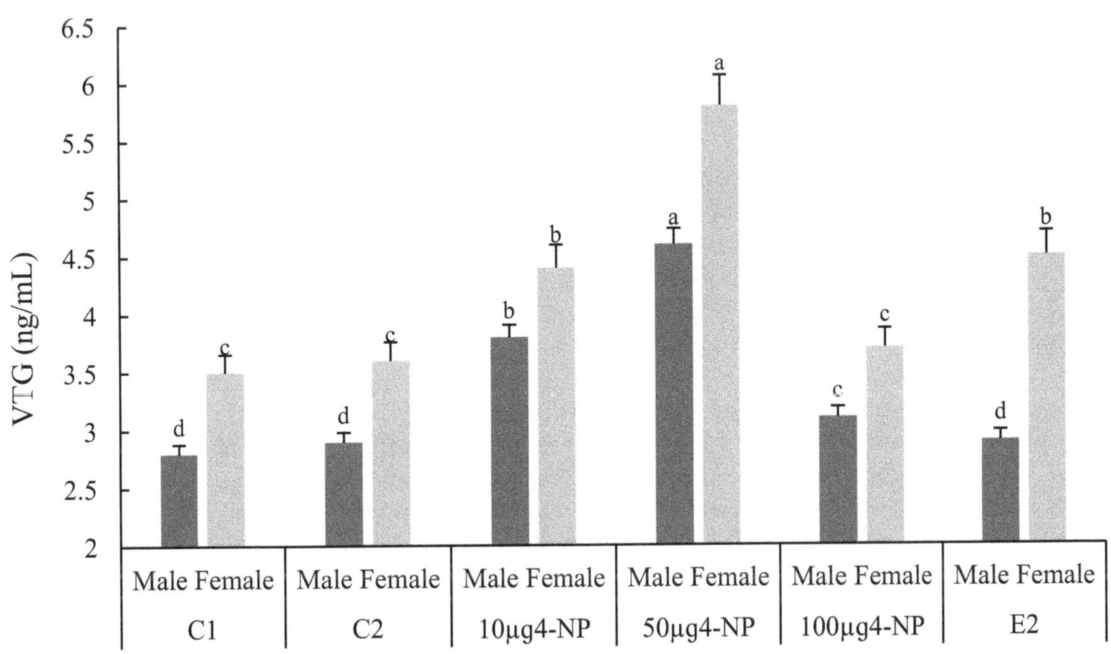

Fig. 2 Vitellogenin (VTG) changes in koi carp after 21 days exposure to different concentrations of 4-nonylphenol. Different letters indicate significant difference between experimental groups ($n = 3$, $P < 0.05$)

Female histological findings

No histological changes were seen in the ovaries of control fish (Fig. 3a, b). The ovary of female fish exposed to 10 and 50 µg 4-NP was loaded with oocytes in different vitelogenic stages and late vitellogenesis (Fig. 3c, d). The vitellogenic oocytes contained normal yolk materials (fatty yolk globules and protein granules) and surrounded by zona radiate and follicular epithelium that consists of granulose cell layer and thecal layer. Fish exposed to 100 µg 4-NP showed a reduction in size of the growing vitellogenic oocytes and ova damage including vacuoles (v), irregular shape of ovum with rupture of follicular layer and atretic (Fig. 3e, f). The number of follicular atretic after exposure to 100 µg 4-NP was significantly increased compared to control fish (arrows). No histological change was seen in ovaries of control fish (Fig. 3a, b).

Necrosis, degeneration, psychosis, and cytoplasmic vacuolization of hepatic parenchyma were seen in the liver samples all treated fish (Fig. 4b, c) but with more severe histological changes observed in fish treated with higher dosage of 4-NP, i.e., 100 µg/g BW (Fig. 4d). Also, liver samples of fish treated with 100 µg 4-NP showed a sedimentation of cytoplasmic and an increase in cytoplasmic vacuolization (Fig. 4e, f).

Ovary TEM

No abnormality was seen in sections obtained from the control fish (Fig. 5a). Ovaries of female fish exposed to 10 µg 4-NP showed an increase in size and number of previtellogenic oocytes and most of ova demonstrated a vitellogenesis stage containing droplets of fat and cortical alveoli. The external membrane presented a normal in shape having a higher amount of yolk particles. Yolk particles with a relatively high volume of pores and channels through the porous passing to the inner membrane and density of porous channels were also seen (Fig. 5b). In fish exposed to 50 µg 4-NP, an increase in vitellogenesis with a remarkable increase in vitellogenin particles was observable. Outer membrane of the yolk particles with a relatively high volume of pore channels in passing to the inner membrane and making micro-pinocytosis cascade was seen (Fig. 5c, d). An abnormal external membrane with a few channels containing a little penetration of vitellogenin inside the internal membrane was also seen in fish exposed to 100 µg 4-NP. Further, a decrease in number of channels with no transport of materials in this layer was observable (Fig. 5e, f).

Male histological findings

The testis of the control fish contained spermatids, spermatozoa, lumen space, seminiferous lobule, and interlobular connective tissue histological structure (Fig. 6a). In testis of fish exposed to 10 µg 4-NP, a large extended sperm duct loaded with spermatozoa and seminiferous lobules with different amounts of spermatozoa, together with wide lumen free of spermatozoa, were detected (Fig. 6b). In fish exposed to 50 µg 4-NP, a discontinuous interlobular connective tissue with a reduction in testis mass was seen. Also, the seminiferous lobules were filled with spermatozoa or late spermatids (Fig. 6c). A decrease in spermatids and spermatozoa, degeneration of Leydig cells, atrophy of testis cells and cellular vacuolation were also observable in sections obtained from fish treated with 100 µg 4-NP (arrows) (Fig. 6d).

Discussion

The results of the present work showed that 4-NP at 10 and 50 µg/g BW was able to increase the concentrations of steroid hormones of E2 and P in both male and female koi carp, while the value of testosterone was reduced. However, these hormones' concentrations were reduced at the higher dosage, i.e., 100 µg/g BW. Hasselberg et al. (2004) showed that there is a reverse relationship between CYP1A and amount of estradiol in fish plasma in a way that CYP1A is effective in metabolizing steroids. In our study, a gradual increase in E2 of fish treated with 10 and 50 µg/g BW could be a reason for a decrease in the activity of metabolizing enzymes. Yang et al. (2008) reported similar results when goldfish (*Carassius auratus*) was exposed to 4-NP at 100 µg 4-NP for 20 days. Also, a significant increase of E2 was seen in female fathead minnows (*Pimephales promelas*) after fish being exposed to 4-NP at 0.09 µg/L (Giesy et al. 2000). Lavado et al. (2004) demonstrated an inhibition in glucuronidation and sulfidation of sexual steroids in common carp exposed to NP. Thus, such

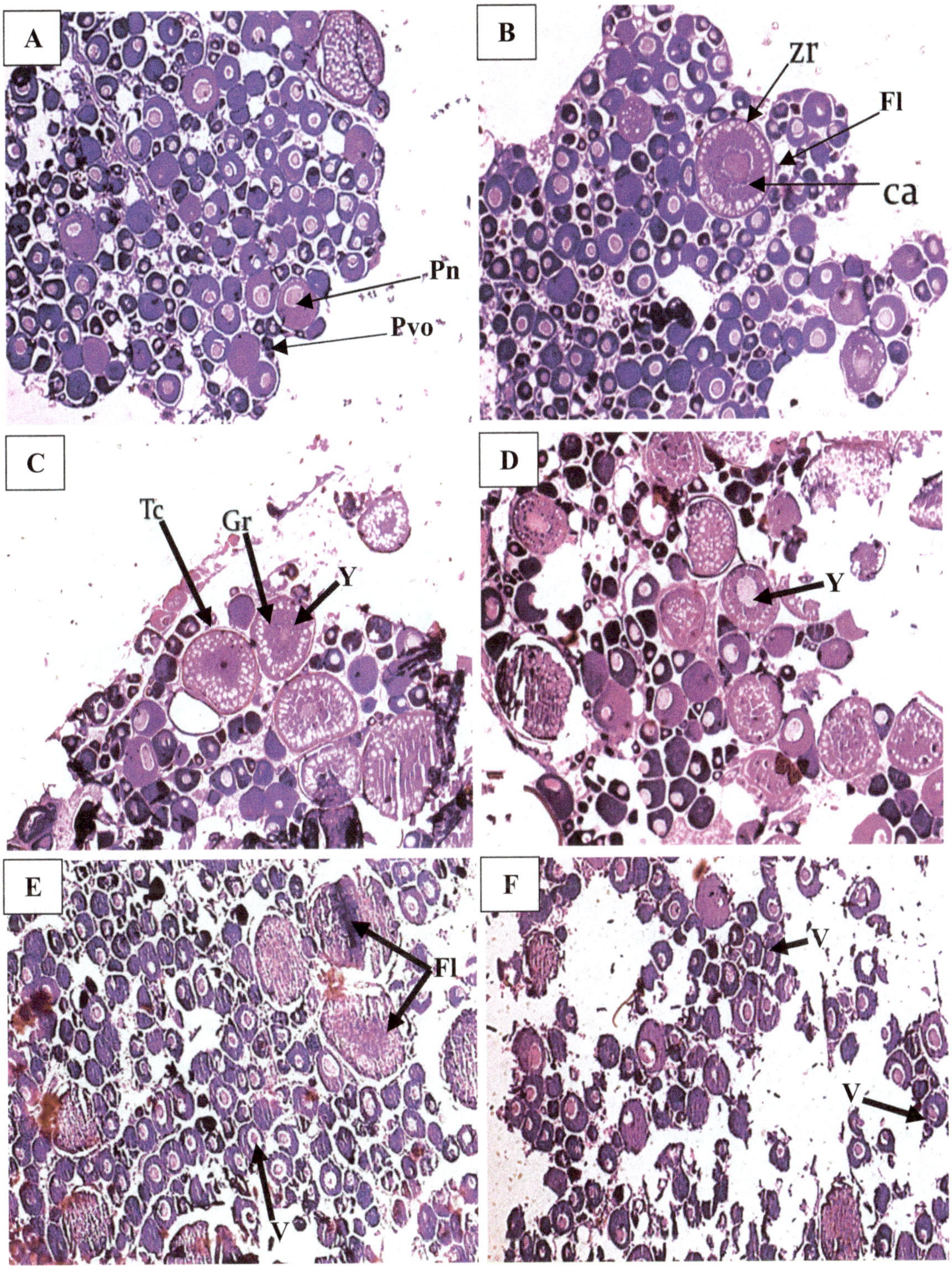

Fig. 3 Sections of koi carp ovary exposed to 4-nonylphenol at different concentrations. **a** control fish showing normal ovarian structures, **b** fish treated with 17β-Estradiol, **c**, **d** fish treated with 10 and 50 μg/g b w, respectively; **e**, **f** fish exposed to 100 μg 4-NP/g BW (H&E, ×400). Pvo, previtellogenic oocytes; Pn, peri-nucleolar; Fl, intact follicular layer; Zr, zona radiate; Ca, cortical alveoli; Gr, granulose cell layer; TC, thecal layer; Y, yolk; V, vacuolization

inhibition of glucuronidation and sulfidation in liver could be the cause of an increase in E2 seen in our work. However, when turbot (*Scophthalmus maximus*) was exposed to nonylphenol at 30 μg/g significantly, a decrease was occurred in the estradiol and testosterone levels as well as a reduction in the ovary aromatase enzyme and glucuronidation of sex steroids (Martin-Skilton et al. 2006).

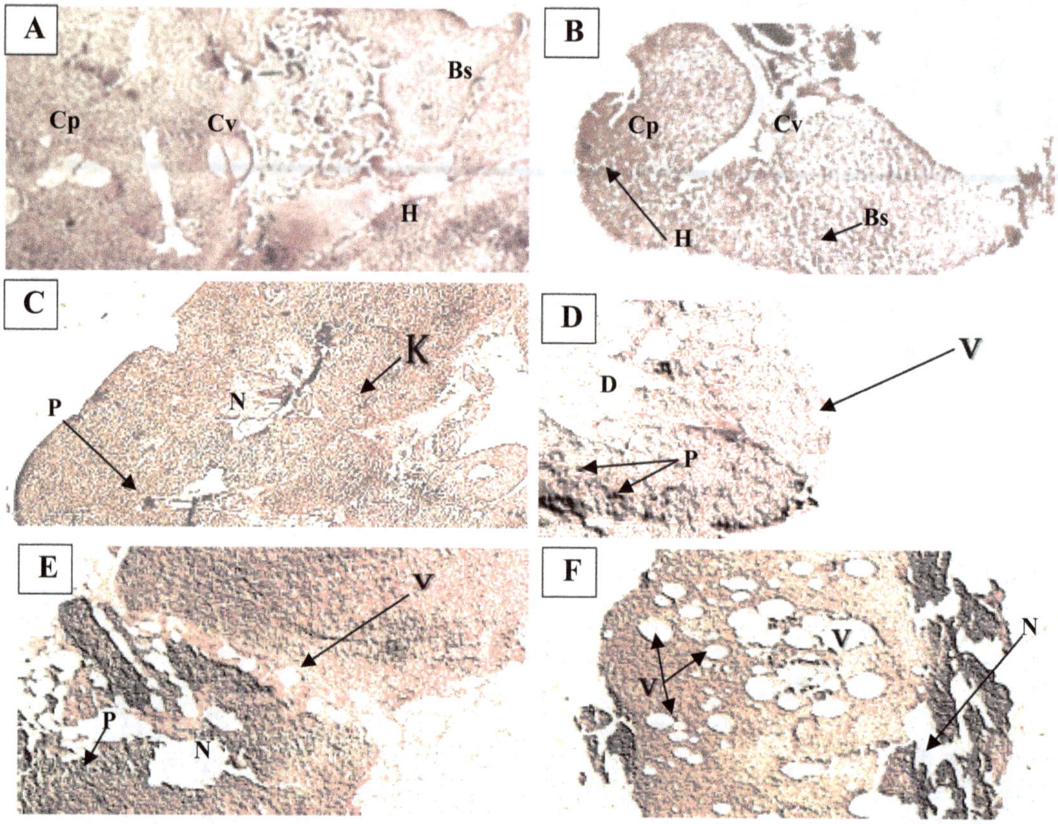

Fig. 4 Sections of koi carp liver exposed to different concentration of 4-nonylphenol. **a** Control fish showing normal histological structures, **b** fish treated with 17β-Estradiol, **c**, **d** fish treated with 10 and 50 μg 4-NP/g b w, respectively; **e**, **f** fish treated with 100 μg 4-NP/g BW (H&E, ×400). Cp, cytoplasm; Cv, central vein; Bs, blood sinusoids; H, hepatocyte; N, necrosis; K, Kupffer cell; D, degeneration; P, Pyknosis; V, vacuolation

Zenoestrogens like 4-NP can be accomplished by inhibitory effects in different sections of hypothalamus–hypophysis–gonads axis, and an effluence of the hormone receivers can cause a change in the secretion of hormones (Folmar et al. 2002; Lavado et al. 2004). Also, 4-NP could inhibit the activity of 17 alpha-hydroxylase enzymes which are involved in testosterone synthesis (Laurenzana et al. 2002). However, Villeneuve et al. (2002) have demonstrated no effect on E2, T and VTG in matured common carp male exposed to 4-NP at 0.05–5.4 μg/L for 28–31 days.

The plasma VTG in both male and female fish treated with 50 μg/g BW was significantly higher than control fish, whereas it significantly decreased at higher concentration, i.e.100 μg/g BW that may be in part due to a higher level of estrogen receptors on hepatocytes. Flouriot et al. (1995) demonstrated an increase in level of estrogen receptors and mRNA vitellogenin in hepatocyte cells of rainbow trout after fish being exposed to 4-NP at 10 μM. Therefore, 4-NP may play a role via estrogen receptors in a competition pathway with a much lower potent than E2 providing an estrogenic effect. Such similar function of 4-NP with E2 mimics to enhance the vitellogenin synthesize resulting in both hyperplasia and hypertrophy of hepatocytes in liver tissue (Arukwe et al. 1997).

Remarkable histological malformations were observed in ovaries of koi carp exposed to 4-NP, especially at the higher dosage. This could be in part due to 4-NP effect on oocytes and intrigue of gene expression of VTG as well as an increase in yolk formation and growth of ovary follicles. An increase in yolk production can cause a gathering vitellogenesis in oocytes with a consequence increasing in GSI index in female fish. In contrast, number of degenerated eggs and asteriated follicles in the ovary fish subjected to 4-NP at 100 μg/g BW showed a significant increase compared to other treatments. Degeneration of the cells of the follicular layer may cause a decrease in the level of synthesis of sexual steroids delaying the maturity of sexual cells and

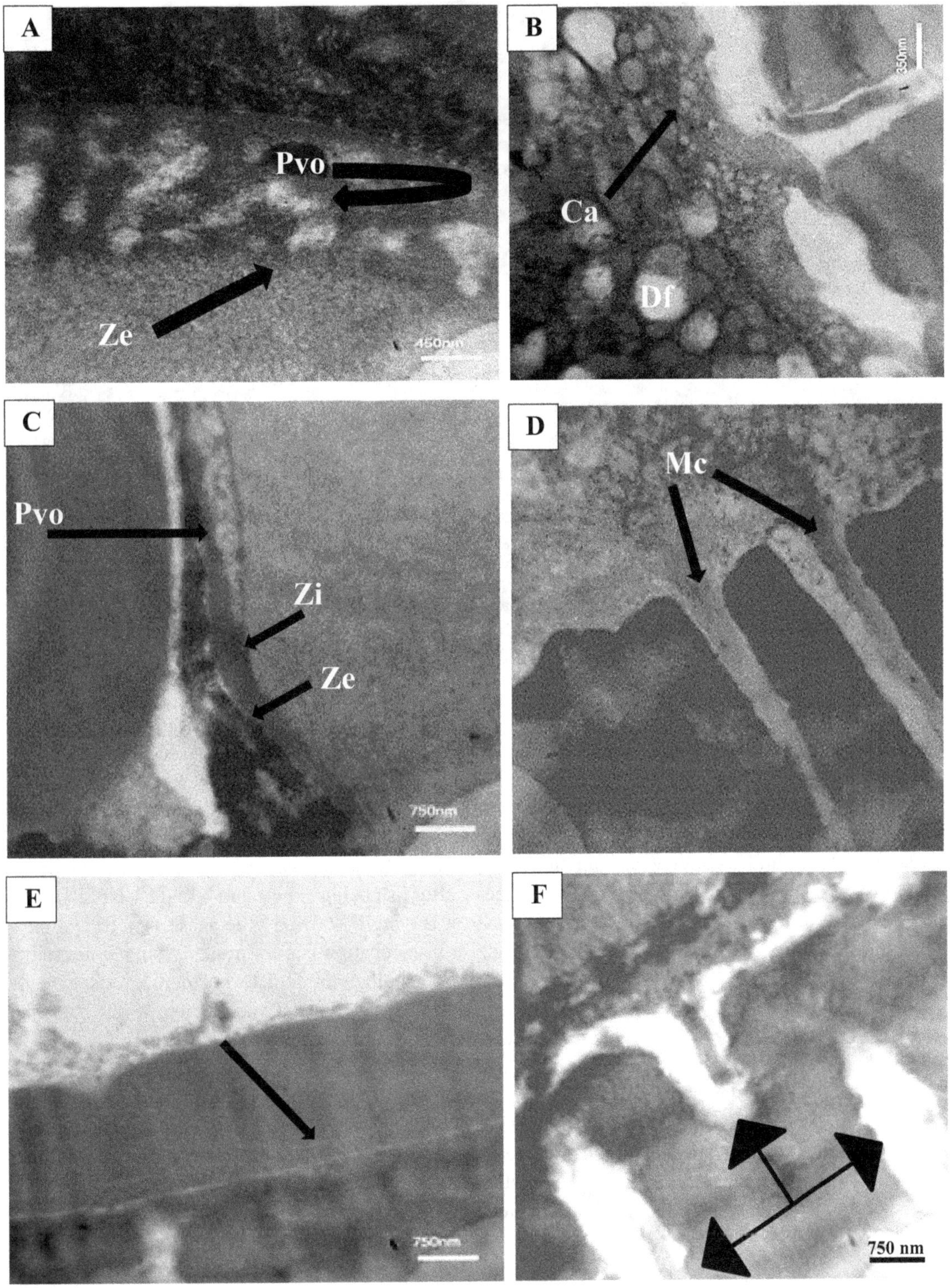

Fig. 5 Transmission electron micrograph of koi carp ovary treated with different concentration of 4-nonylphenol. **a** Control fish showing normal histological structures, **b** fish treated with 17β-Estradiol, **c**, **d** fish treated with 10 and 50 μg/g BW, respectively, **e**, **f** fish treated with 100 μg/g b w showing penetration of vitellogenin inside the zona internal ovary membrane (arrows). Pvo, previtellogenic oocytes; Ze, zona external membrane; Df, droplets of fat; Ca, cortical alveoli; Zi, zona internal membrane; Mc, micro-pinocytosis cascade

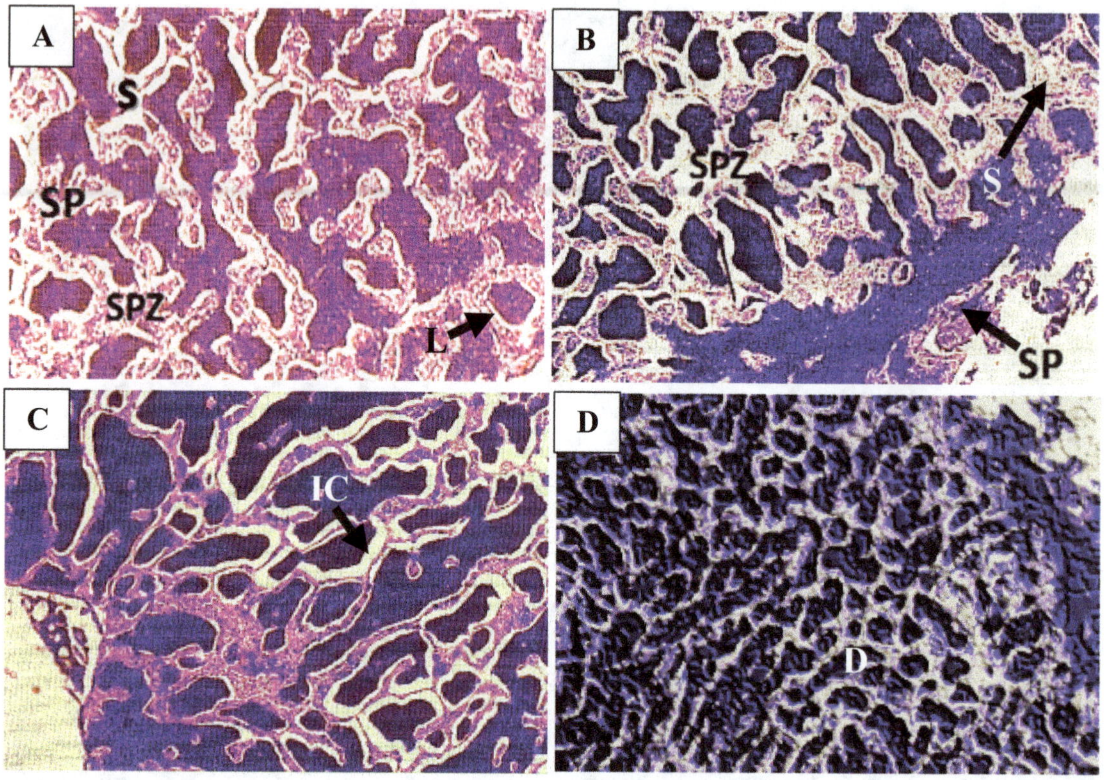

Fig. 6 Sections of koi carp testis treated with different concentrations of 4-nonylphenol. **a** Control fish showing normal histological structures, **b–d** fish treated with 10, 50 and 100 μg/g BW, respectively (H&E, ×400). SP, spermatids; SPZ, spermatozoa; L, lumen space; S, seminiferous lobules; IC, interlobular connective tissue; D, degeneration of Leydig cells

egg expansion. Holdway et al. (2008) reported that exposing fish to EDCs before the spawning can result in a change in reproduction behavior, decrease fecundity and sexual hormone excretion. Also, Sayed et al. (2014) demonstrated a structural pathology and function in Nile tilapia (*Oreochromis niloticus*) with a high decrease in gonadosomatic indexes, fecundity and oocyte diameter after fish being exposed to 100 μg 4-NP. In addition, African catfish (*Clarias gariepinus*) subjected to 4-NP at 0.1 mg/L showed a decrease in fertilization, egg hatching and fecundity as well as a remarkable reduction in incubation period, mortality and abnormality of embryos (Sayed et al. 2012). In the present study, liver histology of female koi carp exposed to 4-NP demonstrated several cellular disorders including degeneration, necrosis and vacuolation. It has been shown by Lee et al. (2003) that 4-NP is able to create such disorders in testicules structures. Also, similar histopathological signs have been also reported in different fish spices subjected to NP at different concentrations (Lee et al. 2004; Bhattacharya et al. 2008; Kaptaner and Ünal 2011). However, lower concentration of 4-NP can promote sexual glands preparatory phase in female catfishe (*Heteropneustes fossilis*) by ovulation and maturation of oocytes (Gautam et al. 2011). Therefore, NP at lower concentration can cause ovulation and maturation of oocytes, but at the higher concentrations a delay in maturation and growth of sexual glands together with a significant increase in atresia oocytes of stages 1 and 2 can be occurred.

Conclusion

In conclusion, the results of this study demonstrated that 4-NP can mimic to play a role as an estrogenic hormone in the sexual development, hepatotoxicity and gonado-toxicity in koi carp. Also, 4-NP demonstrated to induce vitellogenin production and interfere with fish gonadal maturation. Therefore, as a large volume of detergents containing 4-NP are currently discharged into aquatic ecosystems, a particular attention is required to avoid use of contaminated waters with 4-NP for aquaculture activities that can lead to failure in the propagation and rearing of both wild and commercial fish species.

Acknowledgements Authors would like to thank Mr. Hamid Fathalian and Mr. Saeid Totoonchi Mashhor for their valuable assistance.

Author contributions PA conducted the experiment, collection and assembly of data, MS and HHS designed and supervised the study, MS wrote the manuscript and SPHS analyzed the data and assist in writing the manuscript.

Conflict of interest The authors report no conflict of interest.

References

Arukwe A, Förlin L, Goksøyr A (1997) Xenobiotic and steroid biotransformation enzymes in Atlantic salmon *Salmo salar* liver treated with an estrogenic compound, 4-nonylphenol. Environ Toxicol Chem 16:2576–2583

Bancroft JD, Steven A (1982) Theory and practice of histological technique. Churdchil Livingstone, Edinburgh

Bhattacharya H, Xiao Q, Lun L (2008) Toxicity studies of nonylphenol on rosy barb (*Puntius conchonious*): a biochemical and histopathological evaluation. Tissue Cell 40:243–249

Cardinali M, Maradonna F, Olivotto I, Bortoluzzi G, Mosconi G, Polzonetti-Magni AM (2004) Temporary impairment of reproduction in freshwater teleost exposed to nonylphenol. Reprod Toxicol 18:597–604

Carrera EP, García-López A, del Río MDPM, Martínez-Rodríguez G, Solé M, Mancera JM (2007) Effects of 17β-estradiol and 4-nonylphenol on osmoregulation and hepatic enzymes in gilthead sea bream (*Sparus auratus*). Comparative Comp Biochem Physiol C Toxicol Pharmacol 145:210–217

Casini S, Fossi MC, Mori G, Bjornstad A (2002) Vitellogenin induction in *Cyprinus carpio* treated with 17 β-estradiol and 4-nonylphenol. Environ Monit Assess 75:235–239

Choi S, Lee CH, Park W, Kim DJ, Sohn YC (2010) Effects of shortened photoperiod on gonadotropin-releasing hormone, gonadotropin, and vitellogenin gene expression associated with ovarian maturation in rainbow trout. Zool Sci 27:24–32

Christensen LJ, Korsgaard B, Bjerregaard P (1999) The effect of 4-nonylphenol on the synthesis of vitellogenin in the flounder *Platichthys flesus*. Aquat Toxicol 46:211–219

Flouriot G, Pakdel F, Ducouret B, Valotaire Y (1995) Influence of xenobiotics on rainbow trout liver estrogen receptor and vitellogenin gene expression. J Mol Endocrinol 15:143–151

Folmar LC, Hemmer MJ, Denslow ND, Kroll K, Chen J, Cheek A (2002) A comparison of the estrogenic potencies of estradiol, ethynylestradiol, diethylstilbestrol, nonylphenol and methoxychlor in vivo and in vitro. Aquat Toxicol 60:101–110

Gautam GJ, Chaube R, Joy KP (2011) 4-Nonylphenol Impairs Ovarian Recrudescence and Induces Atresia in the Catfish *Heteropneustes fossilis*. Indian J Sci Technol 4:245–246

Giesy JP, Pierens SL, Snyder EM, Miles-Richardson S, Kramer VJ, Snyder SA (2000) Effects of 4-nonylphenol on fecundity and biomarkers of estrogenicity in fathead minnows (Pimephales promelas). Environ Toxicol Chem 19:1368–1377

Hasselberg L, Meier S, Svardal A, Hegelund T, Celander MC (2004) Effects of alkylphenols on CYP1A and CYP3A expression in first spawning Atlantic cod (*Gadus morhua*). Aquat Toxicol 67:303–313

Holdway DA, Hefferman J, Smith A (2008) Multigeneration assessment of nonylphenol and endosulfan using a model Australian freshwater fish, *Melanotaenia fluviatilis*. Environ Toxicol 23:253–262

John DM, House WA, White GF (2000) Environmental fate of nonylphenol ethoxylates: differential adsorption of homologs to components of river sediment. Environ Toxicol Chem 19:293–300

Kaptaner B, Ünal G (2011) Effects of 17α-ethynylestradiol and nonylphenol on liver and gonadal apoptosis and histopathology in *Chalcalburnus tarichi*. Environ Toxicol 26:610–622

Laurenzana EM, Weis CC, Bryant CW, Newbold R, Delclos KB (2002) Effect of dietary administration of genistein, nonylphenol or ethinyl estradiol on hepatic testosterone metabolism, cytochrome P-450 enzymes, and estrogen receptor alpha expression. Food Chem Toxicol 40:53–63

Lavado R, Thibaut R, Raldúa D, Martın R, Porte C (2004) First evidence of endocrine disruption in feral carp from the Ebro River. Toxicol Appl Pharmacol 196:247–257

Lee HJ, Chattopadhyay S, Gong EY, Ahn RS, Lee K (2003) Antiandrogenic effects of bisphenol A and nonylphenol on the function of androgen receptor. Toxicol Sci 75:40–46

Lee HB, Peart TE, Chan J, Gris G (2004) Occurrence of endocrine-disrupting chemicals in sewage and sludge samples in Toronto, Canada. Water Qual Res J Can 39:57–63

Maradonna F, Carnevali O (2007) Vitellogenin, zona radiata protein, cathepsin D and heat shock protein 70 as biomarkers of exposure to xenobiotics. Biomarkers 12:240–255

Martin-Skilton R, Thibaut R, Porte C (2006) Endocrine alteration in juvenile cod and turbot exposed to dispersed crude oil and alkylphenols. Aquat Toxicol 78:S57–S64

Matsumura N, Ishibashi H, Hirano M, Nagao Y, Watanabe N, Shiratsuchi H (2005) Effects of nonylphenol and triclosan on production of plasma vitellogenin and testosterone in male South African clawed frogs (*Xenopus laevis*). Biol Pharm Bull 28:1748–1751

Mitchelmore CL, Rice CP (2006) Correlations of nonylphenol-ethoxylates and nonylphenol with biomarkers of reproductive function in carp (*Cyprinus carpio*) from the Cuyahoga River. Sci Total Environ 371:391–401

Pait AS, Nelson JO (2003) Vitellogenesis in male *Fundulus heteroclitus* (killifish) induced by selected estrogenic compounds. Aquat Toxicol 64:331–342

Rankouhi TR, Sanderson JT, Van Holsteijn I, Van Leeuwen C, Vethaak AD, Van den Berg M (2004) Effects of natural and synthetic estrogens and various environmental contaminants on vitellogenesis in fish primary hepatocytes: comparison of bream (*Abramis brama*) and carp (*Cyprinus carpio*). Toxicol Sci 81:90–102

Rivero CL, Barbosa AC, Ferreira MFN, Dorea JG, Grisolia CK (2008) Evaluation of genotoxicity and effects on reproduction of nonylphenol in *Oreochromis niloticus* (Pisces: Cichlidae). Ecotoxicol 17:732–737

Sayed AEDH, Mahmoud UM, Mekkawy IA (2012) Reproductive biomarkers to identify endocrine disruption in *Clarias gariepinus* exposed to 4-nonylphenol. Ecotoxicol Environ Saf 78:310–319

Sayed AEDH, Abdel-Aziz SH, El-Sayed AF, Zeid S (2014) Structural and functional effects of early exposure to 4-nonylphenol on gonadal development of Nile tilapia (*Oreochromis niloticus*): b-histological alterations in testes. Fish Physiol Biochem 40:1495–1507

Sone K, Hinago M, Kitatama A, Morokuma J, Ueno N, Watanabe H, Iguchia T (2004) Effects of 17β-estradiol, nonylphenol, and bisphenol-A on developing *Xenopus laevis* embryos. Gen Comp Endocr 138:228–236

Uguz C, Iscan M, Ergüven A, Isgor B, Togan I (2003) The bioaccumulation of nonyphenol and its adverse effect on the liver of rainbowtrout (*Onchorynchus mykiss*). Environ Res 92:262–270

Villeneuve DL, Villalobos SA, Keith TL, Snyder EM, Fitzgerald SD, Giesy JP (2002) Effects of waterborne exposure to 4-nonylphenol on plasma sex steroid and vitellogenin concentrations in sexually mature male carp (*Cyprinus carpio*). Chemosphere 47:15–28

Vos JG, Dybing E, Greim HA, Ladefoged O, Lambré C, Tarazona JV, Brandt I, Vethaak AD (2000) Health effects of endocrine-disrupting chemicals on wildlife, with special reference to the European situation. Crit Rev Toxicol 30:71–133

Weber LP, Hill Jr RL, Janz DM (2003) Developmental estrogenic exposure in zebrafish (*Danio rerio*): II. Histological evaluation of gametogenesis and organ toxicity. Aquat Toxicol 63:431–446

Yadetie F, Arukwe A, Goksøyr A, Male R (1999) Induction of hepatic estrogen receptor in juvenile Atlantic salmon in vivo by the environmental estrogen, 4-nonylphenol. Sci Total Environ 233:201–210

Yang L, Lin L, Weng S, Feng Z, Luan T (2008) Sexually disrupting effects of nonylphenol and diethylstilbestrol on male silver carp (*Carassius auratus*) in aquatic microcosms. Ecotoxicol Environ Saf 71:400–411

Yaron Z, Gur G, Melamed P, Rosenfeld H, Elizur A, Levavi-Sivan B (2003) Regulation of fish gonadotropins. Int Rev Cytol 225:131–185

Zaccaroni A, Gamberoni M, Mandrioli L, Sirri R, Mordenti O, Scaravelli D, Sarli G, Parmeggiani A (2009) Thyroid hormones as a potential early biomarker of exposure to 4-nonylphenol in adult male shubunkins (*Carassius auratus*). Sci Total Environ 407:3301–3306

Zha J, Sun L, Spear PA, Wang Z (2008) Comparison of ethinylestradiol and nonylphenol effects on reproduction of Chinese rare minnows (*Gobiocypris rarus*). Ecotoxicol Environ Saf 71:390–399

Permissions

All chapters in this book were first published in IAR, by Springer; hereby published with permission under the Creative Commons Attribution License or equivalent. Every chapter published in this book has been scrutinized by our experts. Their significance has been extensively debated. The topics covered herein carry significant findings which will fuel the growth of the discipline. They may even be implemented as practical applications or may be referred to as a beginning point for another development.

The contributors of this book come from diverse backgrounds, making this book a truly international effort. This book will bring forth new frontiers with its revolutionizing research information and detailed analysis of the nascent developments around the world.

We would like to thank all the contributing authors for lending their expertise to make the book truly unique. They have played a crucial role in the development of this book. Without their invaluable contributions this book wouldn't have been possible. They have made vital efforts to compile up to date information on the varied aspects of this subject to make this book a valuable addition to the collection of many professionals and students.

This book was conceptualized with the vision of imparting up-to-date information and advanced data in this field. To ensure the same, a matchless editorial board was set up. Every individual on the board went through rigorous rounds of assessment to prove their worth. After which they invested a large part of their time researching and compiling the most relevant data for our readers.

The editorial board has been involved in producing this book since its inception. They have spent rigorous hours researching and exploring the diverse topics which have resulted in the successful publishing of this book. They have passed on their knowledge of decades through this book. To expedite this challenging task, the publisher supported the team at every step. A small team of assistant editors was also appointed to further simplify the editing procedure and attain best results for the readers.

Apart from the editorial board, the designing team has also invested a significant amount of their time in understanding the subject and creating the most relevant covers. They scrutinized every image to scout for the most suitable representation of the subject and create an appropriate cover for the book.

The publishing team has been an ardent support to the editorial, designing and production team. Their endless efforts to recruit the best for this project, has resulted in the accomplishment of this book. They are a veteran in the field of academics and their pool of knowledge is as vast as their experience in printing. Their expertise and guidance has proved useful at every step. Their uncompromising quality standards have made this book an exceptional effort. Their encouragement from time to time has been an inspiration for everyone.

The publisher and the editorial board hope that this book will prove to be a valuable piece of knowledge for researchers, students, practitioners and scholars across the globe.

List of Contributors

Imane Afandi
Faculty of Sciences Ain Chock, Hassan II University, Maarif, B. P. 5366, 20100 Casablanca, Morocco
National Institute of Halieutic Research, Street Sidi Abderrahmane Club equestre Ould Jmel, 20100 Casablanca, Morocco

Sophia Talba
National Institute of Halieutic Research, Street Sidi Abderrahmane Club equestre Ould Jmel, 20100 Casablanca, Morocco
Faculty of Sciences Ben M'sik, Avenue du Cdt Driss Lharti, 20700 Casablanca, Morocco

Mohammed Bekkali
Faculty of Sciences Ain Chock, Hassan II University, Maarif, B. P. 5366, 20100 Casablanca, Morocco

Ali Benhra, Samir Benbrahim, Rachid Chfiri, Hicham Masski and Fatima Zohra Bouthir
National Institute of Halieutic Research, Street Sidi Abderrahmane Club equestre Ould Jmel, 20100 Casablanca, Morocco

Maylis Labonne, Raymond Laë and Luis Tito De Morais
UMR LEMAR 6539, IUEM, Place copernic, Plouzané, 29820 Brest, France

Güntekin Doğan
Suleyman Demirel University, Isparta, Turkey

Ömer Osman Ertan
Department of Fisheries Basic Sciences, Egirdir Fisheries Faculty, Suleyman Demirel University, Isparta, Turkey

Behrooz Heidari
Department of Biology, Faculty of Science, University of Guilan, Namjoo St., Rasht 4193833697, Iran
Department of Marine Sciences, Caspian Sea Basin Research Center, University of Guilan, Rasht, Iran

Motahereh Ghadiri, Soodabeh Kazemi and Mehdi Rassa
Department of Biology, Faculty of Science, University of Guilan, Namjoo St., Rasht 4193833697, Iran

Takahiro Ikari, Yoichiro Kitani, Toshio Sekiguchi and Nobuo Suzuki
Noto Marine Laboratory, Division of Marine Environmental Studies, Institute of Nature and Environmental Technology, Kanazawa University, Noto-Cho, Ishikawa 927-0553, Japan

Yuki Kobayashi
Graduate School of Integrated Arts and Sciences, Hiroshima University, Higashi-Hiroshima 739-8521, Hiroshima, Japan

Masato Endo
Department of Marine Biosciences, Division of Marine Science, Tokyo University of Marine Science and Technology, Minato-Ku, Tokyo 108-8477, Japan

Akira Kambegawa
The Kambegawa Laboratory, Komae, Tokyo 201-0013, Japan

K. Asahina
Department of Marine Science and Resources, College of Bioresource Sciences, Nihon University, Fujisawa 252-0813, Kanagawa, Japan

A. Hattori
Department of Biology, College of Liberal Arts and Sciences, Tokyo Medical and Dental University, Ichikawa, Chiba 272-0827, Japan

Y. Tabuchi
Division of Molecular Genetics Research, Life Science Research Center, University of Toyama, Toyama 930-0194, Japan

T. Amornsakun
Fisheries Technology Program, Faculty of Science and Technology, Prince of Songkla University, Pattani 94000, Pattani, Thailand

K. Mizusawa and A. Takahashi
School of Marine Biosciences, Kitasato University, Sagamihara, Kanagawa 252-0373, Japan

Takeo Kurihara, Shoji Nakano, Yukihiko Matsuyama, Kazumasa Hashimoto and Katsumasa Yamada
Seikai National Fisheries Research Institute, Japan Fisheries Research and Education Agency, 1551-8 Taira-machi, Nagasaki 851-2213, Japan

Atsushi Ito and Masaei Kanematsu
National Research Institute of Fisheries and Environment of Inland Sea, Japan Fisheries Research and Education Agency, 2- 17-5 Maruishi, Hatsukaichi, Hiroshima 739-0452, Japan

Lia Kusmita, Erlita Verdia Mutiara, Petrick Ariska Pratama and Awang Surya Wiguna
STIFAR "Yayasan Pharmasi", Letjend Sarwo Edie Wibowo KM 1, Semarang 50193, Indonesia

Handung Nuryadi and Ocky Karna Radjasa
Faculty of Fisheries and Marine Science, Diponegoro University, Semarang, Central Java 50275, Indonesia

Rajib Majumder
Department of Zoology, University of Kalyani, Kalyani, W.B. 741235, India
Department of Zoology, Vivekananda Mahavidyalaya, Hooghly, W.B. 712405, India

Anilava Kaviraj
Department of Zoology, University of Kalyani, Kalyani, W.B. 741235, India

Alexey A. Morozov, Grigorii M. Chuiko and Victoria V. Yurchenko
Laboratory of Physiology and Toxicology of Aquatic Animals, I.D. Papanin Institute for Biology of Inland Waters Russian Academy of Sciences, IBIW RAS, 152742 Borok, Russia

Mansour Torfi Mozanzadeh, Morteza Yaghoubi and Jasem G. Marammazi
Agriculture Research, Education and Extension, Iran Fisheries Research Organization (IFRO), South Iran Aquaculture Research Center, Ahwaz, Iran

Omid Safari
Department of Fisheries, Faculty of Natural Resources and Environment, Ferdowsi University of Mashhad, Khorasan Razavi, Iran

Enric Gisbert
Unitat de Cultius Experimentals, Centre de Sant Carles de la Rápita (IRTA-SCR), IRTA, Crta. del Poble Nou Km 5.5, Sant Carles de la Rápita, Spain

R. P. Pikle, R. M. Jatiger and C. B. Ganesh
Neuroendocrinology Research Lab, Department of Studies in Zoology, Karnatak University, Dharwad 580 003, India

Hiroko Seki, Kaori Nakazato and Naoko Hamada-Sato
Tokyo University of Marine Science and Technology, 4-5-7, Konan, Minato-ku, Tokyo 108-8477, Japan

Carolina Vieira da Silva and Raoul Henry
Departamento de Zoologia, Instituto de Biociências, Universidade Estadual Paulista "Julio de Mesquita Filho" (UNESP), Botucatu, SP, Brazil

M. Sohrabnezhad, M. Sudagar and M. Mazandarani
Department of Fisheries, Gorgan University of Agricultural Sciences and Natural Resource, Gorgan, Iran

Chui-Fen Teoh, Leong-Seng Lim and Gunzo Kawamura
Borneo Marine Research Institute, Universiti Malaysia Sabah, Jalan UMS, 88400 Kota Kinabalu, Sabah, Malaysia

Hao Wu
Key Laboratory of Freshwater Aquatic Genetic Resources, Ministry of Agriculture, Shanghai Ocean University,
College of Fisheries & Life Science, Shanghai Ocean Universities, Shanghai 201306, China

Jianlong Zhang, Jizhou Yan and Min Jiang
College of Fisheries & Life Science, Shanghai Ocean Universities, Shanghai 201306, China

Yongxing He
Agricultural Service Center of Pudong New Area, Shanghai, China

Junfang Zhou
Key and Open Laboratory of Marine and Estuarine Fisheries, Ministry of Agriculture, East China Sea Fisheries Research Institute, Chinese Academy of Fishery Sciences, Shanghai, China

Yi Yang, Yuchun Li and Zhenlin Liang
Marine College, Shandong University, Weihai 264200, China

Mehran Yasemi
Institute of Applied-Scientific Higher Education of Jahad-e-Agriculture, Agricultural Research, Education and Extension Organization, Tehran, Iran

Alireza Nazari Bajgan
Higher Education Center of Jahad-e-Agriculture Hormozgan, Bandar Abbas, Iran

Mehran Parsa
Young Researchers and Elite Club, Bandar Abbas Branch, Islamic Azad University, Bandar Abbas, Iran

Avelyno H. D'Costa, S. K. Shyama, M. K. Praveen Kumar and Trisha M. Fernandes
Genetic Toxicology Laboratory, Department of Zoology, Goa University, Goa 403206, India

Lota Alcantara Creencia
College of Fisheries and Aquatic Sciences, Western Philippines University, Sta. Monica, 5300 Puerto Princesa City, Philippines
Faculty of Fisheries, Kagoshima University, 4-50-20 Shimoarata, Kagoshima 8900056, Japan

Tadahide Noro
Faculty of Fisheries, Kagoshima University, 4-50-20 Shimoarata, Kagoshima 8900056, Japan
Kagoshima Prefectural College, 1-42-1 Shimo-ishiki, Kagoshima City 890-0005, Japan

Seinen Chow, Yoji Nakamura, Motoshige Yasuike, Kenji Saitoh and Takashi Yanagimoto
National Research Institute of Fisheries Science, 2-12-4 Fukuura, Kanagawa 2368648, Japan

Nobuaki Suzuki
National Research Institute of Far Seas Fisheries, 5-7-1 Orido, Shizuoka 424-8633, Japan

Parisa Amaninejad and Seyed Pezhman Hosseini Shekarabi
Department of Fisheries Science, Science and Research Branch, Islamic Azad University, Tehran, Iran

Homayoun Hosseinzadeh Sahafi
Iranian Fisheries Science Research Institute Agriculture Research, Education and Extension Organization, Tehran, Iran

Mehdi Soltani
Department of Aquatic Animal Health, Faculty of Veterinary Medicine, University of Tehran, Tehran, Iran

A. K. Prusty
Project Directorate for Farming System Research (PDFSR), Modipuram, Meerut 250110, UP, India

D. K. Meena and B. K. Behera
Central Inland Fisheries Research Institute (CIFRI), Barrackpore, Kolkata 700120, West Bengal, India

S. Mohapatra
Central Institute of Fisheries Education (CIFE), Versova, Andheri (W), Mumbai 400061, India

P. Panikkar
CIFRI Regional Centre, Hessarghatta Lake Post, Bangalore 560089, Karnataka, India

P. Das
CIFRI Regional Centre, Guwahati 781006, Assam, India

S. K. Gupta
Directorate of Coldwater Fisheries Research (DCFR), Chhirapani Fish Farm, Champawat 262523, Uttarakhand, India

Index

A
Ablabesmyia, 132, 138, 140, 142
Abramis Brama, 85-87, 224
Adenosine Monophosphate, 116
Adrenocorticotropic Hormone, 44
Agar Gel Pellet, 156-158
Anguilla Japonica, 44, 49, 209
Aspartic Acid, 16-18, 25, 27, 94, 96, 165
Atrina Pectinata, 53-55, 63
Austrovenus Stutchburyi, 54, 63
Auxis Thazard, 177, 185

B
Bacterial Symbionts, 40, 64-72
Beluga Sturgeon, 146-147, 149, 151

C
Calmodulin, 206, 211-212, 228
Candida Albicans, 30, 32-33, 40, 42
Candida Glabrata, 30
Candida Parapsilosis, 30, 41
Carassius Auratus, 44-45, 51, 113, 115, 218, 224
Carotenoid, 64-65, 67-72, 176
Chlamys Opercularis, 54, 63
Clupeonella, 95-96
Corynoneura, 138, 140, 142
Cyprinid Fish, 85-86, 92
Cyprinus Carpio, 81, 83-84, 103-104, 194, 213-214, 224

D
Danio Rerio, 186-187, 193-194, 224
Defatted Soybean Meal, 146, 151, 154
Dicentrachus Labrax, 105
Docosahexaenoic Acid, 16-17

E
E. Azurea, 132-136, 138-144
Electron Capture Detector, 76
Endotribelos, 132, 138-140, 142-143
Engraulis Encrasicolus, 1-4, 14, 25
Exiguobacterium, 116, 121, 124-125, 130-131

F
Feed Conversion Ratio, 77, 79, 97, 99-100

G
Goldband Goatfish, 16-17

H
Haliotis Diversicolor, 195, 203-205
Heteropneustes Fossilis, 74, 84, 222-223
Huso Huso, 146-147, 154

L
Labrundinia, 132, 139-140, 142-143
Larsia, 138, 140, 142-143
Lipid Peroxidation, 74, 82, 85-86, 88, 92, 192-194
Litopenaeus Vannamei, 125, 162-163, 168
Longtail Tuna, 177-185
Luteinizing Hormone, 105-106
Luteolibacter Algae, 172

M
M. Barbatus, 9, 21-22, 25
Melanocortin Receptors, 44-45, 48, 51
Multienzyme Kemin, 146, 149, 151
Mytella Charruana, 54, 61, 63

O
Ontogenic Scales, 43-47, 49-50
Oreochromis Mossambicus, 105-106, 114, 194
Oreochromis Niloticus, 25, 28, 73-74, 77-78, 81, 83, 103-104, 154, 222, 224
Organophosphate Pesticides, 186-187, 191-192
Ornithinibacillus Contaminans, 169, 172, 175
Oxidative Stress, 81-83, 85-86, 91-93, 192-193
Oxyeleotris Marmorata, 156, 160-161

P
Palmitoleic Acid, 16, 22, 25, 165
Pelagic Fish, 1-2, 7-10, 13-15, 184
Perna Viridis, 54, 61, 63
Phragmites Australis, 54
Plasma Calcitonin, 43-47, 49-50
Proximal Pars Distalis, 105, 107

Pseudoalteromonas Rubra, 64, 68
Pseudoalteromonas Shioyasakiensis, 64, 68

S

Salvelinus Alpinus, 106, 115
Sardina Pilchardus, 1-4, 7-8, 14
Scombridae, 177, 185
Shewanella, 116, 121, 124-125, 128, 130-131, 175
Sparidentex Hasta, 94-95, 104, 155
Spermatocytes, 105-111, 113
Spermatogonia, 105-111, 113
Streptococcus Mutans, 30-32, 41-42
Superoxide Dismutase, 85-86, 88, 93

T

Terrestrial Runoff, 53-54
Thunnus Tonggol, 177, 184-185
Thymallus Thymallus, 159
Trachurus Trachurus, 1-4, 7-8, 14, 25, 27

U

Ultimobranchial Glands, 43-49
Upeneus Moluccensis, 16-18, 28-29

W

White Spot Syndrome Virus, 162, 165, 167-168

www.ingramcontent.com/pod-product-compliance
Lightning Source LLC
LaVergne TN
LVHW080708210926
848485LV00003B/434